ArcGIS for Desktop Cookbook

Over 60 hands-on recipes to help you become a more productive ArcGIS for Desktop user

Daniela Cristiana Docan

BIRMINGHAM - MUMBAI

ArcGIS for Desktop Cookbook

First published: January 2015

Production reference: 1170115

Published by Packt Publishing Ltd.
Livery Place
35 Livery Street
Birmingham B3 2PB, UK.

ISBN 978-1-78355-950-3

www.packtpub.com

Cover image by Daniela Cristiana Docan (daniela.docan@utcb.ro)

Credits

Author
Daniela Cristiana Docan

Reviewers
Ken Doman
Stefano Iacovella
Tram Vu Khanh Truong
John (Yiguang) Zhang

Commissioning Editor
Pramila Balan

Acquisition Editor
Llewellyn Rozario

Content Development Editor
Mohammed Fahad

Technical Editors
Humera Shaikh
Mohita Vyas

Copy Editor
Sarang Chari

Project Coordinator
Danuta Jones

Proofreaders
Simran Bhogal
Sandra Hopper
Joanna McMahon

Indexer
Mariammal Chettiyar

Production Coordinator
Alwin Roy

Cover Work
Alwin Roy

About the Author

Daniela Cristiana Docan is currently a lecturer in the Department of Topography and Cadastre at the Faculty of Geodesy in Bucharest, Romania. She obtained her PhD in 2009 from Technical University of Civil Engineering Bucharest with the thesis *Contributions to quality improvement of spatial data in GIS*. Formerly, she worked for Esri Romania and the National Agency of Cadastre and Land Registration (ANCPI).

While working for Esri Romania, she trained teams (as an authorized instructor in ArcGIS Desktop by Environmental Systems Research Institute, Inc., USA) from state and privately owned companies, such as Romanian Civil Aeronautical Authority, Agency of Payments and Intervention for Agriculture (APIA), Institute of Hydroelectric Studies and Design, and Petrom. She has also trained and assisted the team in charge of quality data control using ArcGIS for Desktop and PLTS GIS Data ReViewer in the Land Parcels Identification System (LPIS) project, in Romania.

For the ANCPI, in 2009, she created the conceptual, logical, and physical data model for the Romanian National Topographic Dataset at the scale 1:5,000 (TOPRO5). She was a member of the workgroup that elaborated TOPRO5 and metadata technical specifications for the ANCPI and the Member State Report for Infrastructure for Spatial Information in the European Community (INSPIRE) in 2010.

I would like to thank Llewellyn Rozario, Mohammed Fahad, Humera Shaikh, Mohita Vyas, and everyone else from Packt Publishing for all their hard work to get this book published.

I would also like to thank the reviewers for their work and practical advice.

A special thanks goes to my friends for their support.

About the Reviewers

Ken Doman is a GIS developer for Bruce Harris & Associates, Inc., a land-record mapping and software development company that is an Esri Business Partner. Ken has worked both in municipal government GIS and in the private sector. He has experienced many facets of GIS technology, from field data collection and GPS to mapping and data analysis to publishing web maps and applications.

Ken previously reviewed *Building Web and Mobile ArcGIS Server Applications with JavaScript, Eric Pimpler Packt Publishing*. He is also in the middle of writing another book for Packt Publishing on a similar topic.

I would first like to thank my wife, who puts up with my night-owl tendencies when working on these books. I would also like to thank Bruce Harris & Associates, Inc., for giving me opportunities and exposure to learn more about GIS technologies. I would also like to thank the City of Plantation, Florida, and the City of Jacksonville, Texas, for providing career opportunities for me in GIS. Thanks also goes out to Packt Publishing, who found me out of the blue and let me read cool stuff such as this book. Finally, I would like to thank God, without whom I believe none of this would be possible.

Stefano Iacovella is a longtime GIS developer and consultant living in Rome, Italy. He also routinely works as a GIS course instructor.

He obtained a PhD in Geology. Having a very curious mind, he developed a deep knowledge of IT technologies, mainly focusing on GIS software and related standards.

Starting his career as an Esri employee, he was exposed to and became confident with proprietary GIS software, mainly the Esri suite of products.

In the last 14 years, he has become more and more involved with open source software, also integrating it with proprietary software. He loves the open source approach and really trusts collaboration and sharing of knowledge. He strongly believes in the open source idea and constantly manages to spread it out in all sectors, not just the GIS sector.

He has been using GeoServer since release 1.5, configuring, deploying, and hacking it in several projects. Other GFOSS projects he mainly uses and likes are GDAL/OGR, PostGIS, QGIS, and OpenLayers.

He authored two books on GeoServer with Packt Publishing: *GeoServer Cookbook*, a practical set of recipes to get the most out of the software, and *GeoServer Beginner's Guide*, a first approach to GeoServer features.

When not playing with maps and geometric shapes, he loves reading about science, mainly physics and math ematics; riding his bike; and having fun with his wife and his two daughters, Alice and Luisa.

You can contact him at `stefano.iacovella@gmail.com` or follow him on Twitter at `@iacovellas`.

Tram Vu Khanh Truong received her Master's degree in Regional and City Planning from the University of Oklahoma. Currently, she is a transportation planner at the Greensboro Urban Area Metropolitan Planning Organization in Greensboro, North Carolina. Her duties include GIS development, data analysis, and transportation system planning. Tram Truong has a passion for utilizing GIS in transportation planning to support decision making and linking multimodal transportation planning with mixed-use land development planning.

John (Yiguang) Zhang has been in the geospatial industry for over 20 years with a strong background in GIS, software development, database management, photogrammetry, and remote sensing. He has been working as a software developer and GIS analyst for the past 15 years and has experienced various GIS projects from start to finish on GIS application design, development, and implementation as well as GIS analysis and map production. He has also managed complex spatial databases and experienced a lot of spatial data conversion and integration processes. His creative thinking skills have helped him solve problems effectively through the course of his career in the public and private sectors, such as City of Chilliwack and Intergraph Corporation. He is proficient with the Esri ArcGIS family of products, including ArcGIS Desktop and ArcGIS Server, and spatial database management systems, such as Oracle Spatial, SQL Server, and open source PostgreSQL/PostGIS. He is also competent in .NET and Web 2.0 technologies. He holds a master's degree in Digital Photogrammetry and an Advanced Diploma in GIS from British Columbia Institute of Technologies, Canada.

Firstly, I'd like to thank to my wife, Winnie, for dedicating her time in taking care of the family and for her patience with this wonderful book review and other projects. I would also like to thank my son, Sylvester, and daughter, Sylvia, for their bright ideas to the problems I had to solve.

www.PacktPub.com

Support files, eBooks, discount offers, and more

For support files and downloads related to your book, please visit www.PacktPub.com.

Did you know that Packt offers eBook versions of every book published, with PDF and ePub files available? You can upgrade to the eBook version at www.PacktPub.com and as a print book customer, you are entitled to a discount on the eBook copy. Get in touch with us at service@packtpub.com for more details.

At www.PacktPub.com, you can also read a collection of free technical articles, sign up for a range of free newsletters and receive exclusive discounts and offers on Packt books and eBooks.

https://www2.packtpub.com/books/subscription/packtlib

Do you need instant solutions to your IT questions? PacktLib is Packt's online digital book library. Here, you can search, access, and read Packt's entire library of books.

Why subscribe?

- Fully searchable across every book published by Packt
- Copy and paste, print, and bookmark content
- On demand and accessible via a web browser

Free access for Packt account holders

If you have an account with Packt at www.PacktPub.com, you can use this to access PacktLib today and view 9 entirely free books. Simply use your login credentials for immediate access.

Table of Contents

Preface

ArcGIS for Desktop is an important component of the Esri ArcGIS platform. ArcGIS for Desktop allows you to visualize, create, analyze, manage, and distribute geographic data.

ArcGIS for Desktop Cookbook starts with the basics of designing a file geodatabase schema. Using your file geodatabase schema, you will learn to create, edit, and constrain the geometry and attribute values of your data. In this book, you will learn to manage Coordinate Reference System (CRS) issues in the geodatabase context.

This book will also cover the following topics: designing and sharing quality maps, geocoding addresses, creating routes and events, analyzing and visualizing raster data in 3D environments, and exporting/importing different data formats.

Knowing and understanding your data is essential in any spatial analysis and geoprocessing process. Therefore, you will work with two main geodatabase structures that fully support all topics covered by this book's chapters.

ArcGIS for Desktop Cookbook will clearly explain all the basic steps performed in every recipe of the book to help you refine your own workflow.

What this book covers

Chapter 1, Designing Geodatabase, teaches you how to create a file geodatabase for a topographic map. It shows you, step by step, how to create feature datasets, feature classes, subtypes, and domains. Furthermore, you will create relationship classes and test the relationship behavior.

Chapter 2, Editing Data, teaches you how to add data to your file geodatabase created in *Chapter 1, Designing Geodatabase*. In addition to this, you will learn to work with COordinate GeOmetry (COGO) elements, such as bearings, angles, and horizontal distances. You will identify all invalid attribute values for the newly created or loaded features in accordance with the domains created in the first chapter. You will also constrain and administrate the spatial relationships between features with geodatabase topology.

Chapter 3, Working with CRS, explains how to transform a CRS into another CRS using a predefined ArcGIS Project tool. You will also learn how to georeference a scanned topographic map. Furthermore, you will learn to define a custom CRS and a custom transformation.

Chapter 4, Geoprocessing, guides you through the geoprocessing tools for vector data, such as Spatial Join, Spatial Adjustment, Attribute transfer, Buffer, and Intersect. In addition to this, you will learn to work with the Model Builder application. You will build a geoprocessing workflow for a project named VeloGIS. This small project will analyze the possible consequences of creating a cycling infrastructure, taking into account the existing road network.

Chapter 5, Working with Symbology, teaches you how to manage a collection of symbols, colors, and map elements into a style format. The chapter also covers the *Representation* topic that refers to an advanced technique to symbolize geographic features on a map.

Chapter 6, Building Better Maps, teaches you how to create labels with Maplex Label Engine. The chapter also covers the *Annotation* topic, which refers to an advanced technique to label the geographic features. In addition to this, you will learn to create a quantitative bivariate map, which analyzes two variables from census data: level of education and unemployment rate for a country/region.

Chapter 7, Exporting Your Maps, teaches you how to design, prepare, and export quality maps. Finally, you will publish your maps on ArcGIS Online.

Chapter 8, Working with Geocoding and Linear Referencing, teaches you how to convert the address information into spatial data and how to manage and use these geocoded addresses. The second part of this chapter will show you how to build routes and events for three different bus lines and a complex route for delivering a customer service.

Chapter 9, Working with Spatial Analyst, teaches you how to work with and analyze raster data. It shows you, step by step, how to create a terrain surface, reclassify a raster, work with Map Algebra and statistical functions, generalize a raster, generate density surfaces, and perform a least-cost path analysis.

Chapter 10, Working with 3D Analyst, teaches you how to create 3D features from 2D features and how to create TIN and Terrain surfaces. You will analyze the visibility between buildings and three geodetic points from the ground, using the ArcScene application. Moreover, you will create an animated fly-by tour of your 3D buildings and Digital Terrain Model (DTM) in ArcScene and will export it to a video file.

Chapter 11, Working with Data Interoperability, teaches you how to manage different data formats. You will export and import a file geodatabase using the XML interchange format. The second part of this chapter will show you how to import vector data into your geodatabase using the ArcGIS Data Interoperability extension.

What you need for this book

To complete the exercises in this book, you will need to have installed ArcGIS 10.x for Desktop (Advanced) and the ArcGIS Data Interoperability for Desktop extension.

Depending on your software version, please download and install the latest patches (bug fixes) or service packs (compilation of bug fixes) from `http://support.esri.com/en/downloads/patches-servicepacks`.

You need to have access to an Internet connection to publish your map in the *Publishing maps on the Internet* recipe of *Chapter 7, Exporting Your Maps*.

Data used in this book is freely available on the Packt Publishing site.

Who this book is for

ArcGIS for Desktop Cookbook is written for GIS users who already have basic knowledge about ArcGIS, but need to increase their productivity using the ArcGIS for Desktop applications and extensions. Even if you don't have previous experience with ArcGIS, this book is useful for you because it will help you to catch up with acquainted users.

Please insert the following here:

Sections

In this book, you will find several headings that appear frequently (Getting ready, How to do it, How it works, There's more, and See also).
To give clear instructions on how to complete a recipe, we use these sections as follows:

Getting ready

This section tells you what to expect in the recipe, and describes how to set up any software or any preliminary settings required for the recipe.

How to do it...

This section contains the steps required to follow the recipe.

How it works...

This section usually consists of a detailed explanation of what happened in the previous section.

There's more...

This section consists of additional information about the recipe in order to make the reader more knowledgeable about the recipe.

See also

This section provides helpful links to other useful information for the recipe.

Conventions

In this book, you will find a number of styles of text that distinguish between different kinds of information. Here are some examples of these styles, and an explanation of their meaning.

Code words in text, database table names, folder names, filenames, file extensions, pathnames, dummy URLs, user input, and Twitter handles are shown as follows: "In the next step, you will create a file geodatabase named `Topo5k.gdb` for a topographic map corresponding to the scale `1:5,000`."

A block of code is set as follows:

```
HYC[int] = HIC[int]
Name[string] = Name[string]
HType[int] = HAT[int]
```

New terms and **important words** are shown in bold. Words that you see on the screen, for example, in menus or dialog boxes, appear in the text like this: "Select **ArcToolbox** from the **Standard** toolbar. Go to **Data Management Tools** | **Workspace**, and double-click on the **Create File GDB** tool."

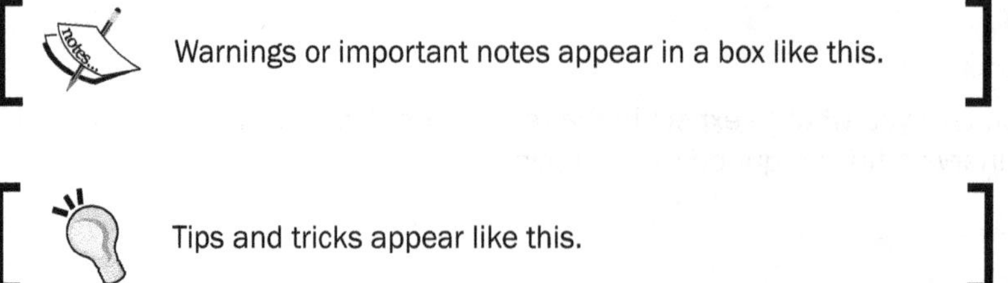

Warnings or important notes appear in a box like this.
Tips and tricks appear like this.

Reader feedback

Feedback from our readers is always welcome. Let us know what you think about this book—what you liked or disliked. Reader feedback is important for us as it helps us develop titles that you really get the most out of.

To send us general feedback, simply e-mail `feedback@packtpub.com`, and mention the book title via the subject of your message.

If there is a topic that you have expertise in and you are interested in either writing or contributing to a book, see our author guide at `www.packtpub.com/authors`.

Customer support

Now that you are the proud owner of a Packt book, we have a number of things to help you to get the most from your purchase.

Downloading the example code

You can download the example code files from your account at `http://www.packtpub.com` for all the Packt Publishing books you have purchased. If you purchased this book elsewhere, you can visit `http://www.packtpub.com/support` and register to have the files e-mailed directly to you.

Downloading the color images of this book

We also provide you with a PDF file that has color images of the screenshots/diagrams used in this book. The color images will help you better understand the changes in the output. You can download this file from: `https://www.packtpub.com/sites/default/files/downloads/B00189_9503OT_Graphics.pdf`.

Errata

Although we have taken every care to ensure the accuracy of our content, mistakes do happen. If you find a mistake in one of our books—maybe a mistake in the text or the code—we would be grateful if you would report this to us. By doing so, you can save other readers from frustration and help us improve subsequent versions of this book. If you find any errata, please report them by visiting `http://www.packtpub.com/submit-errata`, selecting your book, clicking on the **Errata Submission Form** link, and entering the details of your errata. Once your errata are verified, your submission will be accepted and the errata will be uploaded on our website, or added to any list of existing errata, under the Errata section of that title.

To view the previously submitted errata, go to `https://www.packtpub.com/books/content/support` and enter the name of the book in the search field. The required information will appear under the **Errata** section.

Piracy

Piracy of copyrighted material on the Internet is an ongoing problem across all media. At Packt, we take the protection of our copyright and licenses very seriously. If you come across any illegal copies of our works in any form on the Internet, please provide us with the location address or website name immediately so that we can pursue a remedy.

Please contact us at `copyright@packtpub.com` with a link to the suspected pirated material.

We appreciate your help in protecting our authors and our ability to bring you valuable content.

Questions

If you have a problem with any aspect of this book, you can contact us at `questions@packtpub.com`, and we will do our best to address the problem.

1

Designing Geodatabase

In this chapter, we will cover the following recipes:

- Creating a file geodatabase
- Creating a feature dataset
- Creating a feature class
- Creating subtypes
- Creating domains
- Using subtypes and domains together
- Creating a relationship class

Introduction

Real-world objects can be represented in **Geographic Information Systems** (**GIS**) using geographic data. In the context of the **Esri ArcGIS** technology, a database that stores geographic data is a geodatabase. The geodatabase is a native **ArcGIS** format that allows you to store, edit, and manage spatial data and non-spatial data. Before you start adding data to a geodatabase, it is important to think about how your data will be organized in the geodatabase. Another important step is to create an empty but structured schema of your geodatabase (data model). Paul A. Longley (*Geographical Information Systems and Science, 2nd Edition, 2005, John Wiley & Sons, Inc , p.178*) has mentioned three steps in creating a geospatial data model:

1. Conceptual schema
2. Logical schema
3. Physical schema (geodatabase schema)

In this chapter, you will skip steps 1 and 2, and you will manually create the physical schema or the geodatabase structure.

There are three types of geodatabases: personal geodatabase, file geodatabase, and multiuser geodatabase. For more information about geodatabases, please refer to `http://www.esri.com/software/arcgis/geodatabase`.

For comprehensive definitions of specific elements and terms such as geodatabase, file geodatabase, feature dataset, spatial domain, resolution, tolerance, subtype, domain, feature class, relationship, or referential integrity, please refer to the ESRI GIS Dictionary online at `http://support.esri.com/en/knowledgebase/GISDictionary`.

In this chapter, you will work with the most common geodatabase elements, such as feature dataset, feature class, table, and relationship class. You will test the main advantages of using a geodatabase by performing the following actions:

- Define the common spatial reference using feature datasets
- Organize the features with the same geometry and spatial reference in feature classes
- Define attribute constraints to eliminate edit errors using subtypes, domains, and default values
- Define spatial and non-spatial relationships using relationship classes
- Add supplementary spatial and attribute behaviors by defining relationship rules for relationship classes

Creating a file geodatabase

In this chapter, all exercises will refer to the single-user file geodatabase format. A file geodatabase is suited to ArcGIS for Desktop users and is stored in a filesystem folder. The main advantages of a file geodatabase are:

- Editing of different feature classes or tables at the same time by multiple users
- A maximum size of up to 1 terabyte (TB) for the individual datasets stored in a file geodatabase

If you decide to use a single-user geodatabase for your project, but you are still thinking about the two options, personal geodatabase and file geodatabase, then please read the paper *The Top Nine Reasons to Use a File Geodatabase*: `http://www.esri.com/news/arcuser/0309/files/9reasons.pdf`.

Getting ready

In the next step, you will create a file geodatabase named `Topo5k.gdb` for a topographic map corresponding to the scale `1:5,000`.

How to do it...

Follow these steps to create a new file geodatabase in ArcCatalog using the context menu:

1. Start ArcCatalog. Select **Connect To Folder** from the **Standard** toolbar. Go to `<drive>:\PacktPublishing\Data`, and click on **OK**.
2. In **Catalog Tree**, select `<drive>:\PacktPublishing\Data`, and right-click to choose **New** | **Folder**. Rename **New Folder** as `MyGeodatabase`.
3. In **Catalog Tree**, select **MyGeodatabase**, and right-click to choose **New** | **New File Geodatabase**. Rename **New File Geodatabase.gdb** as `Topo5k`.

Downloading the example code

You can download the example code files for all Packt books you have purchased from your account at `http://www.packtpub.com`. If you purchased this book elsewhere, you can visit `http://www.packtpub.com/support` and register to have the files e-mailed directly to you.

How it works...

You have created an empty file geodatabase using the context menu in ArcCatalog. You will use **Topo5k.gdb** in the later steps. Open Windows Explorer to see the structure of the filesystem folder created for your file geodatabase.

There's more...

Follow these steps to create a new file geodatabase in ArcCatalog using **ArcToolbox**:

1. Select **ArcToolbox** from the **Standard** toolbar. Go to **Data Management Tools** | **Workspace**, and double-click on the **Create File GDB** tool.
2. For **File GDB Location**, select **Folder Connections**, go to `<drive>:\PacktPublishing\Data`, and select the **MyGeodatabase** folder. For **File GDB Name**, type `Topo5000`. For **File GDB Version (optional)**, select the **CURRENT** option to create a file geodatabase compatible with ArcGIS Version 10.2, and click on **OK**.

See also

- For more information about creating a file geodatabase, please refer to *Chapter 11, Working with Data Interoperability*
- In the next recipe, *Creating a feature dataset*, you will learn how to organize your spatial datasets using feature datasets in a file geodatabase

Creating a feature dataset

A feature dataset is a container for feature classes that have the same spatial reference, **Coordinate Reference System** (**CRS**), spatial domain, resolution, and tolerance. In the case of importing features with the same CRS, the feature dataset will accept only the features that have their coordinates between the minimum and maximum values defined for the *x*, *y*, *z* and *m* values of spatial domain extent. A feature dataset cannot contain other feature datasets or non-spatial tables.

For more details about the feature dataset, please refer to the *online ArcGIS help (10.2)* by navigating to **Geodata** | **Data types** | **Feature datasets** from `http://resources.arcgis.com/en/help/main/10.2`.

Getting ready

You will create seven feature datasets: `Buildings`, `GeodeticPoints`, `Hydrography`, `LandUse`, `Relief`, `Transportation`, and `Boundaries`. In your file geodatabase, `Topo5k.gdb`, all feature datasets will have the same CRS.

How to do it...

Follow these steps to create feature datasets in a file geodatabase using the ArcCatalog context menu:

1. Start ArcCatalog. In **Catalog Tree**, go to `<drive>:\PacktPublishing\Data\MyGeodatabase`, and select **Topo5k.gdb**.
2. Right-click on **Topo5k.gdb**, and navigate to **New** | **Feature Dataset** for the **Name** type `Buildings`. Click on **Next**. Navigate to **Projected Coordinate Systems** | **National Grids** | **Europe** | **Pulkovo 1942 Adj 1958 Stereo 1970**. Select the active yellow star **Add To Favorites** to add the selected projected coordinate system to the **Favorites** section. You will use this projected coordinate system later in this book.
3. Click on **Next**. Navigate to **Vertical Coordinate Systems** | **Europe** | **Constanta**. Click on **Next**. Accept the **XY**, **Z**, and **M tolerance** values. Keep checked the option **Accept default resolution and domain extent (recommended)**.

4. If you want to see the default values for resolution and domain extent, uncheck them and click on **Next**. Examine the default values, and click on the **Finish** button.
5. If you want to examine the default values for domain, resolution, and tolerance, right-click on `Buildings` feature datasets and navigate to the **Properties** | **Domain, Resolution** and **Tolerance** tabs.
6. Repeat the previous steps to create the following feature datasets: `GeodeticPoints`, `Hydrography`, `LandUse`, `Relief`, and `Transportation`, as shown in the following screenshot:

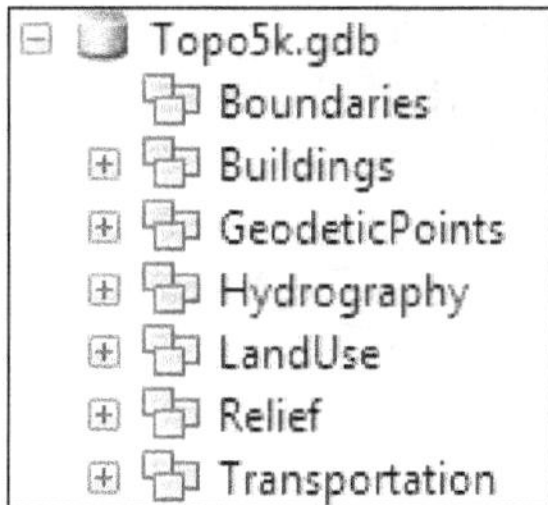

7. To inspect the results, return to the **Feature Dataset Properties** dialog.

How it works...

After you have created a feature dataset, ArcCatalog allows you to change only the *XY* and *Z* coordinate system. If you take the decision to change the *XY* coordinate system for a feature dataset, you should know that you cannot modify the values for domain, resolution and tolerance. If you still need to change the CRS for a feature dataset, it will be more proper to create the feature dataset from the beginning. In conclusion, when you are thinking about a feature dataset, think twice and act once.

There's more...

In ArcCatalog, you can use **Copy** and **Paste** from the context menu in the file geodatabase to duplicate the first feature dataset. In the **Data Transfer** window, you can change the name of the feature dataset by typing in the **Target Name** section. In this case, the newly resulted feature dataset will have the same coordinate reference system as the source feature dataset.

See also

- For more information about CRS, please refer to *Chapter 3, Working with CRS*

Creating a feature class

Features are representations of the real-word objects in a geodatabase. Features are grouped in classes based on their common components: shape (geometry) and attributes. Features from a feature class share the same spatial reference. A feature class stores simple features (for example, point, multipoint, line, and polygon) or non-simple features (for example, dimension and annotation). When a feature class is created, the user can add two optional geometry properties for features: *z* value (3D data) and *m* value (used for linear referencing). The feature class table contains some default attribute fields, which are managed by ArcGIS: **OBJECTID** and **SHAPE**. The **OBJECTID** field has the **Object ID** type property and stores the unique identifier for each feature. The **SHAPE** field refers to the **Geometry** type property and stores the *x* and *y* coordinates of the features and optionally *z* and *m* values if the feature class has the *z* and/or *m* properties enabled. The feature class that stores line or polygon features has two supplementary fields: **SHAPE_Length** (length of the feature line) and **SHAPE_Area** (area of the feature polygon).

For more details about the feature class and field data types, please refer to *Geodata/Data types/Feature Class* and *Geodata/Geodatabases/Defining the properties of data in a geodatabase/Geodatabase table properties/ ArcGIS field data types* from ArcGIS help (10.2) online.

Usually, feature classes are stored inside a feature dataset and automatically inherit the spatial reference of the feature dataset. A feature class that is stored at the root level of a geodatabase is a standalone feature class and has its own spatial reference.

Getting ready

You will continue to work at the geodatabase structure by creating four feature classes: `Buildings` (polygon features), `Watercourse` (polygon features), `WatercourseL` (polyline features), and `LandUse` (polygon features).

How to do it...

Follow these steps to create feature classes using the ArcCatalog context menu:

1. Start ArcCatalog. In **Catalog Tree**, go to `<drive>:\PacktPublishing\Data\MyGeodatabase\Topo5k.gdb`.

2. Right-click on the `Buildings` feature dataset, and choose **New** | **Feature Class.** For **Name**, type `Buildings` and for **Alias**, type `Buildings`. For **Type**, choose **Polygon Feature**. Leave the **Geometry Properties** options unchecked. Click on **Next**. Select **Default** for the **Configuration Keyword** section. Click on **Next**. There are two default attribute fields: **OBJECTID** and **SHAPE**. You will create five new attribute fields: `BD` (**Alias**=`Destination`), `BM` (**Alias**=`Material`), `Stories` (**Alias**=`Number of stories`), `BS` (**Alias**=`State of buildings`) and `LandUseID`. For **Field Name**, in the first empty cell below **SHAPE**, type `BD`, and for **Data Type**, select **Long Integer** from the drop-down list. In the **Field Properties** section, for the **Alias**, type `Destination` and for **Allow NULL values**, select **Yes**. Leave empty the **Default Value** property. You will add the default value later on in this chapter. You should see something similar to the following screenshot:

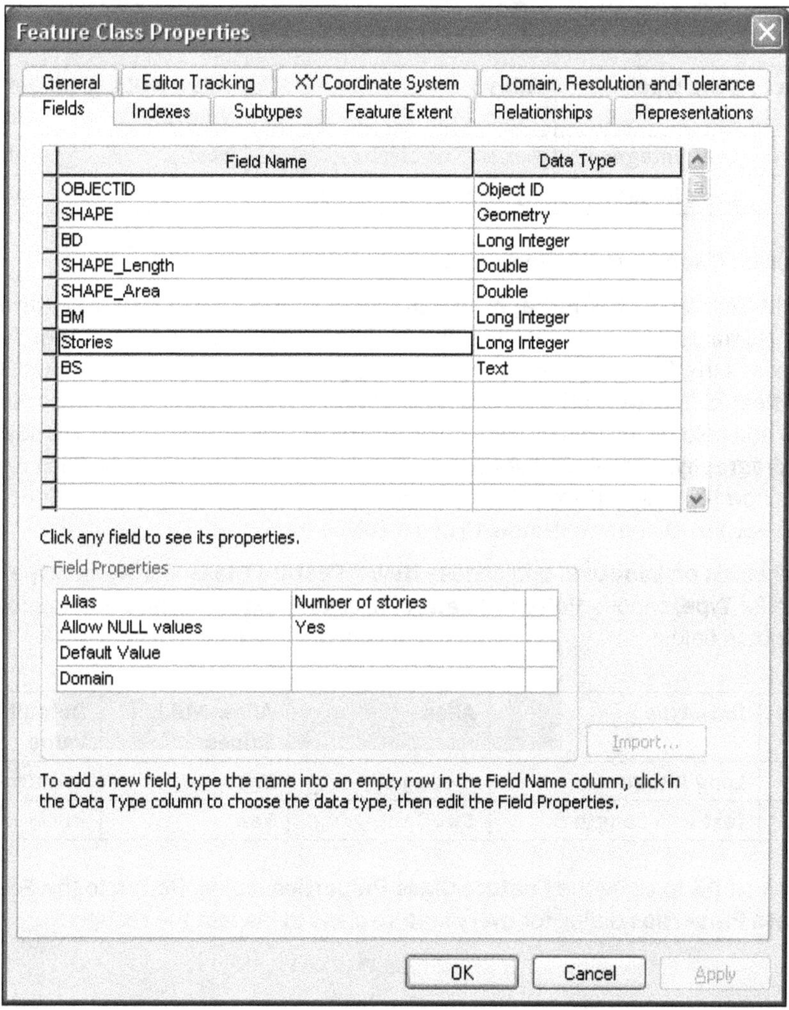

3. Repeat step 2 for `BM` (**Default Value**: `0`), `Stories`, and `LandUseID`.
4. For the `BS` attribute field, choose **Text** from the drop-down list. Notice that the **Field Properties** options have changed according to the data type. For **Alias**, type `State of buildings`, and for the **Length** property, type `4`. Click on **OK** to close the **Feature Class Properties** dialog. To inspect the results, return to the **Feature Class Properties** dialog. Right-click on the **Buildings** feature class, and navigate to **Properties** | **Fields**. Notice the two geometry attribute fields: **SHAPE_Length** and **SHAPE_Area**.
5. Right-click on the **Hydrography** feature dataset, and navigate to **New** | **Feature Class**. For **Name**, type `Watercourse`, and for **Alias**, type `Watercourse`. For **Type**, choose **Polygon Feature**. Leave the **Geometry Properties** options unchecked. Click on **Next**. Select **Default** for the **Configuration Keyword** section. Click on **Next**, and add the following attribute fields:

Name	Data type	Alias	Allow NULL values	Default Value
`HType`	**Long Integer**	`Water category`	**Yes**	`<none>`
`HYC`	**Long Integer**	`Hydrologic category`	**Yes**	`<none>`

6. Click on **OK** to close the **Feature Class Properties** dialog.
7. Right-click on the **Hydrography** feature dataset, and choose **New** | **Feature Class**. For **Name**, type `WatercourseL` and for **Alias**, type `WatercourseL`. For **Type**, choose **Line Feature**. Leave the **Geometry Properties** options unchecked. Click on **Next**. Select **Default** for the **Configuration Keyword** section. Click on **Next**. You will add field definitions by selecting the **Import** button. Navigate to **Topo5k.gdb** | **Hydrography**, and select the `Watercourse` polygon feature class. Click on **Add**. You now have two attribute fields: `HType (Water category)`, `HYC` (`Hydrologic category`), along with **Name**. Click on **OK**.
8. Right-click on **LandUse**, and choose **New** | **Feature Class**. For **Name**, type `LandUse` and for **Type**, choose **Polygon Feature**. Click on **Next** twice, and add the following attribute fields:

Name	Data type	Alias	Allow NULL values	Default Value
`CAT`	**Long Integer**	`Category`	**Yes**	`<none>`
`SCAT`	**Text** with **Length** `5`	`SubCategory`	**Yes**	`<none>`

9. Click on **OK** to close the **Feature Class Properties** dialog. Return to the **Feature Class Properties** dialog for every feature class to inspect the results.

How it works...

For every feature class, you defined different attribute fields. Because the `WatercourseL` feature class has the same attribute fields as `Watercourse`, you imported the field definitions from `Watercourse` at step 7. For all attribute fields, you have selected **Yes** for **Allow NULL values**. When you add a new feature in **ArcMap**, the attribute field will have a `NULL` value or an empty value that will be ignored by the functions in ArcMap. To enforce data integrity, you can specify that a field cannot contain empty or null values by setting **Allow NULL values** to **No**. This condition can be validated in the ArcMap edit session using the **Validate Features** option from **Editor** in **Editor toolbar**. You have another option to enforce field values by setting the **Default Value** property. A default value is automatically assigned to an attribute field when a new feature is being created in the ArcMap edit session.

There's more...

Regarding spatial reference of the feature class, if you want to move `WatercourseL` in the `LandUse` feature dataset, you simply select the feature class and drag and drop it into the `LandUse` feature dataset. You will succeed because both feature datasets have the same spatial reference. If you drag and drop the `WatercourseL` feature class into the root of the geodatabase, you will have a standalone feature class.

See also

- In the following *Creating subtypes* recipe, you will learn how to group features in a feature class based on a subtype field

Creating subtypes

Subtypes are properties of feature classes or non-spatial tables. The subtypes gather the features from a feature class or records from a non-spatial table that share the same attribute values using an attribute field. The attribute field that groups the features must be of the data type **Short** or **Long integer**, and it will be named the subtype field. A feature class can have only one subtype field. You can assign different behaviors to individual subtypes from a feature class/table. Generally speaking, the behavior is defined by the actions or characteristics of features in a geodatabase. A subtype has a code and a description. After you have defined the subtypes for a subtype field, you can change everything related to a subtype: change the value or description of code, add more codes, and delete code. The subtypes help you in the geometry editing process and prevent errors when editing feature attribute values. The subtypes maintain the integrity in a geodatabase.

Getting ready

You will continue to work with geodatabase schema by grouping features from the feature classes that you created in the previous *Creating a feature class* recipe. The integer subtype fields are the following: `BD` for `Buildings`, `HType` for `Watercourse` and `WatercourseL`, and `CAT` for `LandUse`.

How to do it...

Follow these steps to define subtypes using the context menu in ArcCatalog:

1. Start ArcCatalog. In **Catalog Tree**, go to `<drive>:\PacktPublishing\Data\MyGeodatabase\ Topo5k.gdb\Buildings`.
2. Right-click on the `Buildings` feature class, and navigate to **Properties** | **Fields**. You will define five subtypes for the `BD` attribute field. Select the **Subtypes** tab in the **Feature Class Properties** window. For **Subtype Field**, select the drop-down arrow and choose `BD`. In the **Subtypes** section, you have a default subtype with **Code** with the value `0` and **Description** with the value `New Subtype`. Change **Description** of the first subtype by typing `Unknown`.
3. Continue to create subtypes as shown in the following table:

Code	Description	Code	Description
0	Unknown	3	Industrial and municipal
1	Dwelling	4	Historical sites, monuments, statues
2	Administrative buildings and socialcultural	5	Dwelling annex

4. For **Default Subtype**, choose the **Dwelling** subtype. Click on **Apply** and on **OK**. Open the **Feature Class Properties** dialog for the **Buildings** feature class to check the subtypes you just added. Right-click on the **Buildings** feature class, and choose **Properties** | **Subtypes**. You should see something similar to the following screenshot:

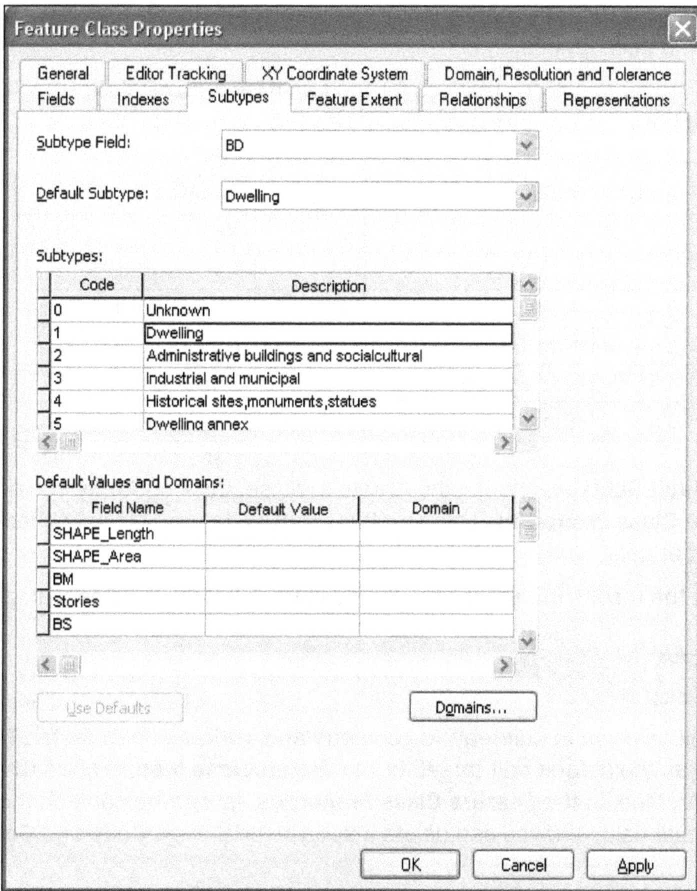

5. In **Catalog Tree**, select the **Hydrography** feature dataset. Right-click on the `Watercourse` feature class, and choose **Properties** | **Fields**. Select the **Subtypes** tab in the **Feature Class Properties** window. You will define five subtypes for the `HType` attribute field. For **Subtype Field** select the drop-down arrow, choose `HType`, and add the following subtypes:

Code	Description	Code	Description
0	Unknown	3	Creek
1	River	4	Canalized stream
2	Stream		

6. For **Default Subtype** choose the **River** subtype. Click on **Apply** and on **OK.** Open the **Feature Class Properties** dialog for the **Watercourse** feature class to check the subtypes you just added.

7. Repeat step 4 to create subtypes for the **WatercourseL** feature class using the same codes and descriptions.
8. Repeat step 4 to create subtypes for the **LandUse** feature class using the **CAT** (**Description**: `Category`) attribute field and the following subtypes:

Code	Description	Code	Description
0	Unknown	21	Forest
11	Arable	31	Hydrography
12	Pasture	41	Transportation
13	Meadow	42	Other terrains
14	Vineyard	51	Unproductive
15	Fruit orchard		

9. For **Default Subtype**, choose the **Arable** subtype. Click on **Apply** and on **OK**. Open the **Feature Class Properties** dialog for the **LandUse** feature class to check the subtypes you just added.
10. Inspect the results in **Subtypes** in the **Feature Class Properties** dialog.

How it works...

The subtypes will help you to edit feature geometry and attribute fields faster. The default subtype is used as the default edit target for the **Watercourse** feature class when you start an edit session in ArcMap. In the **Feature Class Properties** dialog, you can add more subtypes for the **BD** attribute field, and you can delete a subtype or change **Code** and **Description**.

If you delete all code, your feature class will not have subtypes anymore, and all rules and behaviors related to subtypes will be corrupt or lost for the feature class.

Sometimes, when you are establishing codes and descriptions for a subtype field, it is impossible to anticipate correctly all possible values. It is good practice to add code that covers all unforeseen values: **Code**: `42` and **Description**: `Other terrains`. Another good practice is to add code that refers to the situation in which you know the values for all other attribute fields, but you are not sure about the subtype, and it will be defined later: **Code**: `0` and **Description**: `Unknown`.

See also

- For information about how you can enforce attribute values for a feature class table, please refer to the following *Creating domains* recipe

Creating domains

The domains define the valid values for the attribute fields of feature classes or non-spatial tables. The domains are properties of a geodatabase, and obviously are stored at the geodatabase level. In a feature class/table, an attribute field can have one or many domains associated with it. The multiple domains associated with a field are based on the existing subtypes that you have already defined for a feature class. A domain can be associated with one or many attribute fields of one or more feature classes from a geodatabase.

There are two domain types: **Range** and **Coded Values**. A range domain defines minimum and maximum values and can be used only by the numeric and date field types. A coded domain can be used by the numeric, date, and text field types and defines an explicit list of valid values. Every valid value has a code and a description. A domain defines the attribute behavior when a feature is split or merged in ArcMap using the **Split policy** and the **Merge policy**.

For more details about domains, please refer to *Geodata/Data types/Domains* from ArcGIS help (10.2) online.

Getting ready

Let's continue to work with geodatabase schema by defining valid values for the attribute fields of feature classes that you created in the *Creating a feature class* recipe. You will create the domains in the geodatabase **Properties**.

In the **Database Properties** dialog, the **Domains** section has three main sections:

- **Domain Name** | **Description**: This defines domain name and description
- **Domain Properties**: This sets the domain properties, such as the data type, domain type and split/merge policy
- **Coded Values**: This defines the codes and descriptions

How to do it...

Follow these steps to create domains using the ArcCatalog context menu:

1. Start ArcCatalog. In **Catalog Tree**, go to `<drive>:\PacktPublishing\Data\MyGeodatabase\Topo5k.gdb`. Right-click on the **Topo5k.gdb** file geodatabase, and select **Properties** | **Domains**.

2. In the first cell from the **Domain Name** column, type `dBM`. For the **Description** column, type `Building material`. In the second section, **Domain Properties**, for the **Field Type** tab, choose **Long Integer** from the drop-down list. For **Domain Type**, select **Coded Values** from the drop-down list. For **Split policy**, select **Duplicate** and for **Merge policy,** leave **Default Value**. You will establish the default values later in this recipe. In the **Coded Values** section, add the following codes and descriptions:

Code	Description	Code	Description
`0`	`Unknown`	`4`	`Clay`
`1`	`Concrete`	`5`	`Mixture`
`2`	`Brick`	`6`	`Metal`
`3`	`Wood`	`100`	`Other`

3. To save changes, click on **Apply** from time to time. You should see something similar to the following screenshot:

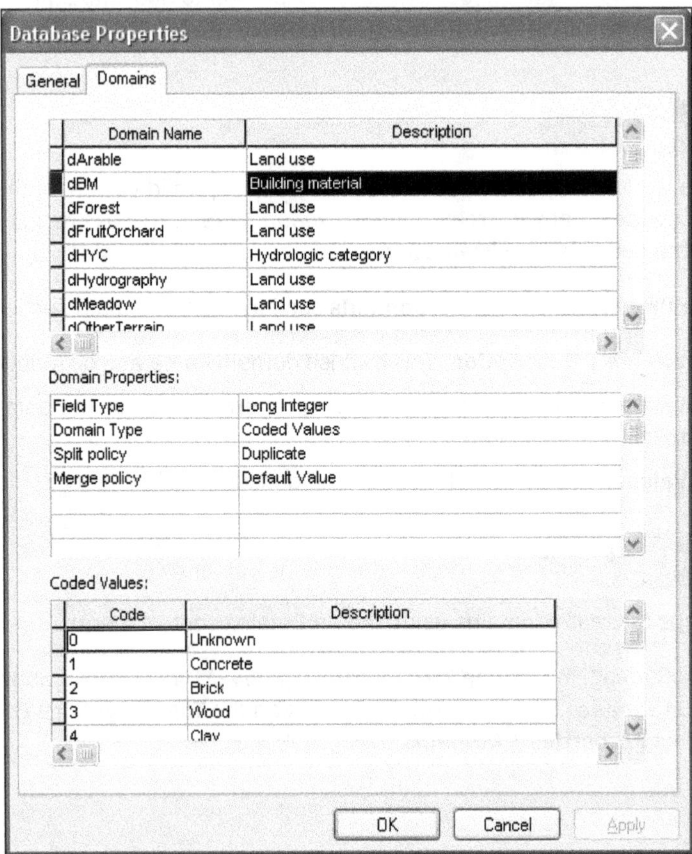

4. Create a new domain named `dStories` with **Description** of `Number of stories.` In the **Domain Properties** section, for the **Field Type** option, choose **Long Integer**. For **Domain Type**, select **Range** from the drop-down list. For **Minimum value**, type `1`, and for **Maximum value**, type `6`. For **Split policy**, select **Duplicate**, and for **Merge policy**, leave **Default Value**.

5. Create a new domain named `dState` with **Description**: `State of building.` In the **Domain Properties** section, for the **Field Type** option, choose **Text**. For **Domain Type**, select **Coded Values** from the drop-down list. For **Split policy**, select **Duplicate**, and for **Merge policy**, leave **Default Value**. You will establish the default values later in this recipe. In the **Coded Values** section, add the following code and description:

Code	Description	Code	Description
fb	Very good	r	Bad condition
b	Good	i	Insanitary
s	Satisfactory	c	Construction

6. To save changes, click on **Apply** from time to time.

7. The next domain is `dHYC`. For the **Description** column, type `Hydrologic category`. In the **Domain Properties** section, for the **Field Type** option, choose **Long Integer**. For **Domain Type**, select **Coded Values** from the drop-down list. For **Split policy**, select **Duplicate**, and for **Merge policy**, leave **Default Value**. In the **Coded Values** section, add the following code and description:

Code	Description	Code	Description
0	Unknown	2	Intermittent
1	Perennial	3	Ephemeral

8. Create a last domain `dUnkn`. For the **Description** column, type `Unknown Text`. In the **Domain Properties** section, for the **Field Type** option, choose **Text**. For **Domain Type**, select **Coded Values** from the drop-down list. For **Split policy**, select **Duplicate**, and for **Merge policy**, leave **Default Value**. In the **Coded Values** section, you will add the code and description. For the **Code** column, type `Unkn`, and for **Description**, type `Unknown`.

9. Click on **Apply** to save changes. Click on **OK** to close the **Database Properties** window.

10. Right-click on **Topo5k.gdb**, and navigate to **Properties** | **Domains** to inspect the results.

How it works...

At step 2, for **Split policy**, you selected **Duplicate**. When you split a building made of `Brick` material (domain code: `2`) during the edit session in ArcMap, you will have two buildings made from bricks. For **Merge policy**, you selected **Default Value**. Let's suppose you already established the **Default Value** option for the `BM (Building material)` field to `0`. When you merge two buildings made of `Brick` material (domain code: `2`) during the edit session in ArcMap, you will have two buildings made of `Unknown` material (domain code: `0`). The range and coded values can be validated in the ArcMap edit session using the **Validate Features** option from **Editor** present in the **Editor** toolbar. The **Validate Features** option helps you to find mistaken attribute values in the sense that it is not a valid value.

There's more...

To create a new coded value domain for a file geodatabase, based on a table that contains the defined codes and descriptions for a domain, you can use the **Table To Domain** tool from **ArcToolbox**. In `...\Data\DesigningGeodatabase\LandUseDomains` folder, you have five dBASE tables corresponding to the following domains: `dArable`, `dPasture`, `dMeadow`, `dVineyard`, and `dFruitOrchard`.

Follow these steps to create more domains for your file geodatabase using **ArcToolbox**:

1. Select **ArcToolbox** from the **Standard** toolbar. Navigate to **Data Management Tools** | **Domains**, and double-click on the **Table To Domain** tool. Set the following parameters:
   - Set the **Input Table** parameter as `...\Data\DesigningGeodatabase\LandUseDomains\Arable.dbf`
   - Set the **Code Field** option as **CodeArable**
   - Set the **Description Field** option as `Descript`
   - Set the **Input Workspace** field as `...\Data\MyGeodatabase\Topo5k.gdb`
   - Set the **Domain Name** field as `dArable`
   - Set the **Domain Description (optional)** field as `Land Use domain from table`
   - Accept default option for the **Update Option (optional)** field
2. Click on **OK** to close the **Table To Domain** dialog.
3. Open the **Database Properties** dialog for the `Topo5k.gdb` feature class to check the subtypes you just added. Repeat step 1 to add the `dPasture`, `dMeadow`, `dVineyard`, and `dFruitOrchard` domains.

See also

- In the *Using subtypes and domains together* recipe, you will learn how to combine subtypes and domains for the feature classes you just created

Using subtypes and domains together

In this recipe, you will use both subtypes and domains to better control, constrain, and partially automate the editing processes of the field values.

Getting ready

In this section, you will assign default values and domains to fields and subtype fields for the following feature classes: `Buildings`, `Watercourse`, `WatercourseL`, and `LandUse`.

The exercise tests three cases:

- A simple domain assigned to the `BM` (`Material`), `Stories` (`Number of stories`), and `BS` (`State of buildings`) attribute fields for the `Buildings` feature class.
- Two feature classes will share the same domain. The `dHYC` domain will be assigned to the `HYC` (`Hydrologic category`) attribute field for the `Watercourse` and `WatercourseL` feature classes.
- Six domains will be assigned to a single attribute field based on the subtypes of the `LandUse` feature class. The `dUnkn`, `dArable`, `dPasture`, `dMeadow`, `dVineyard`, and `dFruitOrchard` domains will assign to the `SCAT` (`SubCategory`) attribute field for the `LandUse` feature class based on the subtypes defined for the `CAT` (`Category`) subtype field.

How to do it...

Follow these steps to assign domains to attribute fields of the feature classes:

1. Start ArcCatalog. In **Catalog Tree**, go to `...\Data\MyGeodatabase\ Topo5k.gdb\Buildings`.
2. Right-click on the `Buildings` feature class, and navigate to **Properties** | **Fields**. Select the `BD` attribute field. In **Field Properties**, the **Default Value** field is `1` because **Default Subtype** is `Dwelling` with code with a value of `1`. To see the subtypes, select the **Subtypes** tab. For every subtype, you will assign three different domains to the following fields: `BM`, `Stories`, and `BS`.

3. In the **Subtypes** section, select the **Unknown** subtype by clicking the small box on the left-hand side of the row with code the value of which is `0`. The rows will become black. In the **Default Values and Domains** section, set the **Default Value** and **Domain** columns, as shown in the following screenshot:

Default Values and Domains:

Field Name	Default Value	Domain
SHAPE_Length		
SHAPE_Area		
BM	0	dBM
Stories	1	dStories
BS	b	dState

4. The default value `0` for the `BM` field corresponds to the `Unknown` description. The default value `1` for the `Stories` field corresponds to a building with one storey. The default value `b` for the `BS` field corresponds to a `Good` state of building. Click on **Apply** to save changes.
5. Select the `Dwelling` subtype. In the **Default Values and Domains** section, set the **Default Value** and **Domain** columns, as shown in the following screenshot:

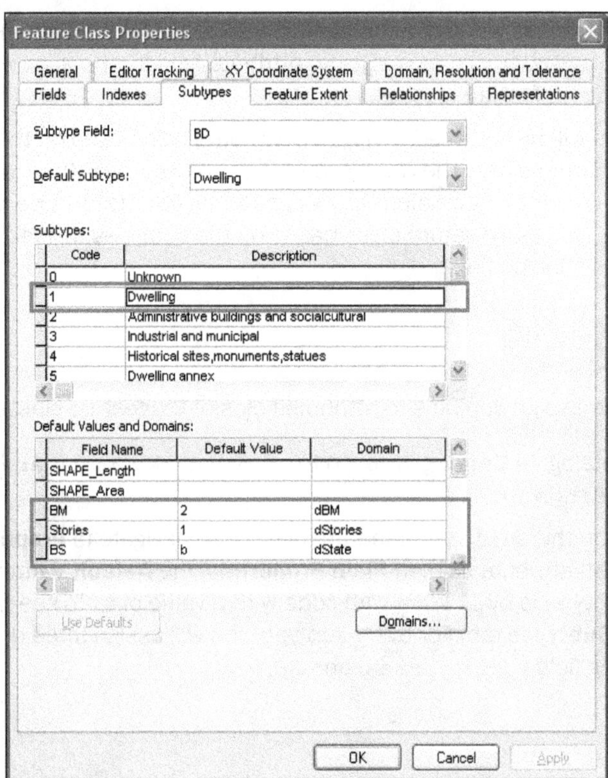

6. When you define values in the **Default Value** field, you must think what the most common value used for the `Dwelling` subtype is. For example, most of the dwellings are made of `brick` (domain code: `2`). During an edit session, when a feature is added, split, or merged, the default value will be assigned to the attribute field value.
7. Continue to assign the same domains to the next four subtypes. Click on **Apply** and on **OK** to save and close the **Feature Class Properties** dialog.
8. In **Catalog Tree**, select the `Hydrography` feature dataset. Right-click on the `Watercourse` feature class, and navigate to **Properties** | **Fields**. Select the **HType** attribute field. In **Field Properties**, the **Default Value** field is `1` because the **Default Subtype** field is `River`, where **Code** is `1`.
9. For every subtype, you will assign a single domain to the **HYC** attribute field. In the **Subtypes** section, select the **Unknown** subtype, and set the **Default Value** and **Domain** columns, as shown in the following screenshot:

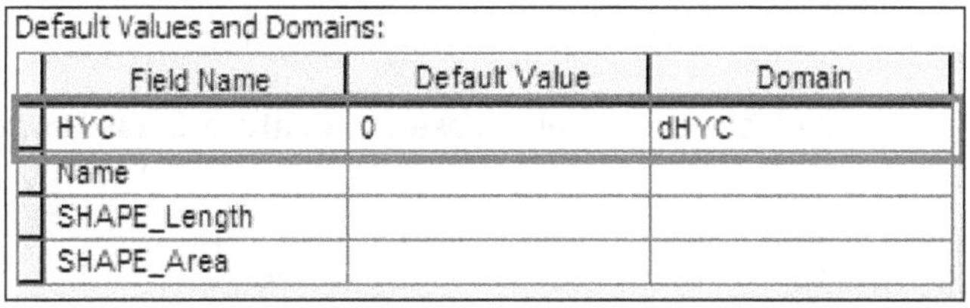

Default Values and Domains:

Field Name	Default Value	Domain
HYC	0	dHYC
Name		
SHAPE_Length		
SHAPE_Area		

10. Continue to assign values to the **Default Values and Domain** section for the rest of the subtypes, as shown in the following table:

Subtype code	Code description	Field	Default Value	Description	Domain
`1`	`River`	`HYC`	`1`	`(Perennial)`	`dHYC`
`2`	`Stream`	`HYC`	`2`	`(Intermittent)`	`dHYC`
`3`	`Creek`	`HYC`	`3`	`(Ephemeral)`	`dHYC`
`4`	`Canalized stream`	`HYC`	`1`	`(Perennial)`	`dHYC`

11. Click on **Apply** and on **OK** to save and close the **Feature Class Properties** dialog.
12. In **Catalog Tree**, select the `LandUse` feature dataset. Right-click on the `LandUse` feature class, and navigate to **Properties** | **Subtypes**. You have 10 subtypes. For every subtype, you will assign a domain and a default value for the `SCAT` field, as shown in the following table:

Subtype code	Description	Field	Default Value	Domain
0	Unknown	SCAT	Unkn	dUnkn
11	Arable		A	dArable
12	Pasture		P	dPasture
13	Meadow		F	dMeadow
14	Vineyard		V	dVineyard
15	Fruit orchard		L	dFruitorchard

13. Click on **Apply** and on **OK** to save and close the **Feature Class Properties** dialog.

How it works...

The domains have been assigned to subtypes in the **Subtypes** section in the **Feature Class Properties** dialog. If your feature class doesn't have subtypes, you will assign the domains in the **Fields** section in the **Feature Class Properties** dialog box.

At step 8, the `Arable` subtype has been assigned the `dArable` domain for the **SCAT** attribute field. A good practice is to have similar names for subtypes and related domains. To be easier to read and make a distinction between the field name, subtype name, and domain name, put a small `d`, such as `dArable`, in front of the domain name. It is intuitive and easy to combine subtypes with different domains.

At step 3, choosing the right domain was a little tricky because the field is **BS** and the domain is `dState`. Another good practice is to keep your own documentation regarding the data type, code, and description for domains and subtypes. Another aspect is the default value for domains.

At step 7, you defined the default value for `BM` (`Material`), `Brick` value. When you add a `Dwelling` feature, the default values of the `Material` field will be `Brick`. Probably this will help the editing process, but sometimes it can be an illusory value. The user will not be more preoccupied with the correctness of this value. It would be a good idea to define as the default value for the `Material` field, code `0` (`Unknown`). This value will warn the user that it should be changed to this value according to the reality. There is no perfect solution for a default value, but your experience and the context of using geodatabase will help you to make the most appropriate decision.

There's more...

In the last step, you assigned only six domains to the `LandUse` feature class only if it defined 11 subtypes. There is another way to complete the domain list–by importing a feature class from another file geodatabase. When a feature class is imported, all the domains used by the imported feature class will also be added. Follow these steps to import the `LandUse` feature class from another file geodatabase, `TOPO5000.gdb`.

1. In **Catalog Tree**, right-click on the `LandUse` feature dataset, and navigate to **Import | Feature Class (single)**. For **Input Features**, go to `...\Data\TOPO5000.gdb\LandUse`, and select the `LandUse` feature class. Click on **Add**. For **Output Feature Class**, type `LandUseImport`.
2. You will keep only two attribute fields: `CAT` (long) and `SCAT` (text). These fields are important because the first is a subtype field and the second field uses the subcategory domains for the `Landuse` feature class. You will delete all other attribute fields from the source feature class. In the **Field Map (optional)** section, select `Zone` (long), and select the **X** button on the right-hand side. Repeat the step for the remaining fields. Click on **OK** to close the **Feature Class to Feature Class** dialog.
3. Inspect the newly added domains in the **Database Properties** dialog for `Topo5k.gdb`. You now have five new domains: `dForest`, `dHydrography`, `dTransportation`, `dOtherTerrains`, and `dUnproductive`. Because you don't want to have a duplicate feature class, delete the `LandUseImport` feature class in ArcCatalog. The added domains still remain as file geodatabase properties.
4. Repeat step 12 from the *How to do it...* section to continue to add domains for the `SCAT` field, as shown in the following table:

Subtype code	Description	Field	Default Value	Domain
21	Forest	SCAT	PD	dForest
31	Hydrography		H	dHydrography
41	Transportation		D	dTransportation
42	Other terrains		Cc	dOtherTerrains
51	Unproductive		N	dUnproductive

You can find the results of this section at `...\Data\DesigningGeodatabase\Topo5k.gdb`.

See also

- In the following *Creating a relationship class* recipe, you will go a step further in defining the behavior of a file geodatabase. You will create relationships rules between different feature classes.

Creating a relationship class

Relationships describe how the spatial/nonspatial objects are linked. In a geodatabase, the relationships are stored in relationship classes. In a relationship class, you can relate only two classes (table/feature class) from a geodatabase. Just as with a feature class, a relationship class can be created at the root level of a geodatabase or within a feature dataset.

In a relationship class, you have to decide the followings properties:

- Name of the relationship class
- Origin table and destination table
- Type of relationship, which can be as follows:
    - **Simple**: The objects can exist independently of each other and can have any type of cardinality.
    - **Composite**: The objects from a destination table/feature class (child) cannot exist without the objects from an origin table/feature class (parent); define the feature following and cascade, deleting behaviors (if an origin object is moved, rotated, or deleted, the destination object suffers the same actions); it can have only one-to-one or one-to-many cardinality.
- Label for origin and destination
- Message propagation (controls relationship behavior) with specific effects on composite relationships:
    - **Forward**: This includes cascade delete, feature following, and annotation update
    - **Backward**: This includes cascade delete and no feature following
    - **Both**: This includes cascade delete, feature following, and annotation update
    - **None**: This includes cascade delete and no feature following
- Cardinality (one-to-one, one-to-many, and many-to-many)
- Add an attribute in a relationship class to describe the relationship between objects
- The origin primary key (PK) and the destination foreign key (FK)
- The primary key and foreign key columns must have the same data type.

For more information about annotation concepts in the geodatabase context, please refer to *Chapter 6, Building Better Maps*.

You cannot modify the properties of an existing relationship class. You can delete and define it once again. If you delete one class (table/feature class) that participates in a relationship class, the relationship class will be deleted too.

A table/feature class can be involved in more than one relationship class. Those multiple relationships for a table/feature class work well as long as they are coherently defined and don't generate contradictory behaviors. The last step in creating a relationship class is to define rules in order to refine the cardinality based on the subtypes of the feature classes/tables. All types of cardinality support the relationship rules. A relationship rule is a property of the relationship class.

Getting ready

In this section, you will create two simple relationship classes and one composite relationship class between the `LandUse` feature class, `BuildingsR` feature class, and `Owners` table. Those relationships compose a stacked relationship because they link three classes in an open loop, as shown in the following screenshot:

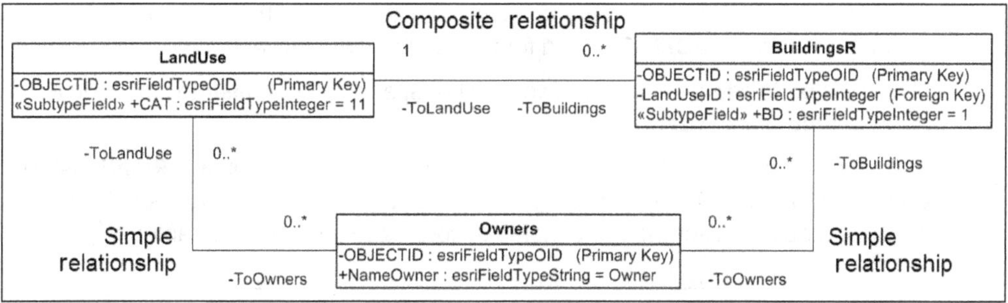

When you define the cardinality, remember that term one can mean zero. For example, in the schema shown in the preceding screenshot, the `Buildings` (origin) are related to `Owners` (destination). The cardinality is many-to-many. The relationship allows the following situations:

- Building with no owner (1..0)
- Building with different owners (1..*)
- Owner with no building (0..1)
- Owner with any number of buildings (*..1)

How to do it...

Follow these steps to create two simple, many-to-many relationship classes:

1. Start ArcCatalog. In **Catalog Tree**, navigate to `...\Data\DesigningGeodatabase\Topo5k.gdb\LandUse`.
2. Right-click on the `LandUse` feature dataset, and select **New** | **Relationship Class**. For **Name of the relationship class**, type `LandUseToOwners`. For **Origin table/feature class**, select the plus sign on the left-hand side of the `LandUse` feature dataset to see the `LandUse` feature class. Select the `LandUse` feature class. For **Destination table/feature class**, select the `Owners` table. Click on **Next**.
3. Select **Simple (peer to peer) relationship**, and click on **Next**. For the first relationship label, type `ToOwners`. For the second relationship label, type `ToLandUse`. Select the **None (no messages propagated)** option, and click on **Next**. Select the cardinality M-N (many-to-many) and click on **Next**. Select **Yes, I would like to add attributes to this relationship class**, and click on **Next**.
4. For **FieldName**, type `OwnerProcent` in the first empty row. Select **Long Integer** from the drop-down list. Select **Yes** for the **Allow NULL values** option, in the **Field Properties** section. As the primary key for the origin table, choose **OBJECTID** from the drop-down list, and as the foreign key, type `LandUseID`. As the primary key for the origin table, choose **OBJECTID**, and for foreign key, type `OwnerID`.
5. Click on **Next** to see a summary of your options, and select **Finish** to close the **New Relationship Class** wizard.
6. Right-click on the `Buildings` feature dataset, and select the **New** | **Relationship Class**. For **Name of the relationship class**, type `BuildingsRToOwners`. Repeat the previous steps, and set the relationship according to the following table:

Relationship	Value
Origin table/feature class	`BuildingsR`
Destination table/feature class	`Owners`
Type of relationship	`Simple`
Origin to destination label	`ToOwners`
Destination to origin label	`ToBuildingsR`
Message propagation	None
Cardinality	M-N (many to many)
Additional attributes	NO
Origin Table/Feature Class	Primary Key: `OBJECTID` Foreign Key: `BuildingsRID`
Destination Table/Feature Class	Primary Key: `OBJECTID` Foreign Key: `OwnerID`

Follow these steps to create a composite one-to-many relationship class between `LandUse` (as parent objects) and `BuildingsR` (as child objects):

7. Right-click on the `LandUse` feature dataset, and select **New** | **Relationship Class**. For **Name of the relationship class**, type `LandUseToBuildings`. For **Origin table/feature class**, select the `LandUse` feature class. For **Destination table/feature class**, select the plus sign on the left-hand side of the `Buildings` feature dataset, and select the `BuildingsR` feature class.
8. Click on **Next**. Select **Composite relationship (peer to peer) relationship** and click on **Next**.
9. For the first relationship label, type `ToBuildingsR`. For the second relationship label, type `ToLandUse`. Select the **Forward (origin to destination)** option and click on **Next**. Select the cardinality 1-M (one-to-many), and click on **Next**.
10. Select **NO, I do not want to add attributes to this relationship class** and click on **Next**.
11. As the primary key, choose **OBJECTID** from the drop-down list, and as the foreign key, choose `LandUseID`.
12. Click on **Next** to see a summary of your options, and select **Finish** to close the **New Relationship Class** wizard.

Follow these steps to add rules for the LandUseToBuildings relationship class:

13. In **ArcCatalog**, double-click on the `LandUseToBuildings` relationship class to open a **Relationship Class Properties** dialog box. Select the **Rules** tab to see the subtypes for the `LandUse` and `BuildingsR` feature classes. You will define relationship rules, as shown in the following table:

BuildingsR subtypes	**Cardinality (M)** `Other terrains: 42`	**Cardinality (M)** `Unproductive: 51`
Unknown	0..10	0..*
Dwelling	0..2	0
Administrative buildings and socialcultural	0..3	0
Industrial and municipal	0..10	0..2
Historical sites, monuments, statues	0..2	0..*
Dwelling annex	0..3	0..1

14. For **Origin Table/Feature class subtypes**, select `Other terrains`. For **Destination Table/Feature class subtypes**, click on the **Code** checkbox on the left-hand side of the `Unknown` subtype. Click on `Unknown` to select the subtype and to enable the **Origin Cardinality** and **Destination Cardinality** sections.

15. In the **Destination Cardinality** section, check the **Specify the range of associated destination objects** option. First, set the **Max** value by typing `10`. Leave the **Min** value at `0`.
16. For **Destination Table/Feature class subtypes**, click on the **Code** checkbox to the left of the `Dwelling` subtype. Click on `Dwelling` to select subtype and to enable the **Origin Cardinality** and **Destination Cardinality** sections.
17. In the **Destination Cardinality** section, check the **Specify the range of associated destination objects** option. First, set the **Max** value by typing `2`. Leave the **Min** value at `0` according to the previous table. The first rule says that an `Other terrains` parcel subtype can have a maximum of two dwellings. You should see something similar to the following screenshot:

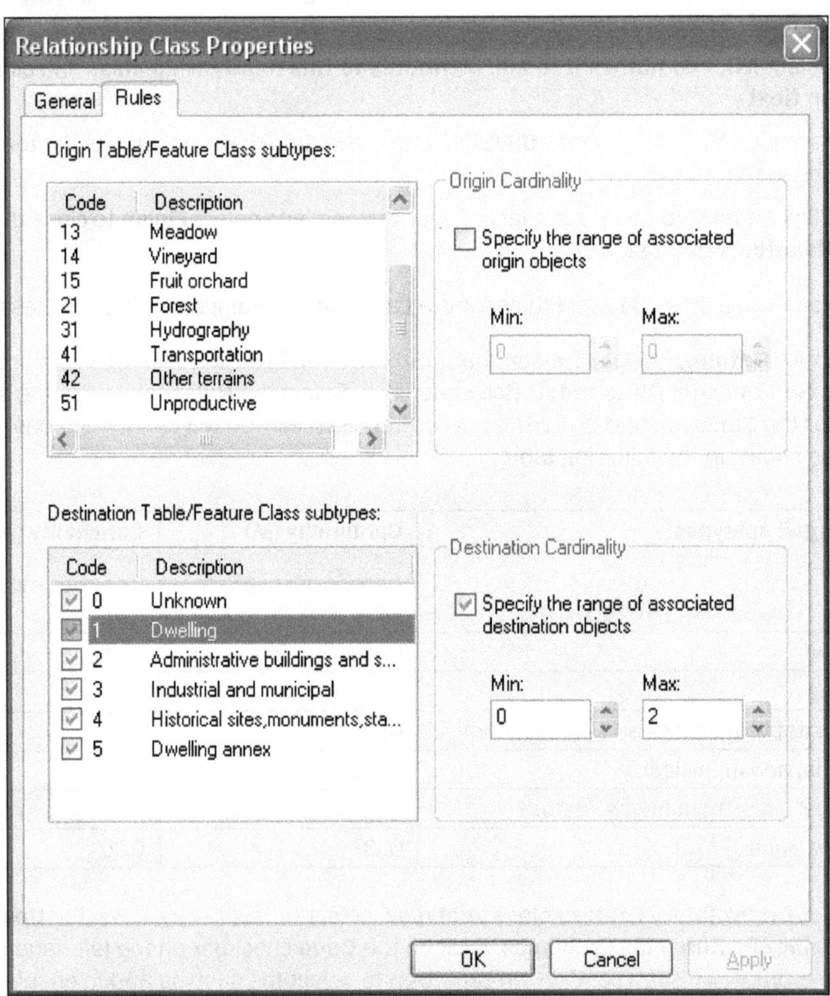

18. Select the **Administrative buildings subtype**, and check the **Specify the range of associated destination objects** option. For the **Max** value, type `3`. For the **Min** value, type `0`. This rule says that there must be between zero and three administrative buildings in a parcel (`Other terrains`).
19. Continue to define the rules in **Destination Table/Feature class subtypes** for all building subtypes. For **Origin Table/Feature class subtypes**, select `Unproductive`. Leave unchecked the `Unknown` subtypes in the **Destination Table/Feature class subtypes** section to allow an unlimited number of `Unknown` buildings.
20. You will permit a maximum of two `Industrial and municipal` buildings and one `Dwelling annex` per `Unproductive` parcel. Select those subtypes from the **Destination Table/Feature class subtypes** section, and set the maximum value according to the previous table.
21. Finally, you will not permit a relationship with the rest of building types, such as `Dwelling`, `Administrative buildings`, and `Historical sites`. Select all those subtypes from the **Destination Table/Feature class subtypes** section, check the **Specify the range of associated destination objects** option, and set the maximum values to `0`.
22. For the rest of the `LandUse` subtypes from **Origin Table/Feature class subtypes**, set the **Max** and **Min** values to `0`. This will not permit `BuildingsR` to associate with parcel subtypes, such as `Arable`, `Pasture`, `Meadow`, `Vineyard`, `Fruit orchard`, `Forest`, `Hydrography`, and `Transportation`.
23. Click on **Apply** and on **OK** to save and close the **Relationship Class Properties** wizard.

You can find the final geodatabase schema at `...\Data\DesigningGeodatabase\MyGeodatabaseResults\Topo5k.gdb`.

How it works...

In the previous exercise, you used `OBJECTID` as the origin primary key field. But you could define your own primary key. The `OBJECTID` field guarantees a unique value for each record because it is managed by ArcGIS. You can't modify the `OBJECTID` values, and this can be inconvenient to maintaining the relationships when you are modifying feature geometry (for example: split, merge) or import features to another feature class.

For more details about the advantages and inconvenience of using the `OBJECTID` field as the primary key, please refer to `Geodata/Data types/Relationships and related objects/Relationship class properties` from ArcGIS Help (10.2) online.

There is a performance cost when you are working with relationship classes. The relationship classes will slow down the edit process because they must maintain the referential integrity in the geodatabase. A relationship class will assure that all changes you have made in the origin feature class/table will be reflected in the destination feature class/table according to the type and rules of the relationship.

At step 3, you created a many-to-many, simple relationship class between the `BuildingsR` feature class and the `Owners` table. A many-to-many relationship class requires an intermediary table that contains two key fields: origin primary key and destination foreign key. Both fields have a foreign key's role. You can see this table in ArcCatalog if you select the `LandUseToOwners` relationship class table. Select the **Preview** tab and the **Table** preview mode from the bottom of the **Preview** panel to examine the attribute fields.

There's more...

Follow these steps to test the relationship behavior:

1. Start ArcMap, and open an existing map document `WorkingRelationships.mxd` from `...\Data\DesigningGeodatabase`. In **Table Of Contents**, you have two layers (`BuildingsR` and `LandUse`), two intermediate tables (`LandUseToOwners` and `BuildingsRToOwners`), and one non-spatial table, `Owners`.
2. Let's validate the relationship rules for the `LandUseToBuildings` composite relationship. Zoom in to the full extent of a map by selecting **Full Extent** in the **Standard** toolbar. Relationships are active and can be validated during an edit session.
3. In the **Editor** toolbar, select **Start Editing**. Select all features from `BuildingsR` and `LandUse` using the **Select Features** tool.
4. In the **Editor** toolbar, navigate to **Editor** | **Validate Features**. A message box will appear: **72 features are invalid**. Click on **OK** to close. All invalid features are selected.
5. Open the **Attributes** window by selecting the **Attributes** button in the **Editor** toolbar. You will see two layers: `BuildingsR` and `LandUse`. Select **Expand All Relationships In Branch** to see all selected features (blue bullets), related tables, and table records.

6. Let's inspect a relationship error. Click on **Bookmarks** and choose **Error2**. With the **Select Features** tool, select the **Others terrains** parcel with **LandUseID=1262**. In the **Attributes** window, expand the `LandUse` layer. You should see something similar to the following screenshot:

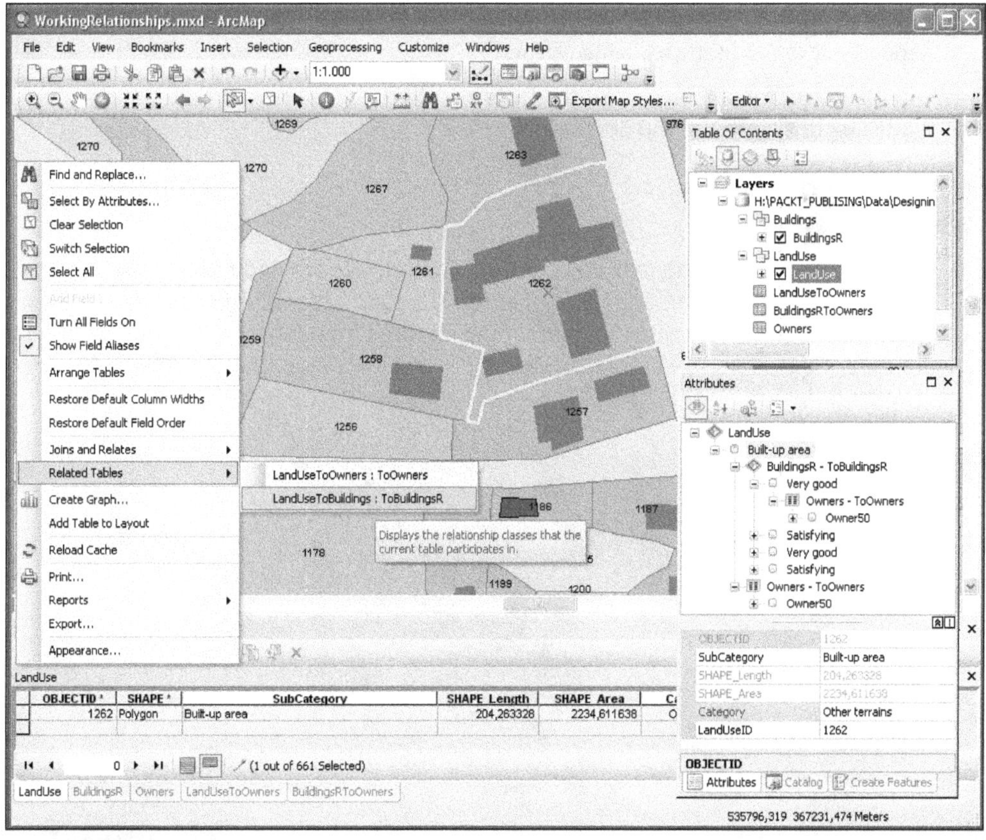

7. Right-click on `LandUse`, and navigate to **Selection** | **Open Table Showing Selected Feature** to see the `LandUse` attribute table in the **Show selected records** mode. Navigate to **Table Options** | **Related Tables**, and choose **LandUseToBuildingsR: ToBuildingsR**. In the **Editor** toolbar, navigate to **Editor** | **Validate Features**. A message box will appear: **The feature LandUse(subtype:Other terrains) has 4 related BuildingsR (subtype:Dwelling) features, thus violating the LandUseToBuildings relationships rule that specifies 0-2**. Click on **OK** to close.
8. To fix the error, delete two buildings. Select the parcel and the remaining buildings and validate features once again. The message box will tell you: **All features are valid.**

9. Let's check the **feature following** and **cascade deleting behaviors** of the `LandUseToBuildings` composite relationship. With the **Select Features** tool, select the `Other terrains` parcel with **LandUseID=1262**. In the **Editor** toolbar, select **Edit Tool**. Drag the parcel away from the previous location. The related buildings should follow the parcel. Select **Rotate Tool**. Rotate the parcel. The building should rotate, too. Delete the parcel using the *Delete* key. The buildings are deleted, too. Restore the features using *Ctrl* + *Z*.

10. In the **Editor** toolbar, select **Editor | Stop Editing**, and save your edit. Select **File | Save as** to save your map document.

See also

- To learn how to edit the features stored in feature classes and how to maintain the spatial relationships among them, please refer to *Chapter 2, Editing Data*. To learn how to create visual representation of the features from different feature classes on a map, please refer to *Chapter 5, Working with Symbology*, and *Chapter 6, Building Better Maps*.

2
Editing Data

In this chapter, we will cover the following topics:

- Editing features in a geodatabase
- Advanced editing in a geodatabase
- Creating geodatabase topology
- Editing geodatabase topology

Introduction

In the first part of this chapter, you will test different ArcGIS tools in order to add different types of feature geometry and their attribute values.

In *Chapter 1, Designing Geodatabase*, you constrained the attribute values using subtypes and domains. With the **Validation** option in an edit session, you can identify the invalid values in order to correct them.

You will continue to use the concept of validation in the feature geometry context in the second part of this chapter. You will constrain and administrate the spatial relationships between features with geodatabase topology. With the geodatabase topology, you will identify and fix the spatial errors from your `Topo5k.gdb` file geodatabase.

Editing features in a geodatabase

In ArcMap, a layer references the spatial datasets (for example, feature class and raster) that contain features with common geometry and attributes. A layer uses the dataset attributes to represent the features by one or more symbols. By default, a layer symbolizes features from a feature class using the subtype field.

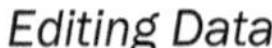

The editing session in ArcGIS for Desktop is based on layers. When you start an edit session, you have to define the **Feature Template** that describes the type of feature that you intend to create. ArcGIS creates feature templates by default, based on the layer symbology. In the **Creating Feature** window, you can see the list of feature templates. You can create new types of features with the **Organize Feature Templates** option. Moreover, you can define a default attribute for one type of feature or for all types of features from a layer. When you create a new feature, the default attribute is automatically inherited. Choosing a default value works for subtypes and domains, too.

After you create a feature, you should edit its attributes using the **Attribute** window. Most of the attribute fields will automatically get completed according to the default subtype and domains you already defined in the *Using subtypes and domains together* recipe of *Chapter 1, Designing Geodatabase*.

Getting ready

In *Chapter 1, Designing Geodatabase*, you learned how to create the `Topo5k.gdb` file geodatabase schema. Now, it's time to add data. There are several ways to add data into a geodatabase:

- Using the **Import/Export** and **Load Data** options in ArcCatalog
- Using the **Load Objects** option in an ArcMap environment
- Creating new features by screen digitizing and by editing existing ones
- Using the **Export Data** option in an ArcMap environment

In this recipe, you will work with the **Editor**, **Snapping**, and **Edit Vertices** toolbars to create new features and to edit the existing features. The `LandUse` and `BuildingsR` feature classes already contain features. The features were digitized based on a scanned topographic map at scale `1:5,000` produced in 1995.

As a base layer for updating features, you will use an orthophotography from 2012 corresponding to a map at scale `1:5,000`. The orthophoto map has a resolution of *1 pixel =0.5 meter* and is stored as **Raster Dataset** in the `Topo5k.gdb` geodatabase.

Even if you assume that you are already comfortable with the elementary edit process in ArcGIS 10, please remember the main steps:

1. Start the edit session.
2. Set the snap environment (optional).
3. Before starting a sketch, choose a feature template and a construction tool.
4. Create the new feature.

5. Add or modify attribute values of the feature.
6. Save edits.
7. Stop the edit session.

How to do it...

You can follow these steps to edit features in a geodatabase:

1. Start ArcMap, and open an existing map document `Editing.mxd` from `<drive>:\PacktPublishing\Data\EditingData`. Notice that layers from the **Table Of Contents** section are symbolized based on their subtypes. The `LandUse` layer has defined a transparency. Right-click on the `LandUse` layer, and navigate to **Properties** | **Display**. You will notice that **Transparent** has a `60` percent value. Let's add a pop-up label with the **OBJECTID** value in the data view. Check **Show MapTips using the display expression**, and choose the desired field. Click on **OK**.
2. From the **Bookmarks** menu, select **Create a Fish Pond**. You will create a new feature in the `LandUse` layer, as shown in the following screenshot:

3. In the **Table Of Contents** section, right-click on the `LandUse` layer, and navigate to **Edit Features** | **Start Editing**. In the **Editor** toolbar, navigate to **Editor** | **Editing Windows**. In the **Create Feature** window, you will see the feature templates for all layers that are visible in the **Table Of Contents** section.
4. In the **Editor** toolbar, navigate to **Editor** | **Snapping**, and check the **Snapping** toolbar. In the **Snapping** toolbar, select only **Vertex Snapping**. Snapping options assure the coincidence of new feature coordinates with the coordinates of existing features.
5. In the **Editor** toolbar, navigate to **Editor** | **Options** | **Attributes**. Navigate to **Display the attributes dialog before storing new features** | **For the following layers options** | **LandUse**. Click on **OK**. Every time you create a new `LandUse` feature, the **Attribute** window will show up after you have finished the sketch.
6. Select the feature with **OBJECTID** as `755` from the `LandUse` layer. In the **Editor** toolbar, select the **Cut Polygons Tool** option to cut the arable polygon.
7. Select the **Trace** tool. Right-click on the polygon, and select **Trace Options**. In the **Trace Options** dialog, type `5` as the **Offset** value, select the **Trace selected features** option, and click on **OK**.
8. Use the screenshot from step 2 to follow the edit steps. Start the sketch from point **(1)** and finish it at point **(2)**. Right-click on point **(2)** to choose a value for the **Length** field, type in `84`, and close the window. Move the segment towards point **(3)**, and add the end of the curve to point **(4)**.

While you are sketching, press *Tab* to display the **Feature Construction** toolbar.

9. Again, go to point **(2)**, and right-click on it to choose a value for the **Length** field, type in `61`, and close the window. Move the segment to point **(5)**, and add the last vertex to point **(6)**. To finish the sketch, use the *F2* shortcut key.

Next, you will edit the attribute fields for the new polygon feature, as shown in the following screenshot:

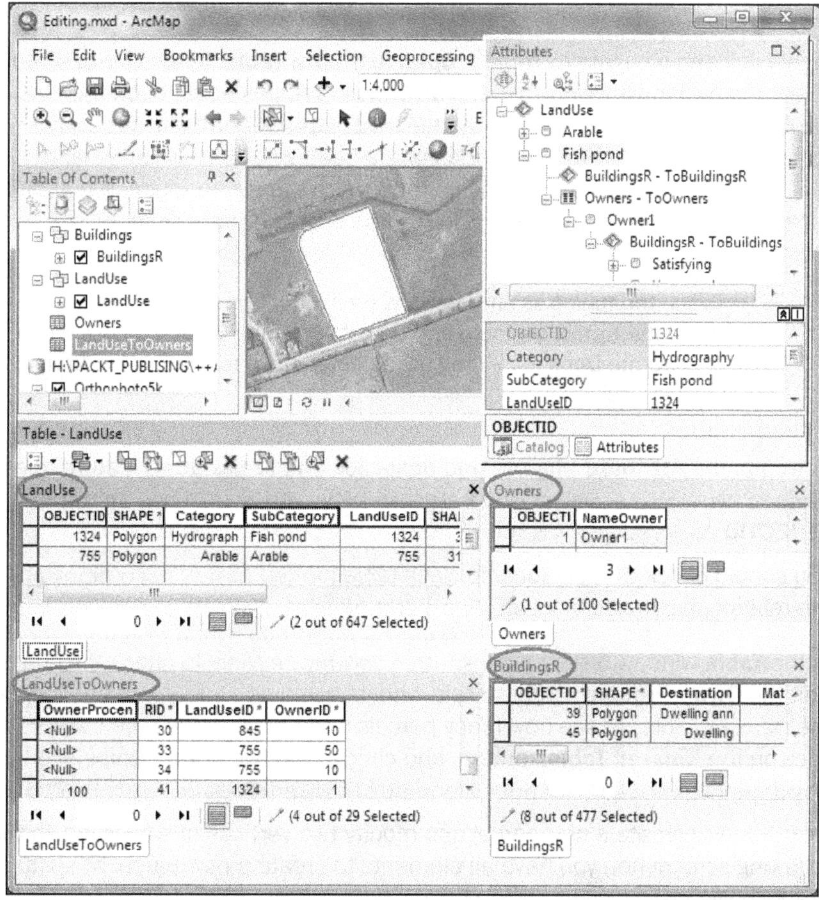

10. In the **Attributes** window that automatically opens, select the new `LandUse` feature you just created. The default value for **Category (CAT)** and **SubCategory (SCAT)** is `Arable`. You defined the default values in the *Using subtypes and domains together* recipe of *Chapter 1, Designing Geodatabase*.
11. Let's change the attribute values. Select the **Category** field, and click on the small button from the right to open the **Choose Symbol Class** window. Select the `Hydrography` value, and click on **OK**. For the **SubCategory** field, select the `Fish pond` value from the drop-down list. For **LandUseID**, type the value of **OBJECTID**. Click on **OK** to close the window.

12. Return to the **Attribute** window, and expand the nodes for the `Fish Pond` feature to see the relationships. The new `LandUse` feature is related to the `BuildingsR` layer (`BuildingsR-ToBuildingsR`) and `Owners` table (`Owners-ToOwners`). Select the `Owners-ToOwners` relationship to define the owners of the parcel.
13. Right-click and navigate to **Table** | **Open Attribute Table**. In the `Owners` table, select the owner with `NameOwners` as `Owner1`. At this point, open the attribute tables for the `LandUse` and `BuildingsR` feature classes and for the `LandUseToOwners` relationship. To see `LandUseToOwners` and `BuildingsRToOwners` relationship tables and the `Owners` nonspatial table, click on the **List By Source** button in the **Table Of Contents** section.

To see the tables as shown in the preceding screenshot, click on the tab of the `LandUse` table and drag it within the **Table** window. When you see the **Dock** icon, drag the selected table to the right or down docked position.

14. Return to the **Attribute** window, and right-click on the `LandUse` feature relationship: `Owners-ToOwners`. Select **Add Selected**. Now, your parcel has an owner with **OBJECTID** as `1`.

    You should check the relationship class table named `LandUseToOwners`; it stores the relationship between `Fish Pond` and `Owner1`.

15. In the **Table** window, select the `LandUseToOwners` table to check the changes. Edit the **OwnerProcent** attribute field with its value as `100`. Select and inspect the `Owners` table. To see how many parcels and buildings own the `Owner1` table, click on the **Related Tables** button, and choose the two relationships one by one: `LandUseToOwners:ToLandUse` and `BuildingsRToOwners:ToBuildingsR`.

    Next, you will create a new parcel and modify two vertices of a dwelling. In the following screenshot, you have all elements to create a new parcel by splitting the arable polygon:

16. Click on **Bookmarks**, and select **Cut a Parcel and modify a Building**.
17. Select the parcel with **OBJECTID** as `755` from the `LandUse` layer.
18. First, go to the **Snapping** toolbox and keep only the **Vertex Snapping** option selected.
19. Second, select the **Cut Polygons Tool** option. Start the sketch from point **(1)** and select the **Distance-Distance** tool. Click on point **(1)** as a center for the first circle, and press the *D* key to type `77` for **Distance**. Click on point **(2)** as a second center, and press the *D* key to type `139` for **Distance.** You have obtained a point as an intersection of two circles. Click on the first intersection to obtain the second vertex of the cutting polyline.
20. Again, select the **Distance-Distance** tool. Click on point **(2)** as a center for the first circle, and press the *D* key to type `128` for **Distance**. Click on point **(3)** as a second center, and press the *D* key to type `135` for **Distance**. Click on the second intersection to obtain the third vertex of the cutting polyline.
21. Select **Straight Segment**, and click on point **(2)**. To finish the cutting sketch, press the *F2* shortcut key.
22. In the **Attributes** window, select the following values: **Category** as `Other terrains`, **SubCategory** as `Built-up area`, and for **LandUseID**, type the value of **OBJECTID**. Click on **OK**.

23. The new `LandUse` feature is related with the `BuildingsR` layer (`BuildingsR-ToBuildingsR`) and the `Owners` table (`Owners-ToOwners`). Try to manage those relationships by yourself. How to start? Firstly, with *Shift* pressed, select the building with the **Select Features** tool. Secondly, select `BuildingsR-ToBuildingsR`, and right-click on it to choose the **Add Selected** option.

    Next, you will correct the position of two vertices of the building according to the orthophoto map used as a background:

24. Keep the building selected and with the *Shift* key pressed, unselect the parcel. Click on **Edit Vertices**, and from the toolbar, choose **Modify Sketch Vertices**. Select the two vertices by drawing a small box around the vertices. Drag the selected segment, and drop it as shown in the preceding screenshot.

    Finally, you will correct a neighboring `LandUse` parcel. You have two options, as shown in the following screenshot:

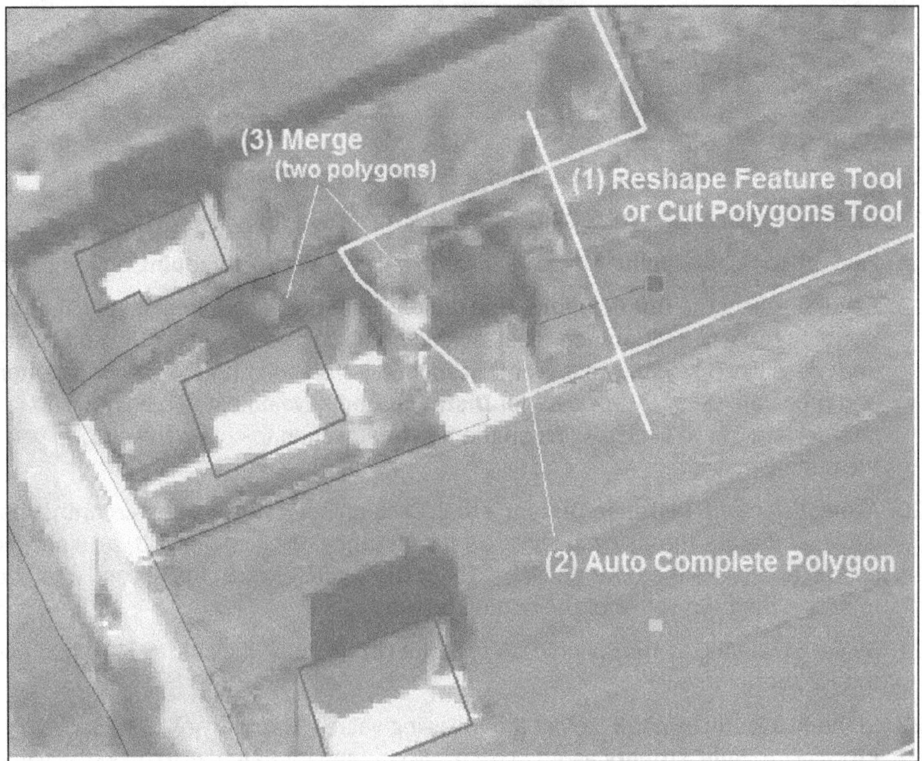

25. Use the **Reshape Feature Tool** option to draw a line as point **(1)** from the preceding screenshot in order to cut the polygon with **OBJECTID** as `755` from the `LandUse` layer. You have created a gap between the parcels. To modify the parcel from the left-hand side, select the **Other Terrains** option with the **Auto Complete Polygon** construction tool from the **Create Feature** window. Put the first vertex in the gap and the second in the neighboring parcel—see point **(2)**. With the **Select Features** tool, select both the parcels, and from the **Editor** toolbar, select **Merge**. In the **Merge** dialog, select **Build-up area (LandUse)**, and click on **OK**.
26. Use the **Cut Polygons Tool** option to split the polygon. With the **Select Features** tool, select the two parcels from point **(3)**. From the **Editor** toolbar, select **Merge**. In the **Merge** dialog, select **Build-up area (LandUse)**, and click on **OK**.
27. In the **Editor** toolbar, navigate to **Editor** | **Stop Editing**, and save your edit. Navigate to **File** | **Save As**, and save your map as `MyEditing.mxd`.

You can find the final results at `<drive>:\PacktPublishing\Data\EditingData\MyEditingResults\Editing.mxd`.

How it works...

To select an existing parcel from the `LandUse` layer, and in order to edit it, first use the **Select Features** tool, and after that, select the specific edit tools from the **Editor** toolbar. If you use an **Edit** tool to select a feature, there is always a risk that you will create a subtle movement of the selected feature. A small shift will create overlaps and gaps between the coincident features. To protect features from this small shift, navigate to **Editor** | **Options** | **General** | **Sticky move tolerance**, and set a minimum distance for cursor movement before the selected feature will be moved too.

At step 23, we tested the multiple vertex editing. We have selected two vertices and moved them at once.

You can change the geometry of a segment by selecting two vertices. Right-click on the segment, select **Change Segment**, and choose one of the options: **Bezier** or **Circular Arc**.

You can also delete multiple vertices at the same time.

In the *Creating a relationship class* recipe of *Chapter 1, Designing Geodatabase*, we created relationship rules between the `LandUse`, `BuildingsR`, and `Owners` tables. We have already noticed at step 12 that relationships are active and must be updated every time we add new features. The relationship rules can be validated during the edit session. Select all features from `BuildingsR` and `LandUse` using the **Select Features** tool. In the **Editor** toolbar, navigate to **Editor | Validate Features**. A message box will appear stating **72 features are invalid**. Click on **OK** to close. All invalid features are selected. In the **Attributes** window, inspect and try to correct the errors yourself.

There's more...

To add more data to your geodatabase schema, we will test two more options: **Load Data** using ArcCatalog and **Load Objects** in the ArcMap environment. Follow these steps to add the data in the geodatabase file using **Load data**:

1. Start ArcMap, and open the existing `Editing.mxd` map document from `<drive>:\PacktPublishing\Data\EditingData`.
2. Open the **Catalog** window from the **Standard** toolbar to have access to your data. In the **Catalog** window, right-click on `WatercourseL` from `...\EditingData\Topo5k.gdb`, and navigate to **Load | Load Data**. Set the following parameters from panels:
    - The **Input data** parameter as the `...\Data\TOPO5000.gdb\Hydrography\ WatercourseL` feature class; click on **Add**
    - Check **I do not want to load all features into a subtype**
    - **Target Field** (destination) and **Matching Source Field** (source):

      ```
      HYC[int] = HIC[int]
      Name[string] = Name[string]
      HType[int] = HAT[int]
      ```

    - Check **Load all of the source data**
3. Click on **Next** and **Finish**. In the **Table Of Contents** section, right-click on `WatercourseL`, and choose **Zoom To Layer** to see the polyline features.

   Continue to follow the steps to populate the `Watercourse` feature class with features from the **LandUse** feature class using **Load Objects** in ArcMap.

4. To add the **Load Objects** tool, click on the drop-down arrow from the right-hand side of the **Editor** toolbar, and select **Customize**. Select **Commands**, and for **Show commands containing**, type `load objects`. From **Commands**, click and drag the **Load Objects** option to the **Editor** toolbar. Click on **Close**. To use this tool, you need to start the editing session.

5. Before starting an edit session, in the **Table Of Contents** section, right-click on `Watercourse`, and navigate to **Edit Features | Organize Feature Templates**. You will define a default value for the **Name** field. All the new features from `Watercourse` that you add in this edit session will have the **Name** field as `River Flake`. In the **Organize Feature Templates** window, select `Watercourse` from the **Layers** section. With the *Ctrl* key pressed, select the **River** and **Stream** templates. Click on **Properties**, and **for Water Name**, type `River Flake`. Click on **OK** and then on **Close**.
6. Start the edit session from the **Editor** toolbar by navigating to **Editor | Start Editing**.
7. In the **Table Of Contents** section, right-click on `Watercourse`, and select **Load Objects**. Set the following parameters from panels:
   - The **Input Data** parameter as the `...\Data\EditingData\Topo5k.gdb\LandUse\LandUse` feature class; click on **Add**
   - The **Target** field as the `Stream` subtype
   - Accept the default matching **Target Field** (destination) and **Matching Source** (source)
   - Check **Load only the features that satisfy a query**
   - Click on **Query Builder** and build the following expression:

     ```
     CAT=31
     ```

   - Accept all the other default parameters
8. Click on **Finish**. In the **Table Of Contents** section, right-click on `Watercourse`, and choose **Zoom To Layer**.
9. Select all vectors using the **Select Features** tool from the **Tools** toolbar and open the **Attributes** window. All nine `Stream` features have the same attribute value for the **HYC** field as `Intermittent`. This is because in the *Using subtypes and domains together* recipe of *Chapter 1, Designing Geodatabase*, we have assigned the domain code 2 (`Intermittent`) as the default value of the **HYC** field for the `Stream` subtype features.

Let's suppose that we mistakenly chose to load all objects in the **Stream** subtype. To change the subtype value to `River` for all the nine features at the same time, select the `Watercourse` layer in the **Attributes** window. Click on the icon from the left `Stream` value to open the **Choose Symbol Class** window. Select the `River` value, and click on **OK**. Read carefully the warning message, and select **Yes** to automatically assign the corresponding `Perennial` default values for **HYC**. The **Name** field already has the `River Flake` value. Notice that every feature has the same **HType** and **HYC** and the **Name** value. In the **Editor** toolbar, navigate to **Editor | Stop Editing**, and save your edit. Navigate to **File | Save As**, and save your map as `Editing.mxd`. You can find the final results in `<drive>:\PacktPublishing\Data\EditingData\ MyEditingResults\Topo5k.gdb`.

Recall that you constrained the attribute values by subtypes and domains in *Chapter 1, Designing Geodatabase*. Every time you load features from other sources, such as existing feature classes or shapefiles, don't forget to validate the attribute values during the edit session. The **Validate Features** option will identify all the invalid values for the attribute fields for which you have already defined the subtypes and domains. All features that have invalid values will be selected. You should inspect and correct the reported errors with the **Attributes** window.

See also

- In the next recipe, *Advanced editing in a geodatabase*, you will continue to work with advanced editing tools and with **COordinate GeOmetry** (**COGO**)

Advanced editing in a geodatabase

The **Advanced Editing** toolbar offers tools to reshape and replace geometry, correct the overshoot or undershoot errors, split line features in intersection, divide a line or curve based on specified lengths or percentages, and edit multipart features.

The **COGO** toolbar allows you to work with **COoordinate GeOmetry**. Coordinate geometry refers to directions or bearings, angles, and horizontal distances that generally came from a topographic survey and that describe a series of consecutive lines. Those lines can be straight lines or simple curves and are named as **courses** or **2-point line**. A course has only two vertices. The survey measurements that describe the courses can be stored in a traverse file in the context of ArcGIS. From the topography point of view, a traverse is a classical method for a detail topographic survey of an area. In ArcGIS, in a closed traverse, the consecutive lines start from a known point and end at the same or another known point. In an open traverse, the consecutive lines start from a known point and do not end at a known point. By known point or control point, we mean a fixed (ground) point of known *X* and *Y* coordinates.

With the **Create COGO Fields**, **Update COGO Attributes**, and **Construct 2-Point Line** tools from the **COGO** toolbar, you can add, calculate, and update the COGO attributes for an existing feature class.

For theory and worked examples about azimuth, bearings, horizontal distances, types of curves, and simple two-dimensional traverse computation, please refer to:

Principles of Geospatial Surveying, Artur L.Allan, Whittles Publishing, 2007, Technical Procedures, pp.7-18, Survey Computations, pp.145-150, and *Construction and Curves, pp.240-244.*

Getting ready

In this recipe, we will work with the `RoadL` layer to create new features using the **Midpoint** tool and to edit the existing features using the **Line Intersection** and **Extent** tools. You will add new attribute fields to the `RoadL` layer using the **Create COGO Fields** tool in ArcCatalog. Finally, you will update some parcels from the `LandUse` layer based on the coordinate geometry coming from a computed traverse. You will use a 2D closed-loop traverse. The line segments are described by direction, which is expressed relative to **North Azimuth**, and by horizontal distance, as shown in the following screenshot:

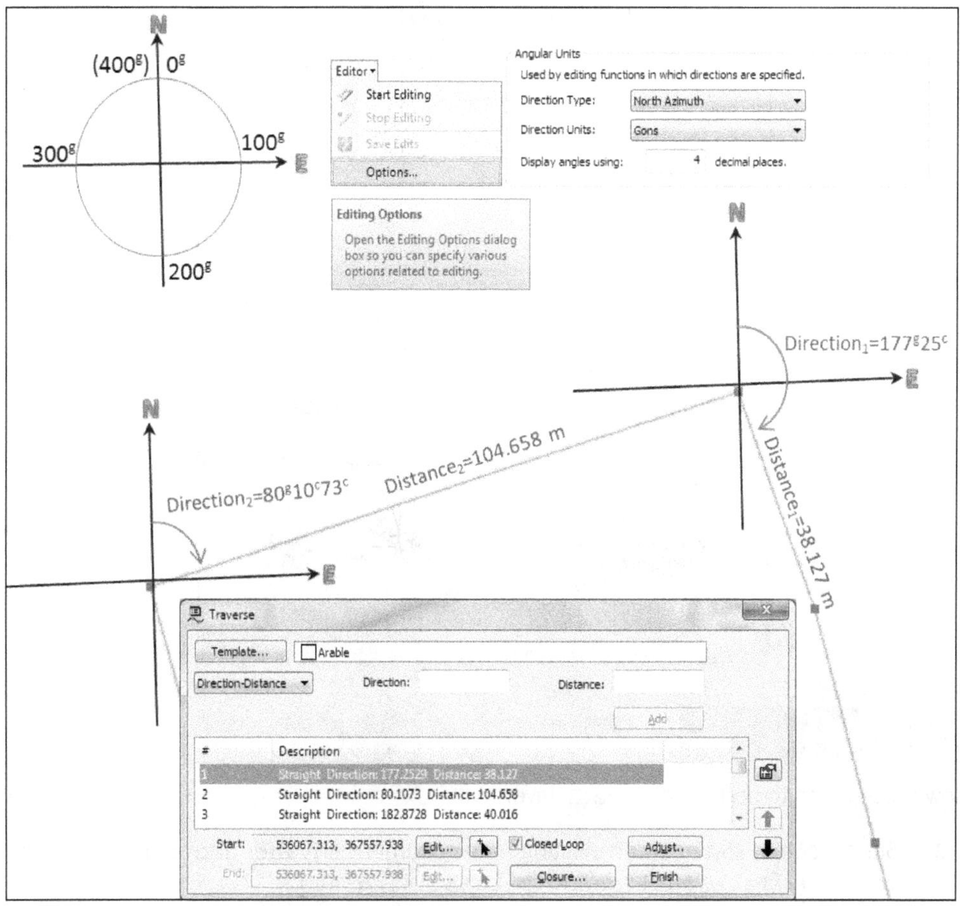

The curves from your traverse are circular curves with constant radius. In the context of ArcGIS, curves are defined by arc, radius, chord direction, and side (**Curve 6**) or by arc, radius, and side when you already know the start point and direction for a tangent and don't need to specify the chord direction and tangent (**Curve 10**). You can calculate the rest of the elements that describe a simple curve, such as **Angle**, **Chord Distance**, or **Tangent Length**, using the **Curve Calculator** from the **COGO** toolbar, as shown in the following screenshot:

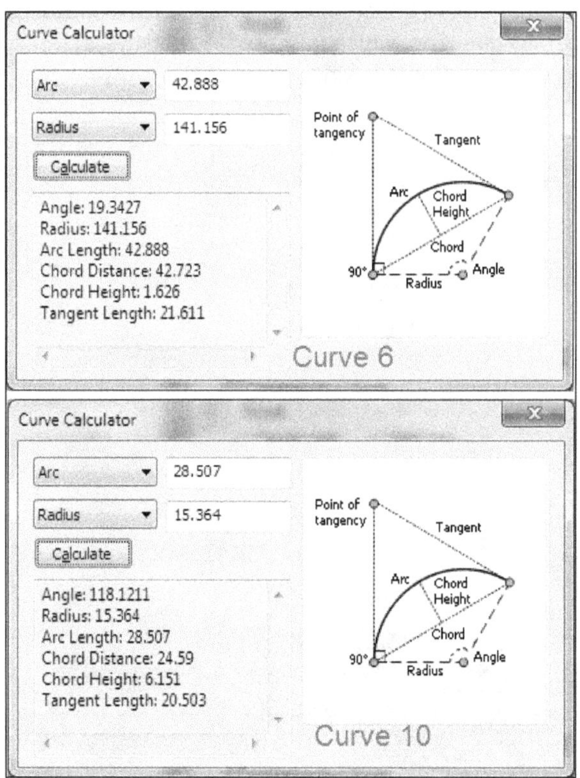

How to do it...

Follow these steps to update the `RoadL` layer:

1. Start ArcMap, and open an existing map document `AdvancedEditing.mxd` from `<drive>:\PacktPublishing\Data\EditingData`.
2. From the **Bookmarks** menu, select **Create Field road**. In the **Table Of Contents** section, right-click on the `RoadL` layer, and navigate to **Edit Features** | **Start Editing**. Add the **Advanced Editing** and **COGO** toolbars. In the **Create Feature** window, select the **Field road** template. In the **Snapping** toolbar, select only **Vertex** and **Edge Snapping**. In the **Editor** toolbar, select the **Midpoint** tool to create a centerline based on the **Transportation** subtype from `LandUse`, as shown in the following screenshot:

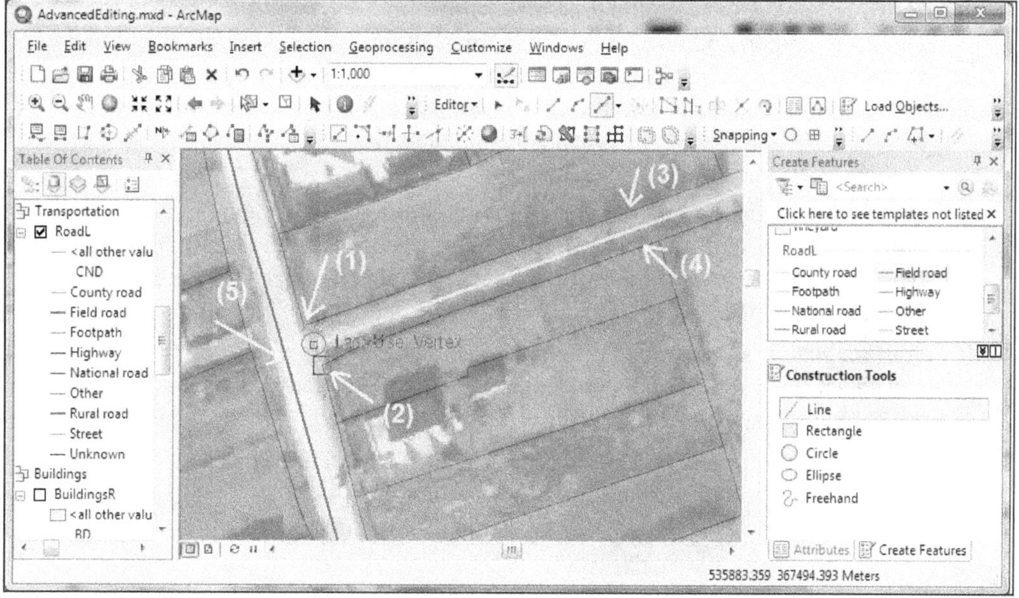

3. Click on point **(1)**, and then click on point **(2)**, to add the first vertex of the sketch at the midpoint between point **(1)** and point **(2)**. To add another vertex, click on points **(3)** and **(4)**. Those points are on the edge of the `LandUse` polygon. Continue to add vertices to the end of the road.
4. To connect `Field road` with `Rural road` (point **(5)**), and to split `Rural road` at the intersection, let's use the **Line Intersection** tool from the **Advanced Editing** toolbar. Select `Field road` and the **Line Intersection** tool. First, click on the end of the selected feature (red circle from the preceding screenshot), and secondly click on point **(5)**. Press *O* to choose the **Extend existing feature** option, and click on **OK**. Finally, click again on point **(5)**. Click on **Undo Line intersection** to undo the last action.
5. You will test another way to split `Rural road` and to correct undershoots. First, to specify the feature that you want to connect to, select the `Rural road`. Secondly, select the **Extent** tool. Finally, click on `Field road` to specify what vector you want to extend to `Rural road` (red circle from the preceding screenshot). To split `Rural road` at the intersection, select the **Split** tool from the **Edit** toolbar, and click on point **(5)**.
6. Save and stop the edit session. Save the map document at `...\Data\EditingData as MyAdvancedEditing.mxd`.

   Let's add new features to the `RoadL` layer using COGO:

7. Open ArcCatalog. Click on the drop-down arrow from the right-hand side of the **Standard** toolbar, and navigate to **Customize** | **Commands**. In **Show commands containing**, type `cogo`. Select the **Create COGO Fields** tool, drag-and-drop it on the **Standard** toolbar. Select the `RoadL` feature class, and click on the **Create COGO Fields** tool to add the specific COGO fields. Close ArcCatalog, and open `MyAdvancedEditing.mxd`.
8. It's time to specify the angular units. In the **Editor** toolbar, navigate to **Editor** | **Options** | **Units**. In the **Angular Units** section, choose **North Azimuth** for **Direction Type** and **Gons** for **Direction Units**. In **Display angles using**, type `4` for decimal places. Click on **OK**.
9. Start an edit session. In the **Table Of Contents** section, right-click on the `RoadL` layer, and select **Open Attribute Table**. Inspect the COGO fields that were created. Leave the **Table** window open.
10. We will transform **polyline 1** and **polyline 2** into COGO lines and add a new **2-Point Line** as shown in the following screenshot:

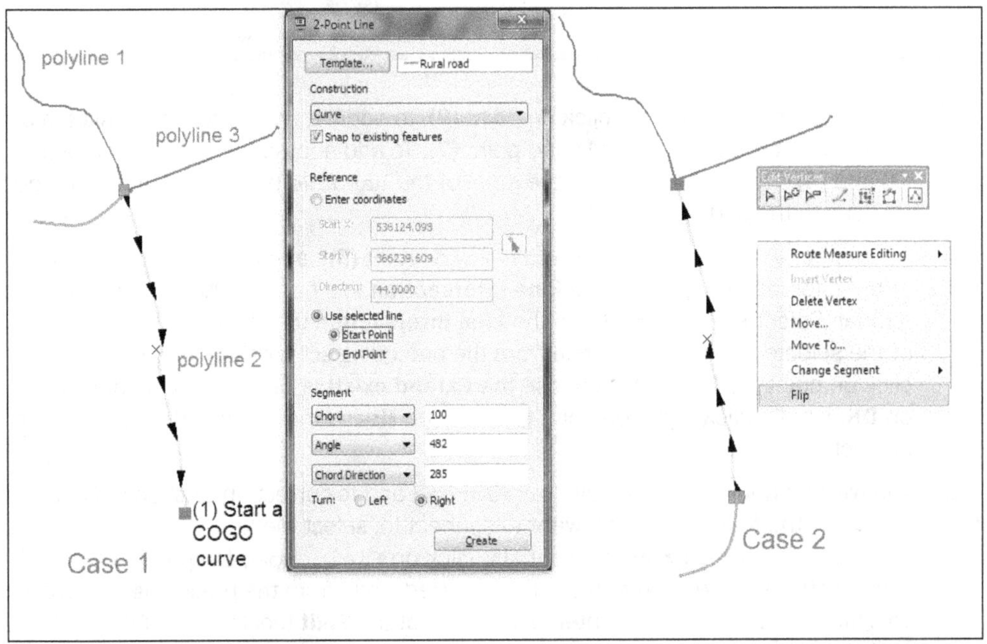

11. Let's consider **polyline 1** coming from an old survey taken in the field and **polyline 3** coming from screen digitizing. To differentiate those two polylines, we will transform **polyline 1** into COGO lines. Select **polyline 1**, and click on the **Split into COGO lines** tool from the **COGO** toolbar. Click on **Template**, choose `Field road` as the feature template, and click on **OK**. To calculate the coordinate geometry for the newly created lines, select the **Update COGO Attributes** tool. Inspect the changes in the **Table** window.

12. Let's add a COGO curve starting from point **(1)** as shown in the preceding screenshot. From the **Bookmarks** menu, select **Create a COGO curve**, then select **Construct 2-Point Line**, and finally select **polyline 2**. The black arrows indicate the direction of the vector. Choose the curve with the **Chord** field of `100` meters, the **Angle** field of `482` gons, the **Chord Direction** field of `285` gons, and the **Turn** field of `Right` to obtain a curve. In the **Reference** section, navigate to **Use selected line** | **Start Point**. Click on **Create** to obtain **Case 1**. We expected to obtain a curve from point **(1)**, but the line started from **polyline 3**.
13. Select the **Edit Vertices** option from the **Edit** tool to see the direction of the vector. The red vertex indicates the end of the vector. To reverse the direction, right-click on the selected vector, and choose **Flip**.
14. Repeat step 7 to obtain **Case 2**. Click on **Create** and close the window.
15. To calculate the coordinate geometry for the newly created curve, select the **Update COGO Attributes** tool, inspect the changes in the **Table** window, and save the edits.

    Let's update a `LandUse` parcel using the **Traverse** tool:

16. From the **Bookmarks** menu, return to **Create Field road**. In the **Table Of Contents** section, check the `LandUse` layer to see the feature templates in the **Create Feature** window. In the **List by Selection** mode, set only the `LandUse` layer as the selectable layer and return to the **List by Source** mode. Select the parcel as shown in the following screenshot:

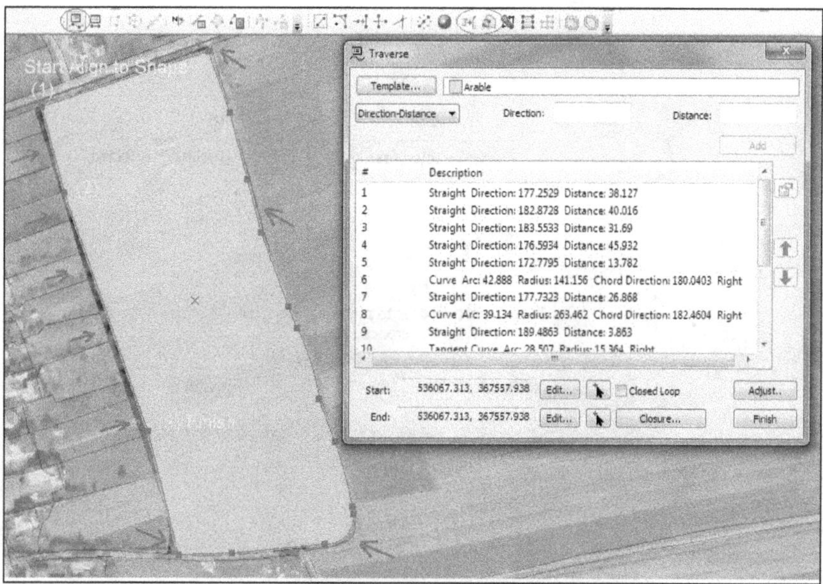

17. We will replace the existing parcel with a new one using the survey recently taken in the field. We will consider that traverse from the `Parcel_Owner1.txt` file was already adjusted for a closure error. Select the **Replace Geometry Tool** option from the **Advanced Editing** toolbar. Select the **Traverse** tool from the **COGO** toolbar. Right-click on the **Traverse** window, and select **Load Traverse**. Go to `...\Data\EditingData`, and select the `Parcel_Owner1.txt` file. Click on **Finish**, and close the window.
18. Examine the spatial errors. Notice there are some overlaps and gaps between the parcels. Let's try to correct some errors. Select the **Align To Shape** tool from the **Advanced Editing** toolbar. Follow the options and steps from the following screenshot:

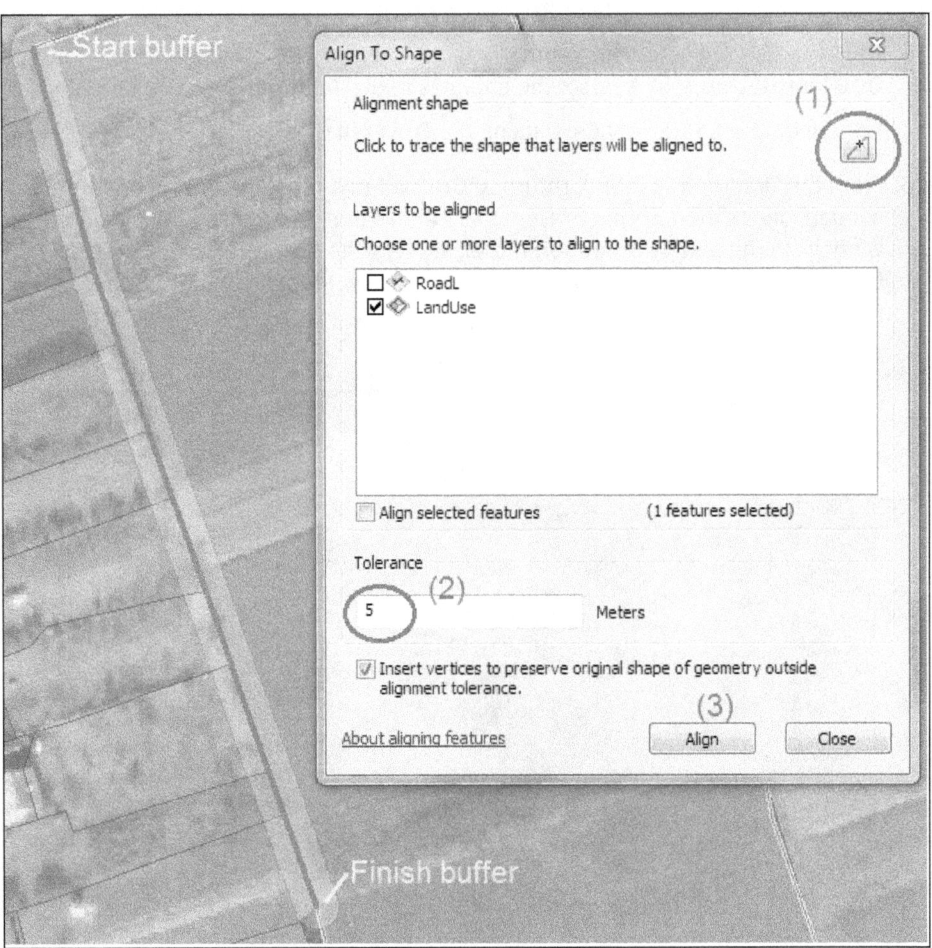

19. With the trace tool, start from the first point (**Start buffer**), and define a short segment on the edge of the updated parcel. To finish the trace and return to the **Align To Shape** window, double-click on the last point (**Finish buffer**). Try to change the value of **Tolerance** to see how the buffer changes. Next, click on **Align** to modify the edges of the neighboring parcels. Then, continue to align the parcel by updating the **Tolerance** value. Finally, click on **Close** to close the window. Inspect the attributes of the new parcel. The attributes remain unchanged because of the **Replace Geometry Tool** option.
20. Save the edits, and stop the edit session. Save the map document to `...\Data\EditingData` as `MyAdvancedEditing.mxd`.

You can find the final results at `<drive>:\PacktPublishing\Data\EditingData\MyEditingResults`.

How it works...

Regarding step 3, it is not necessary to have an existing feature to define the line segment. We just indicate the two points directly on the orthophoto map.

To store the COGO in the attribute table of the `RoadL` layer, we created empty COGO fields in ArcCatalog with the **Create COGO Fields** tool. By adding a COGO curve, we have emphasized the importance of digitized direction.

We used the **Replace Geometry Tool** option from the **Advanced Editing** toolbar in order to keep the attributes of the `LandUse` parcel and its relationship with the `Owner` table.

After we updated the parcel at step 11, we corrected the edges of the neighboring parcels in order to maintain the adjacency between the surrounding parcels. Using the **Align To Shape** tool is a right decision because we assumed that horizontal positions and dimensions of the new parcel are more accurate than manual digitized parcels.

See also

- In the next recipe, *Creating geodatabase topology*, we will learn how to create and validate the spatial relationships between features in a file geodatabase

Creating geodatabase topology

In a map, topology refers to spatial relationships between features, such as:

- **Connectivity**: This indicates that rivers from the `WatercourseL` feature class or roads from the `RoadL` feature class should be connected

- **Coincidence**: This indicates that the water and roads from the `Watercourse` and `Road` polygon feature classes should be covered by the `Hydrography` and `Transportation` subtypes from the `LandUse` feature class
- **Adjacency**: This indicates that parcels from `LandUse` have common edges and should not overlap or have gaps
- **Containment**: This indicates that buildings from the `Buildings` feature class should be within the `Other Terrain (Built-up area)` subtype from the `LandUse` feature class

In the ArcGIS context, a topology is a collection of rules that define the spatial relationship between features within a feature class or between features of two or more feature classes belonging to the same feature dataset.

A topology is stored in a file geodatabase as an individual `File Geodatabase Topology` element. In a topology, we can apply one or more rules to a feature class. We can apply different rules to the feature subtypes. We can create different and complicated spatial relationships between subtype features of one, two, or more feature classes in a single topology.

In a topology, if we want to define topological rules between features of two or more feature classes belonging to different feature datasets, we have to move them in the same feature dataset. We can have more topology elements in a feature dataset, but a feature class can be implicated in only one topology at a time.

For more details regarding topology and topology rules, please refer the online *ArcGIS Help (10.2)* by navigating to **Geodata** | **Data types** | **Topologies** at `http://resources.arcgis.com/en/help/main/10.2`.

Getting ready

The main steps involved in creating a topology are:

1. Name the topology.
2. Set **Cluster Tolerance**.
3. Add one or more simple feature classes.
4. Define the **Ranks** (importance).
5. Add one or more **Rules**.
6. Create the topology in a feature dataset.
7. Validate the topology.

Cluster tolerance is the distance range where the feature vertices are considered coincident. Starting from classical cartographic theory, let's consider that absolute accuracy of our dataset at `scale 1:5,000` is `0.5` meter. We will use a cluster tolerance of about *0.5/10* = *0.05* meter.

Ranks define what feature vertices will be moved during the validation process.

After we define the topology rules between four feature classes from the `Topo5k.gdb` geodatabase, we will validate the topology in order to apply the rules to our feature classes.

How to do it...

Follow these steps to create `File Geodatabase Topology` in a file geodatabase using the ArcCatalog context menu:

1. Start ArcCatalog. In the **Catalog Tree** section, go to `<drive>:\PacktPublishing\Data\EditingData`, and select `Topo5k.gdb`. Expand the `Buildings` and `LandUse` feature datasets.
2. To create a topology for the `BuildingsR` and `LandUse` feature classes, we have to add them in the same feature dataset—`LandUse`—as shown in the following screenshot:

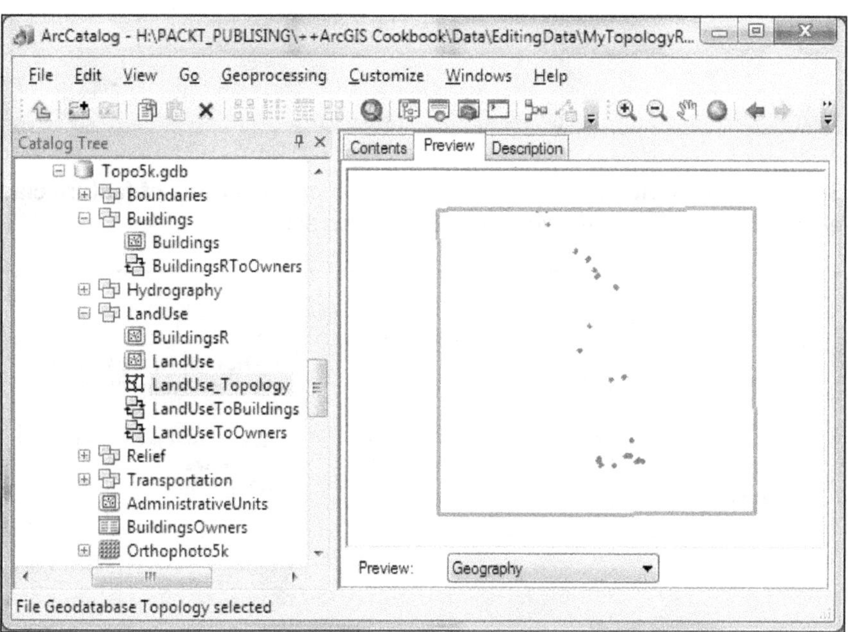

3. Select the `BuildingsR` feature class from the `Buildings` feature dataset, and drag-and-drop it in the `LandUse` feature dataset. Right-click on the `LandUse` feature dataset, navigate to **New** | **Topology**, and click on **Next**. Leave the default value for **Enter a name for your topology**. For **Enter a cluster tolerance**, type `0.05`, and click on **Next**. Click on **Select All** to check all feature classes, and click on **Next**. Define the relative importance of the feature classes by selecting the following ranks from the **Rank** drop-down list: `BuildingsR` = `1` and `LandUse` = `2`. Click on **Next** to see the following panel:

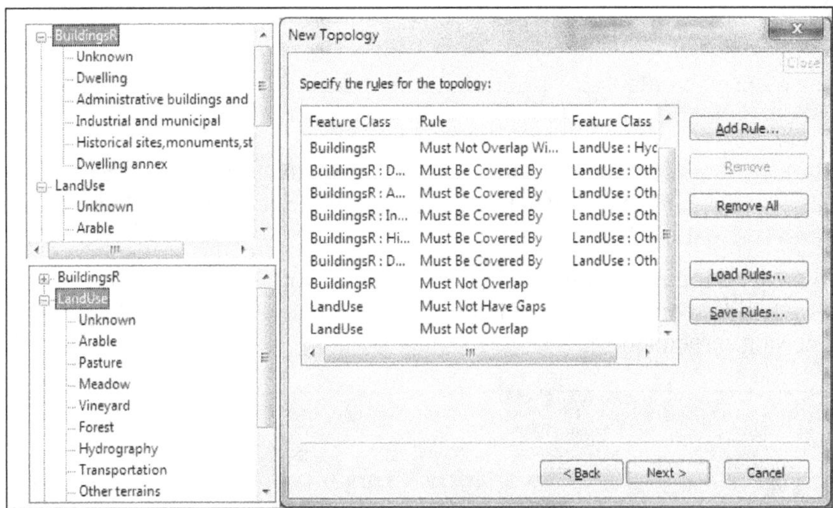

4. Select the **Add Rule** button. We have three sections: **Features of feature class**, **Rule**, and **Feature class**, as shown in the following screenshot:

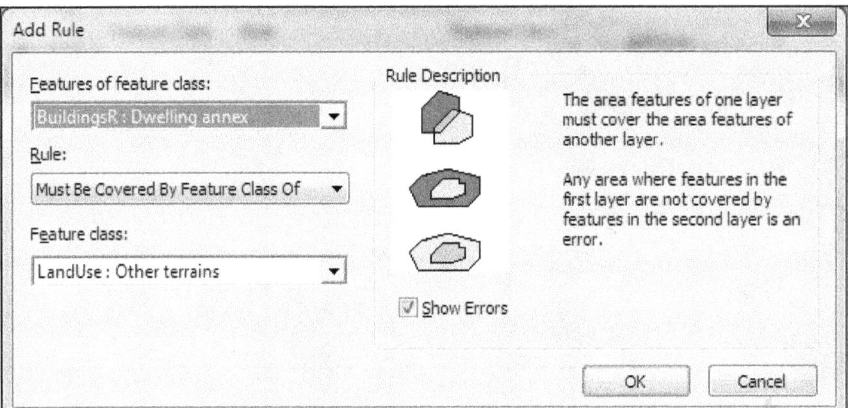

5. In the first and third sections, we can see the feature classes and their subtypes. This will allow us to define rules at feature subtype levels. In the second section, we have a drop-down list with the topology rules corresponding with the type of geometry.
   In the **Add Rule** window, we can define topology rules for a single feature class or between two feature classes. Click on **Cancel** to return to the **New Topology** window.
6. Select the **Load Rules** button to add a set of rules: `Rule Set` (`*rul` format). Go to `...\Data\EditingData`, select the `LandUse_Topology.rul` file, and click on **Open** and on **OK** to load the rules.
7. Inspect the rules. We don't have a general rule between all subtypes from `Buildings` and `LandUse`. As we can see, the rules refer to the `Buildings` and `LandUse` subtypes. In this way, we can exclude some feature subtypes from following a rule that applies to all other subtypes from the parent feature class.
   To exclude a feature subtype, just avoid mentioning it in the list (for example, the `BuildingsR:Unknown` subtype). Click on **Next** and on **Finish**. Select **No** in the pop-up window that appears. You will validate the topology in the next step.
8. In ArcCatalog, select `LandUse_Topology`, and choose the **Preview** mode to see the topology errors. You will see a blue hatched rectangle. This means that the whole area is not validated and requires topology validation in order to find the errors. This area is called a *dirty area*.
9. We can check the topology status in ArcCatalog. Right-click on the `LandUse_Topology` element, and navigate to the **Properties** | **General** tab. The topology status is `Not Validated`.
10. Let's validate the topology to identify any errors. Right-click on the `LandUse_Topology` element, select **Validate**, and from the **View** menu, select **Refresh** to see changes in the **Preview** mode. Now the topology status is `Validated-Errors Exist`.
11. Right-click on the `LandUse_Topology` element, navigate to the **Properties** | **Errors** tab, and click on the **Generate Summary** button to inspect the errors. You can save the report in a `.txt` file.

The **Must be larger than cluster tolerance** rule is added by default. We cannot remove this rule as it is applied automatically to all features participating in a topology.

Every time we make changes in a topology (add/delete rules or cluster tolerance), the entire area covered by our features will become a *dirty area*. When we edit the features (for example, change subtypes and add/delete features or vertices), the area surrounding the features will need additional topology validation and will also be considered as a *dirty area*.

12. You can change the value of **Cluster Tolerance** to see changes in topology status and in the **Preview** mode.
13. If you want to add more rules, open the **Properties** tab of `LandUse_Topology`, and navigate to the **Rules** | **Add Rule** button. Add the rules and validate the topology. Close ArcCatalog.

You can find the final results at `...\Data\EditingData\MyTopologyResults`.

How it works...

A topology has one of the following statuses:

- `Not Validated`
- `Validated-Errors Exist`
- `Validated-No Errors`
- If you edit the features, `Not Validated` with *dirty area*

A topology is not applied to our data until we validate it. Validation helps us to find errors. Validation generates two processes: cracking and clustering. The connected features are disconnected and connected again by adding/deleting vertices and slightly moving feature vertices based on the cluster tolerance and ranks. In the example from the following screenshot, we exaggerated the value of **Cluster Tolerance** to `1` meter for the `LandUse` topology before the validation process:

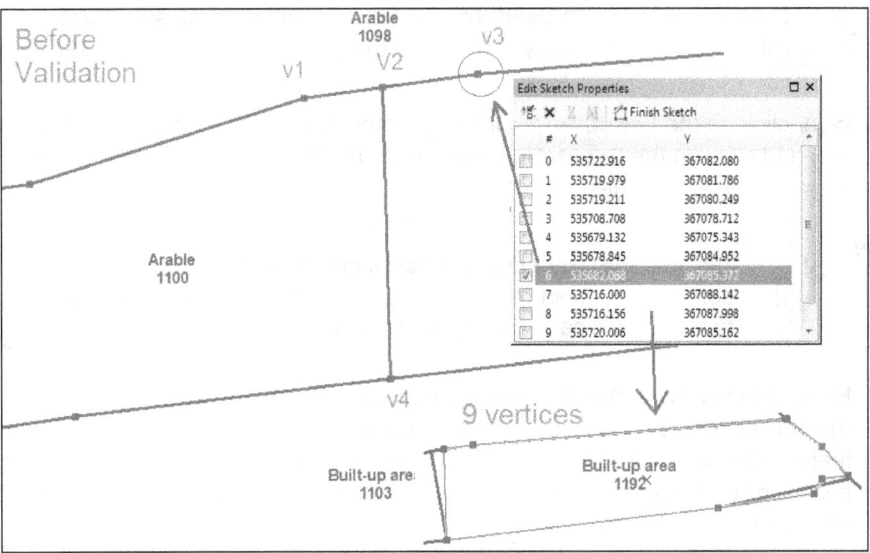

Notice in the following screenshot how the vertices were deleted (**v1**) and moved (**v2**) after the topology validation:

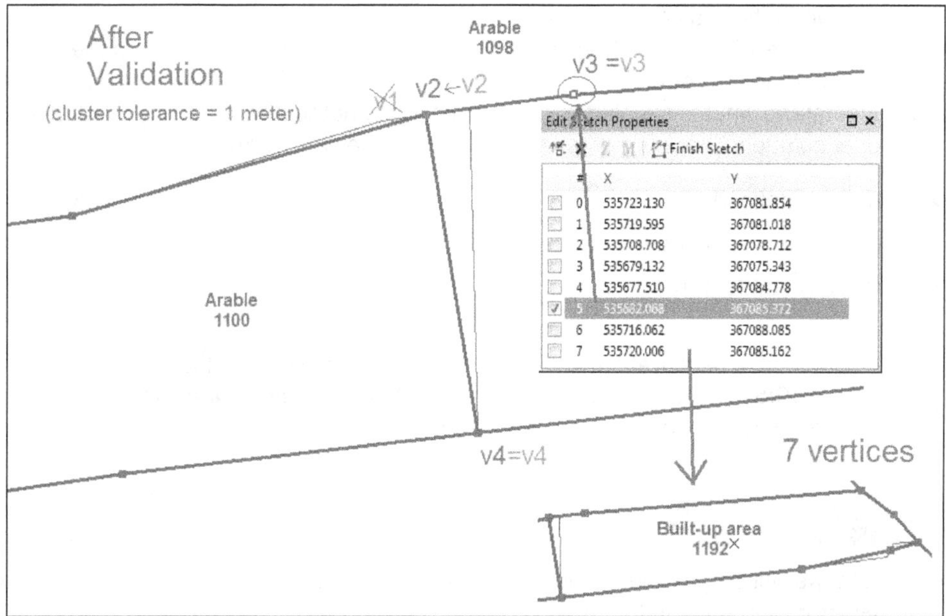

For more details about the cracking and clustering processes, please refer to the Esri Technical Paper (2010): *Understanding Geometric Processing in ArcGIS*.

> If you validate a topology in ArcCatalog, you cannot reverse the effects of the validation process.
>
> If you validate a topology in ArcMap, you can undo the changes made through the validation process.

See also

- In the next recipe, *Editing geodatabase topology*, we will display, analyze, and fix the spatial errors in ArcMap

Editing geodatabase topology

It's very important to have a comprehensive understanding of our dataset and the spatial relationships between features before choosing the right rules for a topology and the right solution to correct the errors. When you inspect the error, firstly identify the features involved, secondly understand what is causing the error from a geometrical point of view, and thirdly use the proper solution. There are different solutions to the same error:

- Use a predefined fix
- Combine a predefined fix with manual edits
- Manually edit features

For more details about the editing topology, please refer to the online *ArcGIS Help (10.2)* by navigating to **Desktop** | **Editing** | **Editing topology** at `http://resources.arcgis.com/en/help/main/10.2`.

Getting ready

In this section, we will display the topology layer in ArcMap, and test different methods to fix the topology errors identified in the previous recipe.

The main steps in editing a topology are:

1. Open the topology in ArcMap.
2. Symbolize your topology by the error type.
3. Examine the list of errors with **Error Inspector**.
4. Use **Fix Topology Error** to examine errors individually.
5. Use predefined fixes to correct an error / manually correct an error by using ArcMap's editing tools or **Topology Edit Tool**.
6. Define exceptions to the rule.
7. Display the *dirty areas* and validate the topology.
8. Save the edits.

How to do it...

Follow these steps to edit `LandUse_Topology` in ArcMap:

1. Start ArcMap, and open an existing map document `EditingTopology.mxd` from `<drive>:\PacktPublishing\Data\EditingData`. Click on the **Add Data** button, and load `LandUse_Topology` from `...\Data\TOPO5000.gdb\LandUse`. Select `Yes` in the pop-up window to add all feature classes that participate in `LandUse_Topology`.
2. In the **Table Of Contents** section, drag the `BuildingsR` layer above the `LandUse` layer. Set the transparency for `BuildingsR` and `LandUse` at `60` percent.
3. Right-click on `LandUse_Topology`, and navigate to **Properties** | **Symbology**. Notice that the error features may be points, lines, or polygons depending on the error types. Inspect all other tabs to review the ranks, rules, and errors from the previous recipe, *Creating geodatabase topology*. Click on **OK** to close the window.
4. Start an edit session. Click on the drop-down arrow from the right-hand side of the **Editor** toolbar, and select **Customize**. Check the **Topology** toolbar, and click on **Close**. Select the **Error Inspector** tool from the **Topology** toolbar and dock the window at the bottom of the ArcMap window, as shown in the following screenshot:

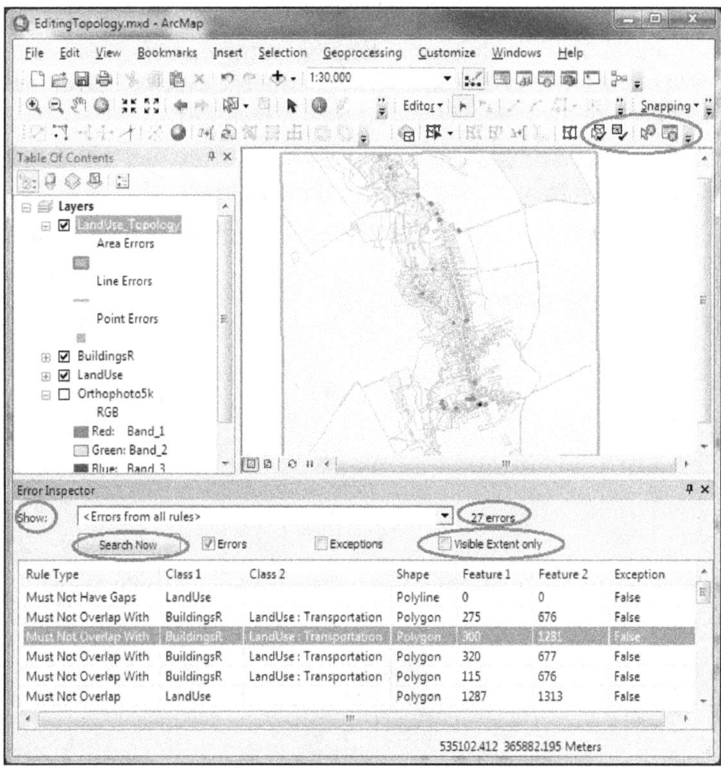

5. Uncheck **Visible Extent Only**, select **<Errors from all rules>** in the **Show** drop-down list, and click on the **Search Now** button. Inspect the rules and the columns in the **Error Inspector** window. If we click on any of the column headings, we change the order of the errors accordingly. If we select a row, we see the selected error on the map. The **Shape** column tells you about the error geometry: polygon error, polyline error, or point error.
6. Let's examine and correct some errors with the **Fix Topology Error** tool. First select **<BuildingsR: Dwelling-Must Be Covered By-LandUse:Other terrains>** in the **Show** drop-down list, and click on the **Search Now** button.
7. Select the first error, shown as **Must Be Covered By** | **BuildingsR:Dwelling** | **LandUse:Other terrains** | **Polygon** | **72** | **0** | **False**. Right-click on the error, and choose **Zoom To**. The `Dwelling` subtype is on the `Arable` parcel. This is a violation of the rule that says that the `Other terrains` subtype parcel must contain only the `Dwelling` subtype buildings. Because the orthophoto map is from 2012, and some changes could have happened, we can assume that this error could be a false error, and we will declare it as an exception to the rule. With the **Fix Topology Error** tool, select the error, right-click on the map, and select **Mark as Exception.** To see this exception, check **Exceptions** in the **Error Inspector** window. Click on **Search Now** to refresh the rows.
8. Select the next error, shown as **Must Be Covered By** | **BuildingsR:Dwelling** | **LandUse:Other terrains** | **Polygon** | **77** | **0** | **False**. Right-click on the error, and choose **Zoom To**. The dwelling is overlapping on two parcels. This is a real error, and you should correct it. How do you do it? With the **Fix Topology Error** tool, select the error, right-click on the map, and select **Select Features.**
9. There are three main possible solutions: delete a vertex, use **Topology Edit Tool**, or use **Reshape Edge Tool** and slightly shift the buildings.
    1. **First solution**: Click on **Edit Vertices** from the **Edit** toolbar. With the pointer on vertex 4, right-click on and choose **Delete Vertex**, as shown in the following screenshot:

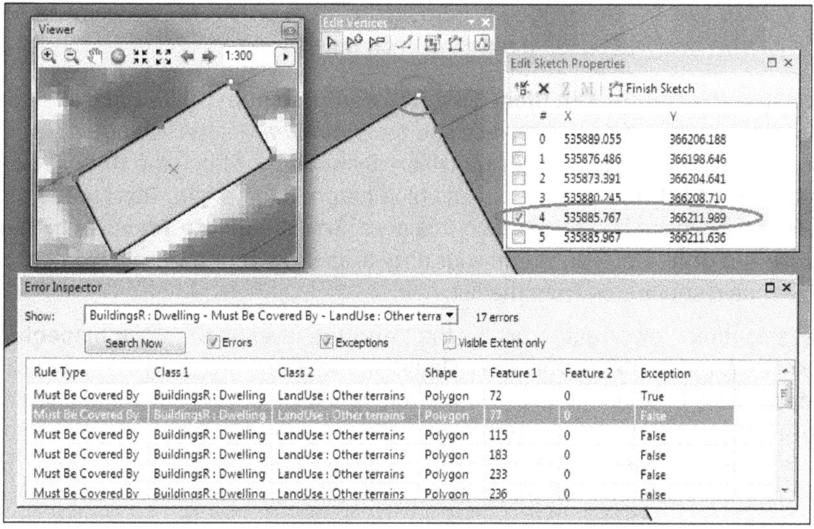

2. **Second solution**: Click on **Topology Edit Tool**, and select edge **(1)**. Select the **Reshape Edge Tool** option, and draw a line between points **(2)** and **(3)**, as shown in the following screenshot. To unselect the edge, click on **Reshape Edge Tool**, and click outside the building feature.

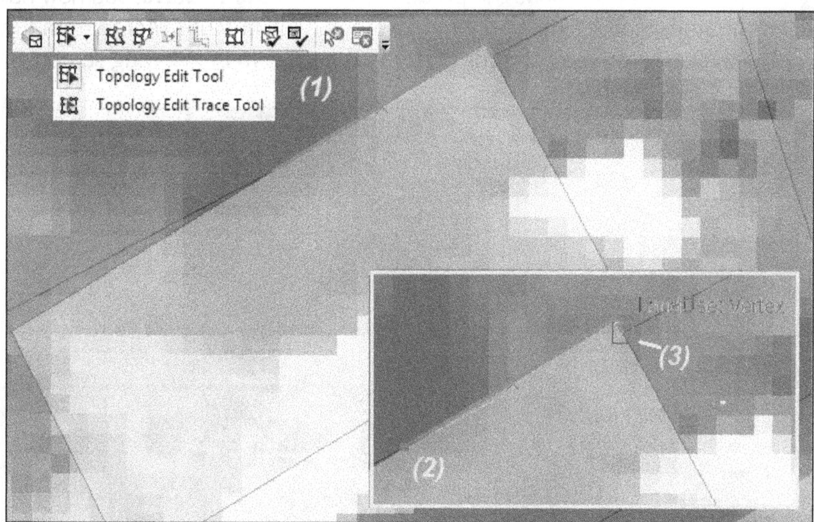

3. **Third solution**: With the building selected, click on **Rotate Tool** from the **Editor** toolbar, press the *A* key to set the rotate angle at `4` gons. Type `4`, and press the *Enter* key. Click on **Edit Vertices** from the **Edit** toolbar, and delete the cracking points that were added on both the edges of the building in the validation process.

As we can see, the last solution seems to be more proper for the building because the geometry of the building remains unchanged.

10. Unselect the feature with **Clear Selected Features** from the **Standard** toolbar. Because we edited a feature, let's see the *dirty area*. Right-click on `LandUse_Topology`, and navigate to **Properties** | **Symbology**. Check the **Dirty Areas** option, and click on **OK**. You will see a small blue hatched rectangle. This means that your topology needs to be validated on this area. Select **Validate Topology in Specified Area**, and draw a bigger box on your dirty area. Now your area is validated, and the error has disappeared from the list.
11. Let's continue to correct errors for the `LandUse` layer. In the **Error Inspector** window, select **< LandUse-Must Not Have Gaps>** in the **Show** drop-down list, and click on the **Search Now** button. Select the first error, right-click on the error, and choose **Zoom To**. The error delineates the entire area. This is a false error, which you will declare as an exception to the rule. With the **Fix Topology Error** tool, select the error, right-click on the map, and select **Mark as Exception**. Click on **Search Now** to refresh the rows.
12. Select the second error in the **Error Inspector** window, right-click on the error, and choose **Zoom To**. With **Fix Topology Error**, right-click on the map, and choose **Select Features.** There are two possible solutions: use a predefined fix or edit the vertices, which are explained in detail here:
    1. **First solution**: With the **Fix Topology Error** tool, select the error, right-click on the map, and choose the predefined fix: **Create Feature**. You now have a new feature that fills the gap. Select both parcels and use **Merge** from the **Editor** toolbar to merge polygons **(1)** and **(2)** as shown in following screenshot:

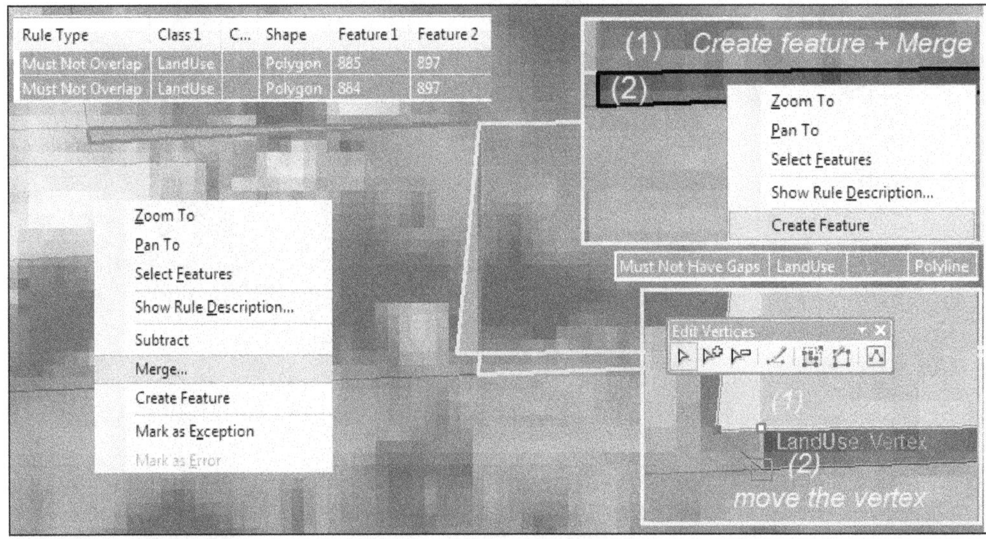

   2. **Second solution**: Select the parcel, and click on **Edit Vertices** from the **Edit** toolbar. With the pointer on the vertex, drag the vertex from point **(1)** to point **(2)**.

13. Use the **Validate Topology In Current Extent** tool to validate the changes you have made.
14. In the **Error Inspector** window, select **< LandUse-Must Not Overlap>** in the **Show** drop-down list, and click on the **Search Now** button. Select the first polygon error, right-click on the error, and choose **Zoom To**. With the **Fix Topology Error** tool, select the error, right-click on the map, and select **Merge**, as shown in the preceding screenshot. In the **Merge** window, select `Built-up area (LandUse)`, and click on **OK**. Click on **Search Now** to refresh the rows. Correct the second error yourself using the same steps. Consider as exceptions the last four errors from the **< LandUse-Must Not Overlap>** list. Select all four errors at once, and right-click on and choose **Mark as Exception**. If you change your mind, just right-click on and choose **Mark as Error**.
15. Zoom to **Full Extent**, validate using the **Validate Topology In Current Extent** tool, and refresh the error list in the **Error Inspector** window.
16. Save your edits and stop the edit session. Save the map document to `<drive>:\PacktPublishing\Data\EditingData` as `MyEditingTopology.mxd`.

You can find the results at `...\Data\EditingData\MyTopologyResults\EditingTopology.mxd`.

How it works...

The validation of spatial data and correction of the topological errors improve the quality of our data but do not increase the accuracy of our spatial data.

The small overshoot and undershoot errors can be fixed through the validation process as shown in the following screenshot:

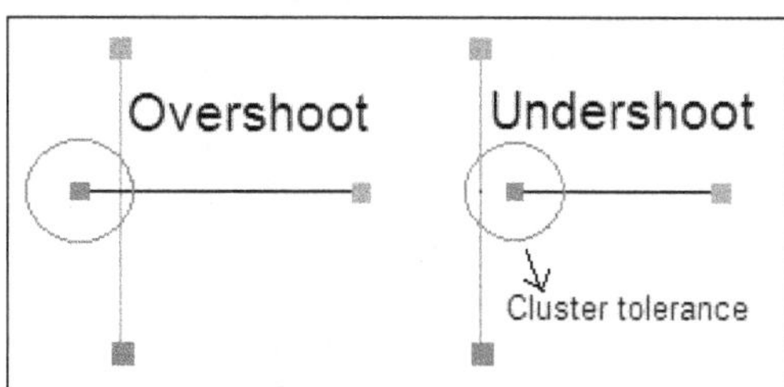

Do not increase the value of cluster tolerance to automatically fix this kind of error in your spatial dataset.

The topology validation is an iterative process. The simultaneous use of different topological rules should be performed moderately. Try to keep the number of topology rules as small as possible.

We have noticed that a single mistake in feature geometry might violate multiple rules and generate different geometry errors (for example, polygon and polyline errors at the same time). The correction of some topological errors might generate new topological errors notified by other topological rules applied to the datasets. In order to avoid the recurrent correction of the errors on the same vectors, it is recommended that you create, for example, a topology only for features from `LandUse`. After we have corrected all errors between features from the same feature class, we should delete the topology. After that, we can create another topology between the `BuildingsR` and `LandUse` feature classes.

Another good idea is that we should first create a topology, and after that, we should create relationships and define relationship rules between the `BuildingsR` and `LandUse` feature classes (the *Creating a relationship class* recipe of *Chapter 1, Designing Geodatabase*). Otherwise, we have to pay attention to those relationships while correcting the topology errors.

The **Must be larger than cluster tolerance** rule helps us to identify the small residual polygons or polylines resulted in the edit process or from different geoprocessing tasks. As a general practice, there is an order of correcting errors from a topology as follows: correct the polygon errors, correct the polyline errors, and finally correct the point errors.

There's more...

Some of the topology errors can be considered *false errors* , which means that involved features will not obey the topology rules. Most of the time, those kinds of errors have to be identified by the operator and be declared as *exceptions* to the topological rules.

In ArcCatalog, inspect the properties of `WatercourseL_Topology` from `...\Data\EditingData\MyTopologyResults\TOPO5000.gdb\Hydrography`. For all 23 errors, refer to the **Must not have dangle** rule that constrains the vectors from being connected to other vectors. This significant number of errors could suggest that rivers from the hydrography network are disconnected. These point errors are false. Only one error has been really identified, the rest of the errors having been declared as exceptions:

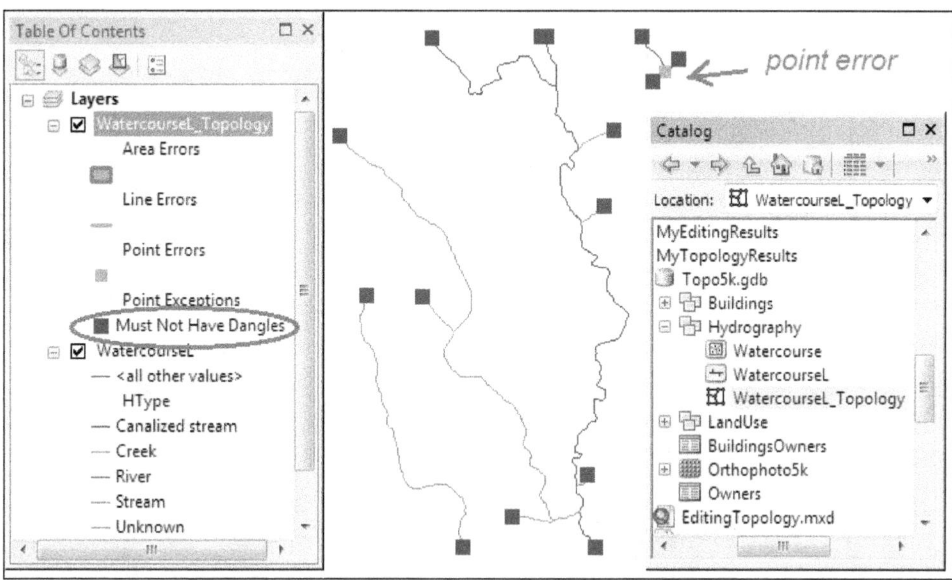

In the quality control report, we should mention the number of identified and corrected errors. It is also necessary to specify the number of exceptions to prevent the following overall assessment: *your spatial data has been created with a doubtful quality and significant corrections were applied.*

See also

- In the next chapter, we will work with different coordinate reference systems in a file geodatabase

3
Working with CRS

In this chapter, we will cover the following topics:

- Understanding projections
- Projecting vector data
- Georeferencing raster data
- Setting a custom coordinate reference system

Introduction

In accordance with the international standard ISO 19111:2007, geographic information spatial referencing by coordinates, the following definitions are universally agreed upon:

- A **Coordinate Reference System** (**CRS**) is a coordinate system that is related to an object through a datum. For geodetic and vertical datums, the object is the Earth.
- A datum is a set of parameters that defines the position of the origin, the scale, and the orientation of a coordinate system.
- A geodetic datum describes the relationship of a 2D or 3D coordinate system to the Earth.

- A geoid is the equipotential surface of the Earth's gravity field that is everywhere perpendicular to the direction of gravity. The geoid best fits mean sea level either locally or globally. A geoid is considering the true shape of the Earth.
- A projected coordinate reference system is a coordinate reference system derived from a two-dimensional geodetic coordinate reference system by applying a map projection.
- A vertical coordinate reference system is a one-dimensional coordinate reference system that is based on a vertical datum. A vertical datum describes the relation of the gravity-related heights to the Earth. A vertical datum is related to a geoid.

A geocentric coordinate reference system is a geodetic coordinate reference system that has the origin in the center of the mass of the Earth (geocenter) and usually is expressed in rectangular Cartesian coordinates (*X*, *Y*, *Z*).

An ellipsoid models the size and shape of the Earth. The ellipsoid (ellipsoid of reference) is formed by rotating the meridian ellipse about its minor axis (oblate ellipsoid). The ellipsoid generated by rotating the meridian ellipse about its major axis is a prolate ellipsoid. You need at least two parameters to define the size and shape of an ellipsoid (one parameter must be the semi-axis).

Generally speaking, the difference between the major and minor axes is small, and the shape is close to a sphere. Because it approximates the shape of a sphere, we call it a spheroid. For example, the **Geodetic Reference System 1980** (**GRS80**) has the semi-major axis length of 6,378,137 meters and the semi-minor axis length of 6,356,752.31414 meters. The terms ellipsoid and spheroid will be used interchangeably.

An ellipsoidal coordinate system works with geodetic latitude (B or φ), geodetic longitude (L or λ), and ellipsoidal height (h).

When the Earth is considered a sphere, we use a spherical coordinate system with its origin in the center of the sphere (spherical latitude φ and spherical longitude λ).

In ArcGIS, you will find both the ellipsoidal and spherical coordinate system in the `geographic coordinate systems` folder. In the ArcGIS context, a geographic coordinate system has three components: angular units, prime meridian, and datum (spheroid).

When you are working with **Vertical Coordinate Systems**, please take into account the difference between the ellipsoidal and gravity-related heights. Ellipsoidal heights (h) use the ellipsoid as the reference surface and represent the vertical component a of 3D ellipsoidal coordinate system. Gravity-related heights (H) use as reference surface the geoid that is the component of the one-dimensional vertical coordinate system. The relationship between the ellipsoid, geoid, and topographic surfaces is as shown in the following screenshot:

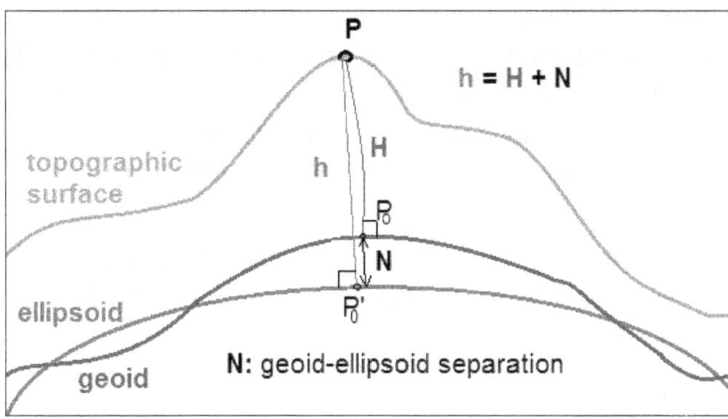

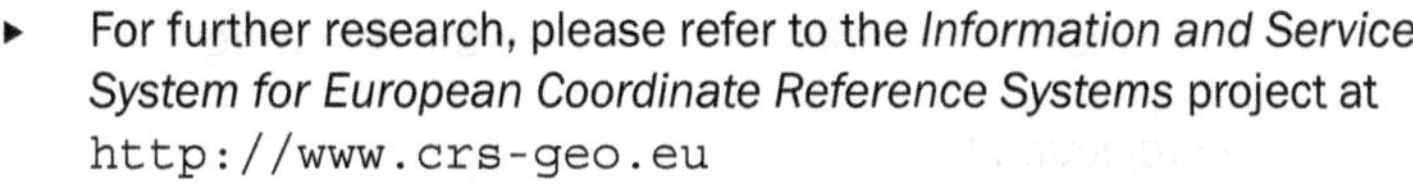

- For further research, please refer to the *Information and Service System for European Coordinate Reference Systems* project at `http://www.crs-geo.eu`

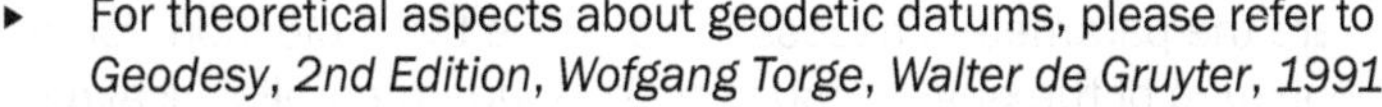

- For theoretical aspects about geodetic datums, please refer to *Geodesy, 2nd Edition, Wofgang Torge, Walter de Gruyter, 1991*

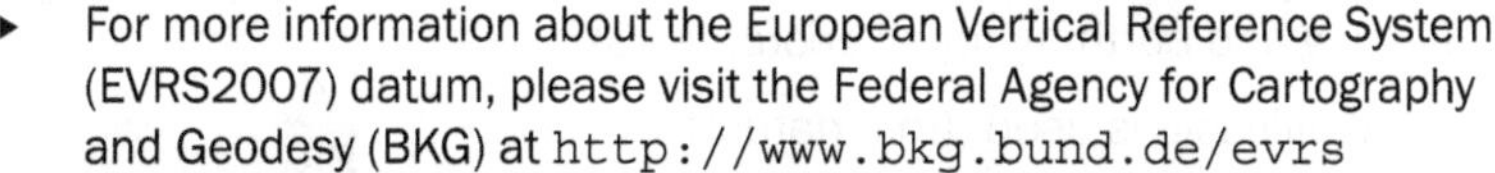

- For more information about the European Vertical Reference System (EVRS2007) datum, please visit the Federal Agency for Cartography and Geodesy (BKG) at `http://www.bkg.bund.de/evrs`

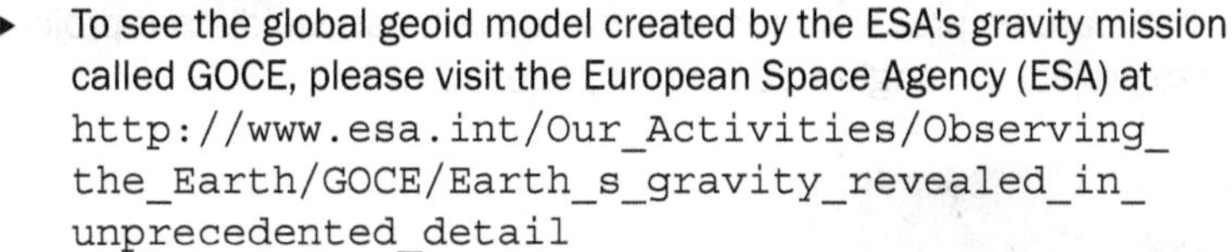

- To see the global geoid model created by the ESA's gravity mission called GOCE, please visit the European Space Agency (ESA) at `http://www.esa.int/Our_Activities/Observing_the_Earth/GOCE/Earth_s_gravity_revealed_in_unprecedented_detail`

As a final remark, please be aware about the difference between the transformation and conversion processes. A transformation involves a change of datum (for example, from GRS80 to Krassowsky 1940), and a conversion does not involve a change of datum (for example, geographic coordinates φ and λ to projected coordinates *N* and *E*).

Understanding projections

In this recipe, you will self-study different map projections in three types of projections: Azimuthal, Cylindrical, and Conical.

For theory regarding the map projections, please refer to:

- *Datums and Map Projections 2nd Edition Jonathan Iliffe* and *Roger Lott, Whittles Publishing, 2012*
- *Understanding Map Projections; Melita Kennedy* and *Steve Kopp , Environmental Systems Research Institute, Inc., 2000*

In the map document `CRS.mxd`, there are marked with a red rectangle the main map projections for Europe as recommended by **INSPIRE Directive** in the *Data Specification on Coordinate Reference Systems* document.

Getting ready

While you inspect every map projection, try to identify the following:

- What is the projection type?
- What are the parameters necessary to define the projection (for example, latitude and longitude of the origin, first/second standard parallel, scaling factor, and false easting and northing for all coordinates)?
- What ellipsoid is used (from datum)?
- What features preserve (for example, distances: equidistant projection; areas: equal area projection; angles: conformal projection)?

How to do it...

Follow these steps to study what Europe looks like in different map projections using ArcMap:

1. Start **ArcMap**, and open the existing map document `CRS.mxd` from: `<drive>:\PacktPublishing\Data\CRS`, as shown in the following screenshot:

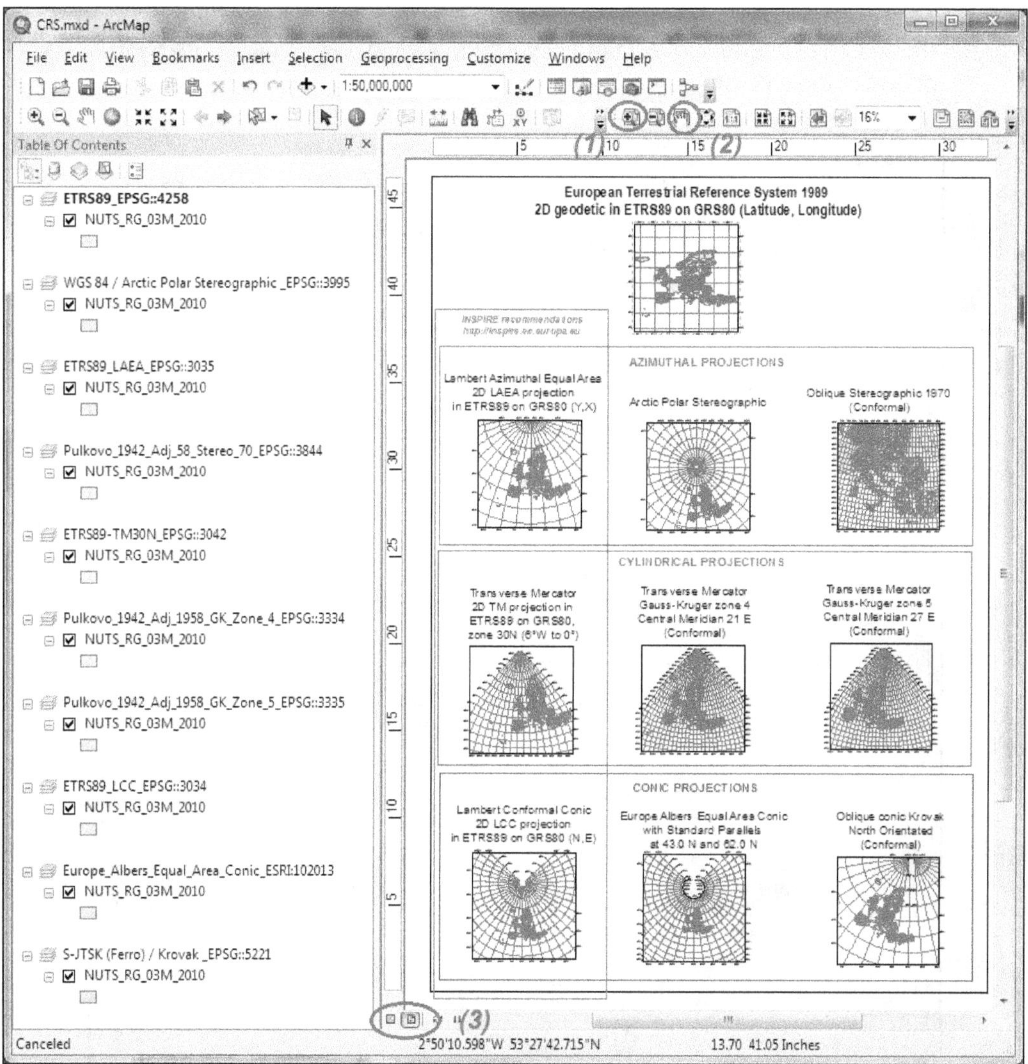

2. The **Table Of Contents** section contains ten **Data Frames** in **Layout View** with different **Coordinate Systems**. Every data frame contains the same `NUTS_RG_03M_2010` layer from the `NUTS_2010_03M.mdb` personal geodatabase. In **Layout View**, you can see all data frames at once. **Data View** allows you to view only one data frame at a time. You can switch between **Data View** and **Layout View** if you click on the small buttons from the lower-left part of the map display window **(3)**. Keep **Layout View** activated. In **Layout View**, use the **Zoom In** (**1**), **Zoom Out**, and **Pan** (**2**) tools from the **Layout** toolbar to study the aspect of the graticules for every data frame's projection.

3. Let's check the first data frame's coordinate system. Right-click on the first data frame **ETRS89_EPSG::4258**, and navigate to **Properties** | **Coordinate System**, as shown in the following screenshot:

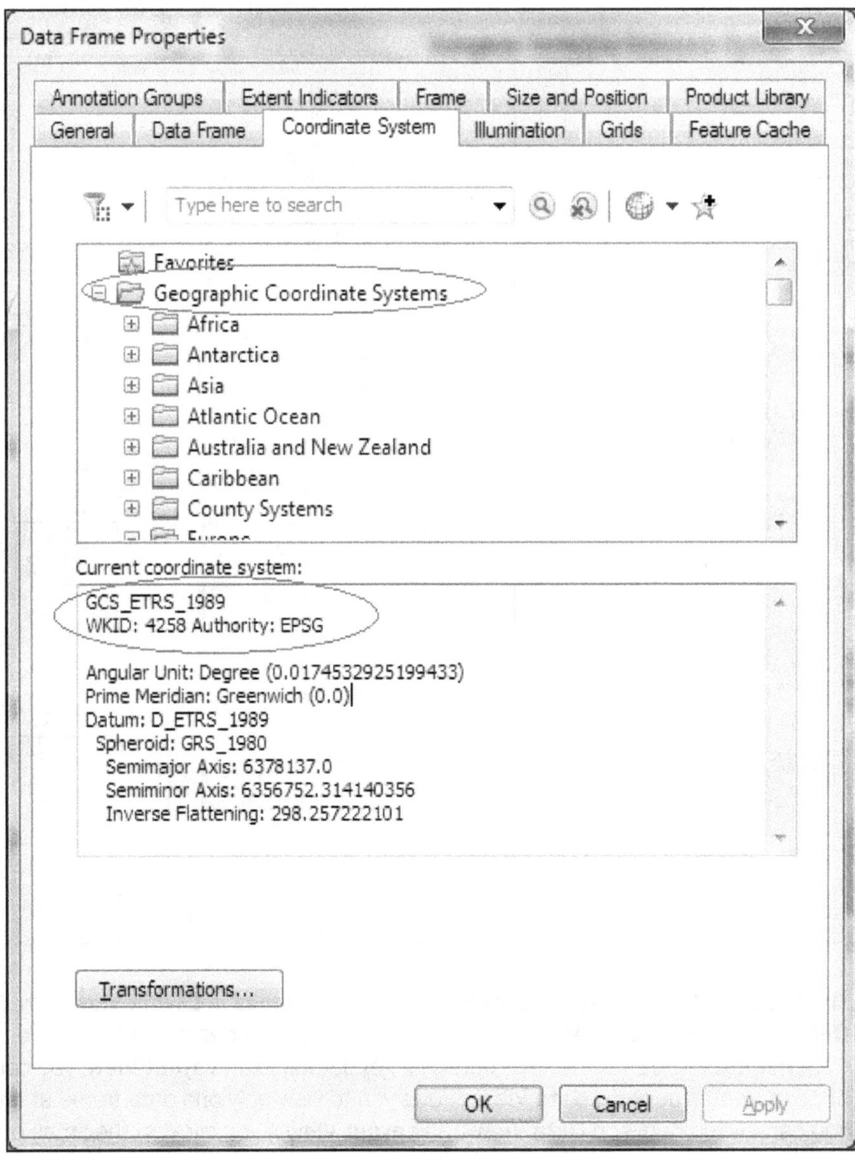

4. Notice that the coordinate system is set to the **GCS_ETRS_1989** coordinate system with the EPSG geodetic code **4258**. This is a 2D geographic coordinate system. To read more about it, please check the website `http://www.epsg-registry.org`. **EPSG** is the acronym for **European Petroleum Survey Group**. As shown in the following screenshot, select **retrieve by code** (**1**), type `4258` (**2**), and click on **Retrieve** (**3**) to find the ETRS89 coordinate reference system characteristics, as shown in the following screenshot:

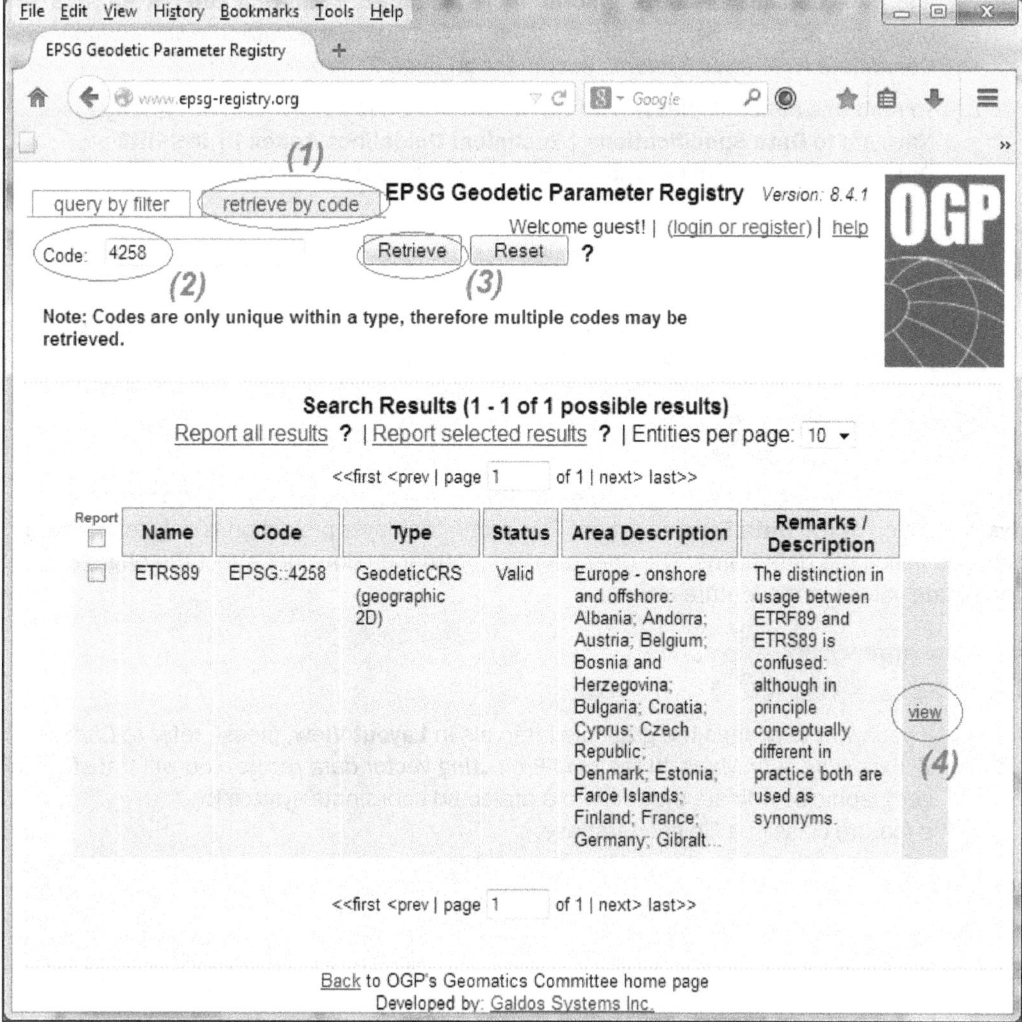

5. To see more details and the geodetic parameters, click on **view** (**4**).
6. Let's inspect the second data frame: **ETRS89_LAEA_EPSG::3035**. To activate the data frame, press the *Alt* key, and select the data frame in the **Table Of Contents** (**TOC**) section. The name of the activated data frame is indicated in bold letters in TOC. You can activate a data frame by right-clicking on **ETRS89_LAEA_EPSG::3035** and selecting **Activate**. The activated data frame will have a gray dashed border in the map display window.
7. It is a **Projected Coordinate System** named **Lambert Azimuthal Equal Area**. This projection is recommended by the INSPIRE Directive in the *Data Specification on Coordinate Reference Systems* document on *page 7*.
8. To read this material, please visit the website `http://inspire.ec.europa.eu`. Navigate to **Data Specifications** | **Technical Guidelines Annex I** | **INSPIRE Data Specification on Coordinate Reference Systems –Technical Guidelines 3.2 17.04. 2014**.
9. Repeat step 5 to see more details on `http://www.epsg-registry.org` using the code **3035**.
10. Continue to inspect every data frame from `CRS.mxd` and compare the projections (for example, graticules and distortions of the feature shapes).

How it works...

The `NUTS_RG_03M_2010` layer is projected on the fly in accordance with the **Coordinate System** option of the **Data Frame** section. This mean that layer projection is only temporarily changed inside the data frame. The on-the-fly transformation does not affect the original coordinate values in the feature class.

See also

- If you want to change the graticule intervals in **Layout View**, please refer to *Chapter 7, Exporting Your Maps*. In the next *Projecting vector data* recipe, you will transform a geographic coordinate system into a projected coordinate system for a feature class in a file geodatabase.

Projecting vector data

In the previous recipe, you saw the same `NUTS_RG_03M_2010` feature class that has the ETRS89-GRS80 geographic coordinate system in nine data frames that have different projected coordinate systems. The projection of the feature class was made on the fly by ArcMap based on every data frame's projected coordinate system.

To permanently transform the coordinate reference system of a feature class, use the **Project** tool from **ArcToolbox**. This will create a new feature class with the new coordinates for the features. ArcGIS offers multiple versions of a transformation between different datums. It's important to choose the optimal transformation because it will affect the coordinates' accuracy.

At `<ARCGISHOME>:\Desktop10.2\Documentation`, you will find a PDF file called `geographic_transformations.pdf` that lists all supported datum transformations and the areas for which they are suited.

For theory and worked examples about coordinate system transformations, please refer to *Datums and Map Projections 2nd Edition Jonathan Iliffe* and *Roger Lott, Whittles Publishing, 2012, chapter Map Projection, p.39-89; Appendix E.3.3, Subsets of the seven-parameter geocentric transformation, p.196-199*

Getting ready

Let's follow the steps to transform the `AdministrativeUnits` standalone feature class from the geographic coordinate system (GCS) `ETRS_1989` to the projected coordinate system `Pulkovo_1942_Adj_58_Stereo_70` using the predefined transformation from ArcGIS:

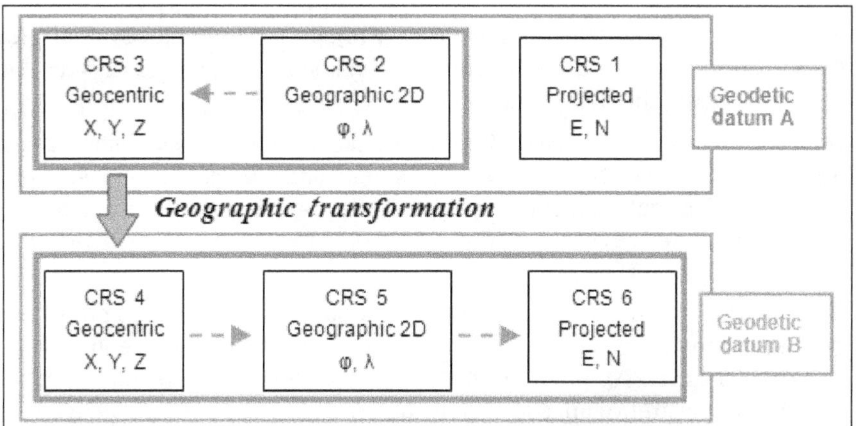

How to do it...

Follow these steps to transform the coordinates for a feature class in the ArcCatalog application:

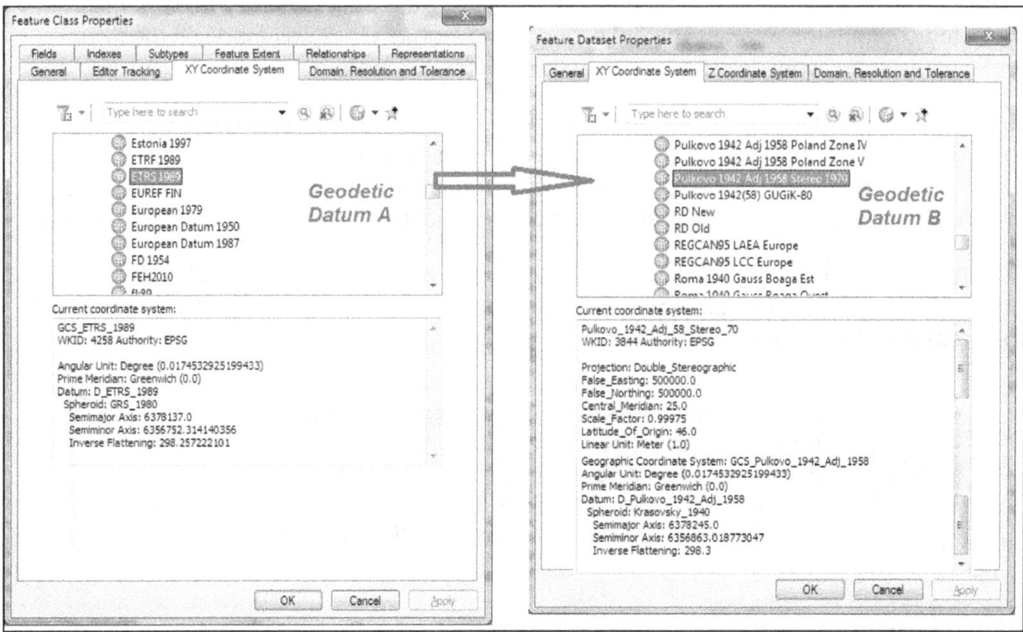

1. Start **ArcCatalog.** In the **Catalog Tree** section, go to `<drive>:\PacktPublishing\Data\CRS` and select `Topo5k.gdb`.
2. The file geodatabase has an empty feature dataset. Right-click on the **Boundaries** feature dataset, and navigate to **Properties** | **XY Coordinate System**. To go to the current coordinate system, navigate to the **Projected Coordinate System** | **National Grids** | **Europe** projected coordinate system named `Pulkovo_1942_Adj_58_Stereo_70` with the `EPSG` code `3844`.
3. Let's add some data. Search on Google for Eurostat website with following key words: `eurostat statistical units`. You should find the following link Administrative units and Statistical units Eurostat: `http://ec.europa.eu/eurostat/web/gisco/geodata/reference-data/administrative-units-statistical-units`.
4. Read carefully the **Copyright** notice, and download the `NUTS 2010` dataset as a personal geodatabase: `NUTS_2010_3M.zip`. Unzip the archive file in `<drive>:\PacktPublishing\Data\CRC\NUTS2010`. You now have a personal geodatabase with the administrative units for Europe at `1:3,000,000` scale.

5. Import the `NUTS_RG_03M_2010` feature class into `Topo5k.gdb`, as a standalone feature class:

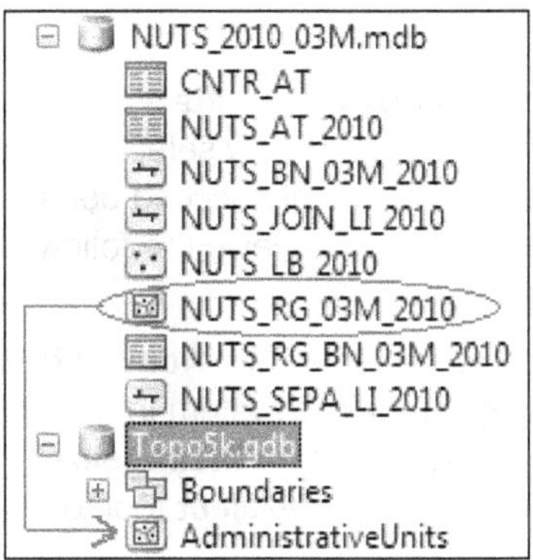

6. In the **Catalog Tree** section, right-click on `Topo5k.gdb` and navigate to **Import** | **Feature Class (single)**. Specify **Input Feature**, **Output Location**, and **Output Feature Class** (as `AdministrativeUnits`), and click on **OK**, as shown in the following screenshot:

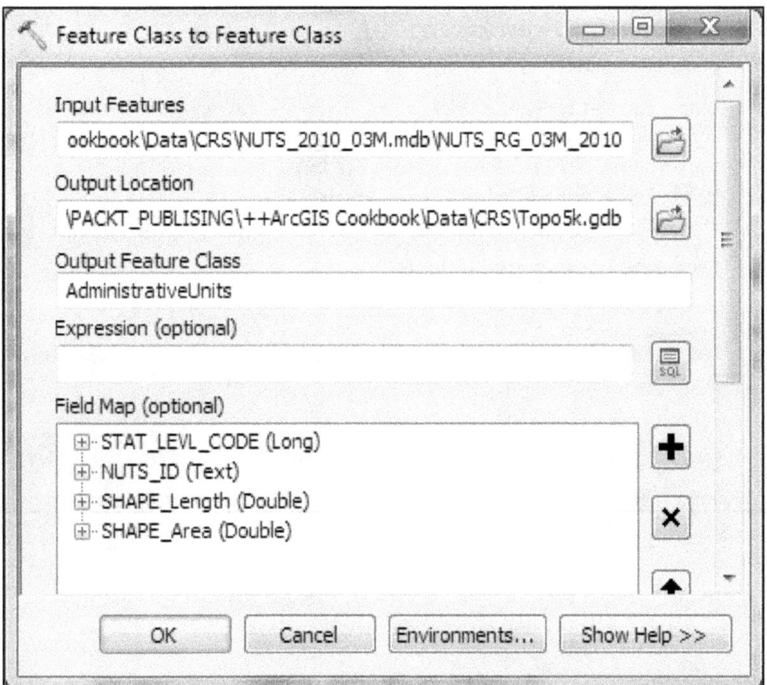

7. Right-click on the **AdministrativeUnits** feature class, and navigate to **Properties** | **XY Coordinate System**. To go back to the current coordinate system, navigate to the **Geographic Coordinate System** | **Europe** GCS named `ETRS 1989` with the `EPSG` code `4258`.
8. If you want to move the **AdministrativeUnits** standalone feature class in the `Boundaries` feature dataset, you simply select the feature class and drag and drop it into the feature dataset. You will not succeed because the feature datasets have a different spatial reference. You will receive the following error message: **The spatial reference does not match**.
9. Let's use the **Project** tool to transform the **AdministrativeUnits** standalone feature class from the geographic coordinate system in a projected coordinate system. Open **ArcToolbox**, and navigate to **Data Management Tools** | **Projections and Transformation**. Double-click on the **Project** tool to open the dialog box shown in the following screenshot:

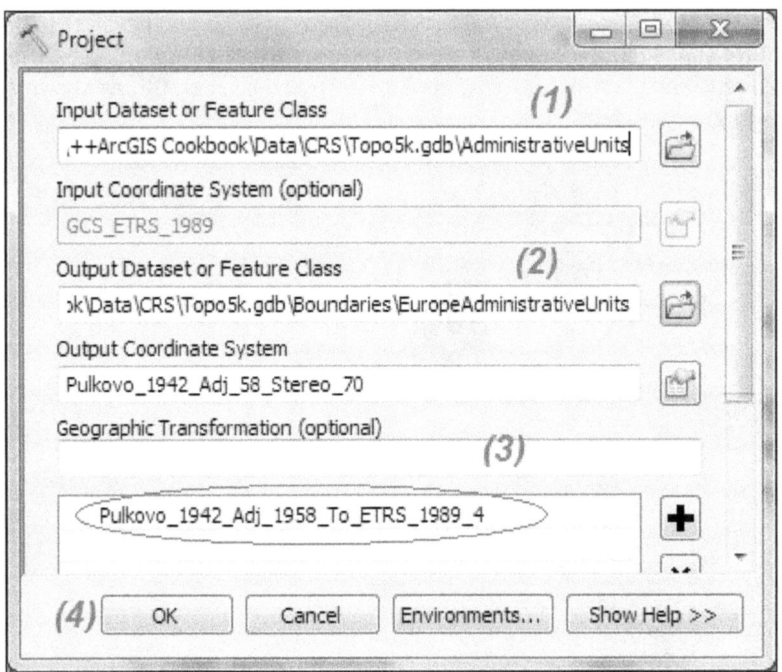

10. Specify the **Input Dataset or Feature Class** step (**1**) of `Topo5k.gdb\AdministrativeUnits`. Specify the **Output Dataset or Feature Class** (step **2**) of `Topo5k.gdb\Boundaries`, and type the name of the feature class as `EuropeAdministrativeUnits`.
11. The **Output Coordinate System** was automatically added because the newly created feature class will inherit the coordinate system of the feature dataset of `Boundaries`.
12. At step (**3**) **Geographic Transformation**, choose the `Pulkovo_1942_Adj_1958_To_ETRS_1989_4` geographic transformation. Click on **OK**.
13. This is the best transformation for your feature class, also take into account the input and output CRSs and the covered area. To choose a proper datum transformation please go through the following steps:
    1. Open the `geographic_transformations.pdf` document from `<ARCGISHOME>:\Desktop10.2\Documentation` and go to page 67 to read the following row: `Pulkovo_1942_Adj_1958_To_ETRS_1989_4 15994 CF 2.3287 -147.0425 -92.0802 0.3092483 -0.32482185 -0.49729934 5.68906266`. Read the details from page 75 regarding the geographic transformation method called `CF` or `Coordinate_Frame` that is a seven-parameter geocentric transformation.
    2. Check the parameters of the transformation **EPSG::15994** at `http://www.epsg-registry.org`. Using the **EPSG::15994** datum transformation the accuracy of the horizontal position will range from 1.5 meters to 3 meters.
14. Use ArcMap to see the **EuropeAdministrativeUnits** feature class.

There's more...

Let's define two custom geographic transformations between the ETRS89 datum with the ellipsoid GRS 1980 (EPSG: 4258) and the Pulkovo 1942 datum with the ellipsoid Krassowsky 1980 (EPSG: 4179) using a seven-parameter transformation.

Follow these steps to define two custom geographic transformations in ArcCatalog using **ArcToolbox**:

1. Open **ArcToolbox**, and navigate to **Data Management Tools** | **Projections and Transformation**. Double-click on the **Create Custom Geographic Transformation** tool to open the dialog box, as shown in the following screenshot:

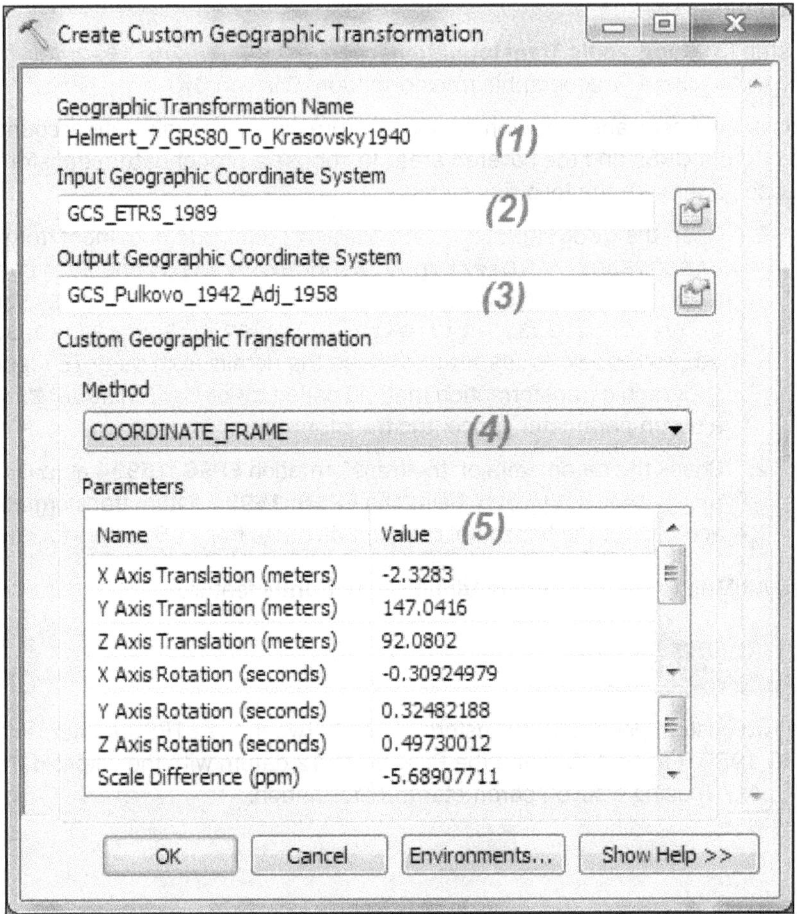

2. For **Geographic Transformation Name**, type `Helmert_7_GRS80_To_Krassowsky1940` (step **1**) to define the name of your transformation.

3. For **Input Geographic Coordinate System** (step **2**), select the icon from the right-hand side to open the **Spatial Reference Properties** dialog. From the **Geographic Coordinate Systems** folder, navigate to **Europe** | **ETRS 1989**. Select the **Z Coordinate System** tab, and navigate to **Vertical Coordinate Systems** | **Europe** | **ETRS 1989**. Click on **OK**.

4. For **Output Geographic Coordinate System** (step **2**), select icon from the right-hand side to open the **Spatial Reference Properties** dialog. From the **Geographic Coordinate Systems** folder, navigate to **Europe** | **Pulkovo 1942 Adj 1958**. Select the **Z Coordinate System** tab, and navigate to **Vertical Coordinate Systems** | **Europe** | **Constanta**. Click on **OK**.
5. In the **Custom Geographic Transformation** section, select as **Method** the **COORDINATE_FRAME** option. You will have to specify all seven parameters (step **4**), as shown the preceding screenshot. Those parameters are from the official site of the National Agency of Cadastre and Land Registration from Romania: `www.ancpi.ro`. Click on **OK**.
6. Open again the **Create Custom Geographic Transformation** tool to create the second transformation named `Helmert_7_ Krassowsky1940_To_ GRS80`, as shown in the following screenshot:

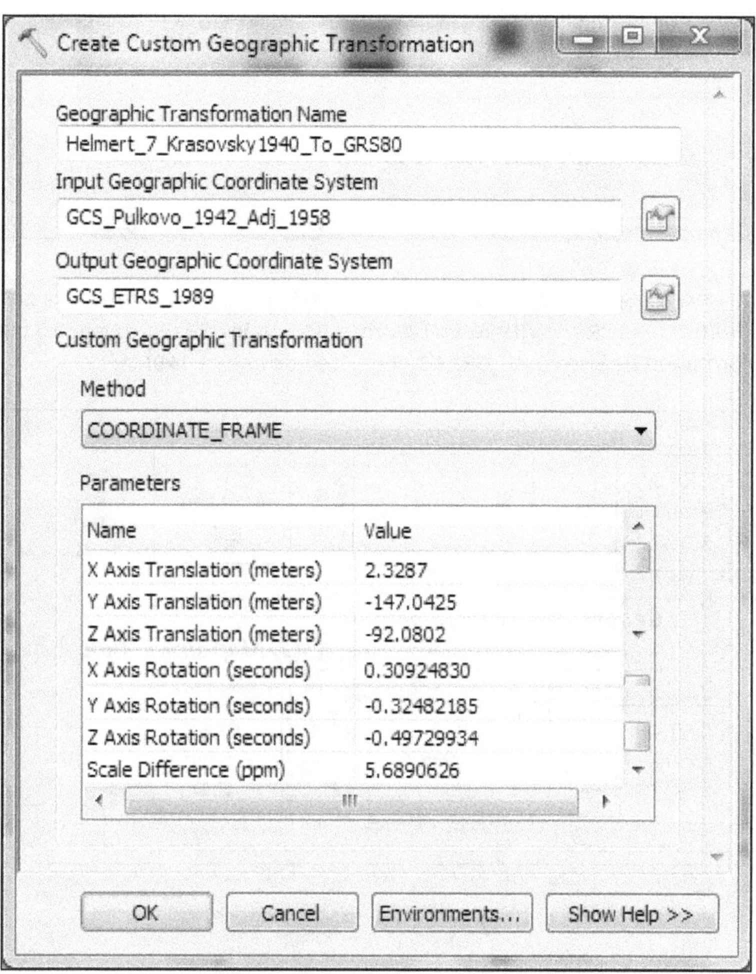

7. You can see the custom geographic transformations in the .gtf format at `C:\Users\<user>\AppData\Roaming\ESRI\Desktop10.2\ArcToolbox\CustomTransformations`, as shown in the following screenshot:

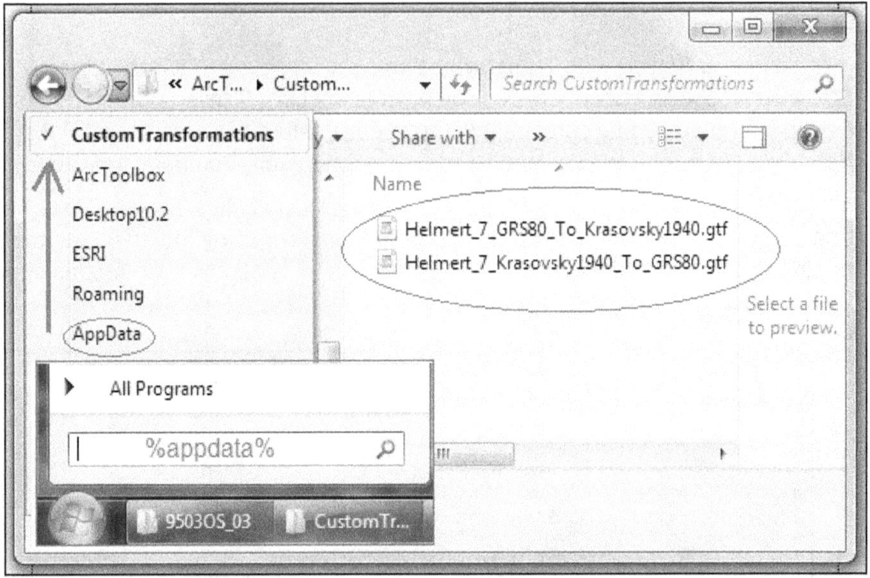

Let's follow the next image to project once again the `AdministrativeUnits` standalone feature class from `GCS_ETRS_1989` to `Pulkovo_1942_Adj_58_Stereo_70` using your custom transformation `Helmert_7_GRS80_To_Krassowsky1940`:

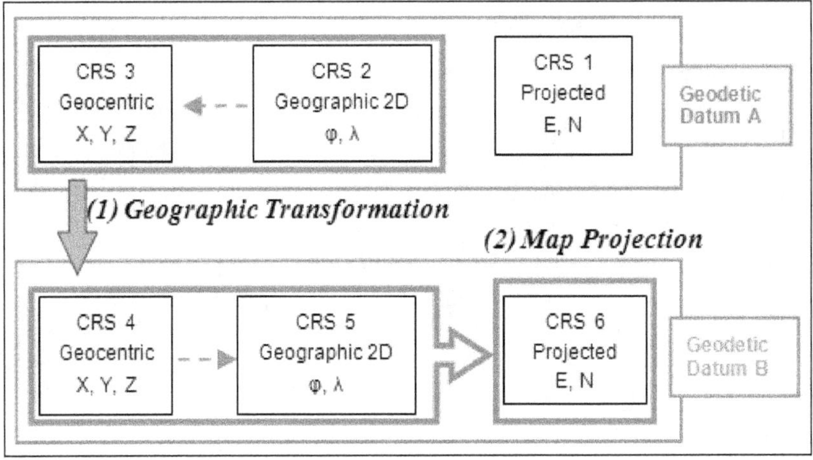

You will transform your feature class using the first two steps:

1. **Geographic transformation or change of datum**: Open **ArcToolbox**, navigate to **Data Management Tools** | **Projections and Transformation**, and double-click on the **Project** tool. Set the following parameters and click on **OK**:
   - **Input Dataset or Feature Class**: `Topo5k.gdb\ AdministrativeUnits`.
   - **Output Dataset or Feature Class**: `Topo5k.gdb\ Krassowsky_Europe`.
   - **Output Coordinate System**: **Geographic Coordinate System | Europe**: `Pulkovo 1942 Adj 1958`.
2. **Map the projection or coordinate conversion from geographic coordinates to Cartesian coordinates**: To add your standalone feature class in the projected coordinate system of the **Boundaries** feature dataset, use the **Project** tool again, and set the **conversion** parameters, as shown in the following screenshot:

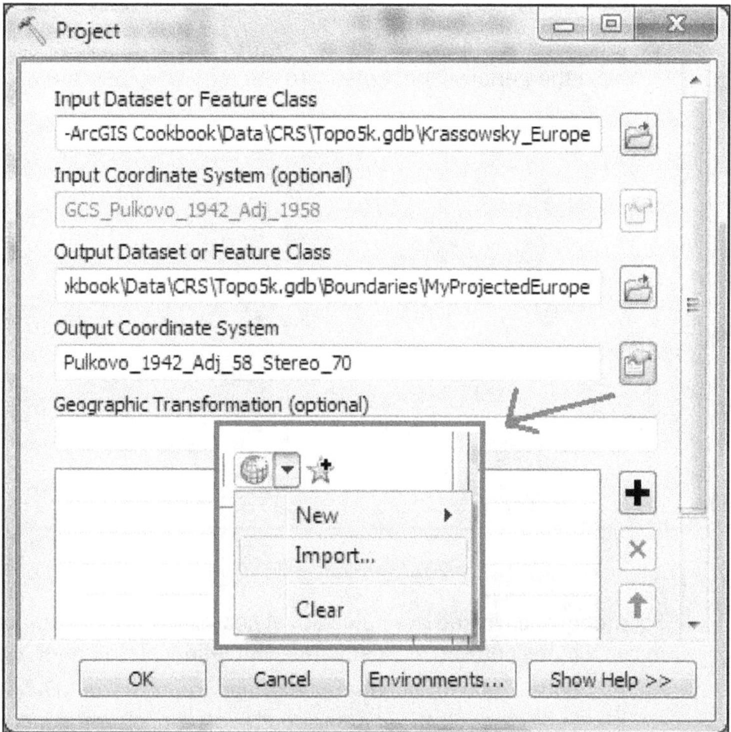

3. For **Output Coordinate System**, open the **Spatial Reference System** dialog. Click on the small globe on the right-hand side and navigate to **Add Coordinate System** | **Import** to go through the coordinate system from the `Boundaries` feature dataset. Click on **Add** and on **OK** to return to the **Project** window. Click on **OK**.

4. Use ArcMap to see the `EuropeAdministrativeUnits` option and `MyProjected_Europe` feature classes. Inspect the two layers with the **Edit Vertices** (start the Edit session) and **Sketch Properties** tools to see whether there are any differences in the *X* and *Y* coordinates of feature vertices. If the transformation and conversion succeed, there will be no difference.

You can find the results of this section at `<drive>:\PacktPublishing\Data\CRS\MyCRSResults`.

Georeferencing raster data

According to the online ESRI GIS Dictionary, the georeferencing process converts the coordinates of a raster to a known coordinate system in order to be viewed, queried, and analyzed with other spatial datasets you might have. The raster's Cartesian coordinate system is considered as a source (arbitrary) system The known or reference Cartesian coordinate system is considered as a destination (fixed) system.

In this recipe, you will follow the general concepts from the following diagram:

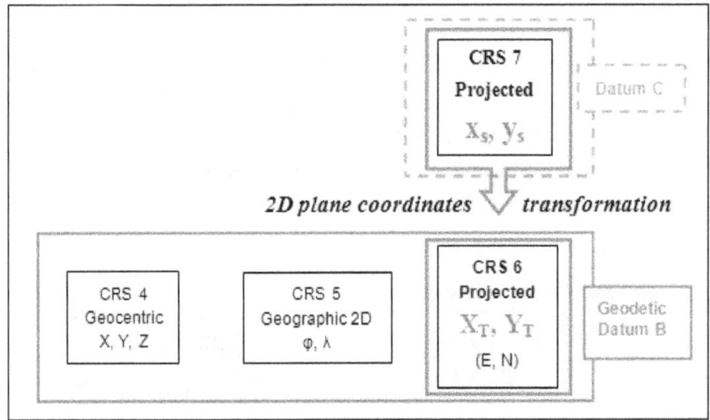

For theory and worked examples about coordinate transformations in two dimensions and least squares estimation, please refer to:

- *Principles of Geospatial Surveying, Artur L. Allan, Whittles Publishing, 2007; chapter Coordinate Transformations, p.41-58* and *chapter Least Squares Estimation, p.88-115*
- *Observations and Least Squares, Eduard M. Mikhail, IEP-A Dun-Donnelley Publisher, 1976; subchapter Coordinate transformations, p.184-212*

Getting ready

You will georeference a raster in the TIF format that is a scanned topographic map corresponding to scale `1:5,000`. The raster will represent the source coordinate system (image system).

As a reference layer, you will have a vector dataset in the `dxf` format (CAD data file format) that contains the grid for `1:5,000` scale maps in oblique stereographic projection with the EPSG code `3844` and the name `Pulkovo 1942(58) / Stereo70`. The reference layer will represent the target coordinate system (fixed).

You will use the affine transformation method or six-parameter transformation. The affine method assumes the following adjustments: translations on *x* and *y*, scale changes on *x* and *y*, and rotation on *x* and *y* (the axes are not orthogonal). The following equations express the affine transformation in 2D space:

XT= AxS + ByS + C

YT= DxS + EyS + F

A, *B*, *C*, *D*, *E*, and *F* are the parameters of the transformation, *xS* and *yS* are the source coordinates (image system), and *XT* and *YT* are the target coordinates (map reference system). To estimate the six parameters using the least squares method, you need at least four points of known coordinates in both coordinate systems (source and target).

After you specify at least four links, ArcGIS will calculate the residuals on *X* and *Y* as the difference between the calculated map coordinates and the known map coordinates. The resultant *Residuals = SQRT (Residual_X2 + Residual_Y2)*.RMS Error represents the root mean square error of an observed point. *RMSE = ± SQRT(Sum(Residual_X2 + Residual_Y2)/ number of equations-number of parameters)*

You should pay attention to the RMSE values. Usually, the values should not be greater than 0.5 pixels in dimensions. After the affine transformation (scale changes on *x* and *y*), you will have a pixel value of around 0.4 m. Trying to obtain a RMS value of 0.2 m in this case is not quite relevant for the scanned map at the scale `1:5,000`.

Taking into account the classical theory of cartography, 1 mm on the map represents 5 m on the ground. The planimetric precision for a map should be between 0.1 mm and 0.3 mm.

Following this idea, let's propose for RMSE a value equal or less than 0.5 m corresponding to 0.1 mm planimetric precision on a map at the scale `1:5,000`.

How to do it...

Follow these steps to georeference a scanned map (raster) in ArcMap using the coordinates from the following screenshot:

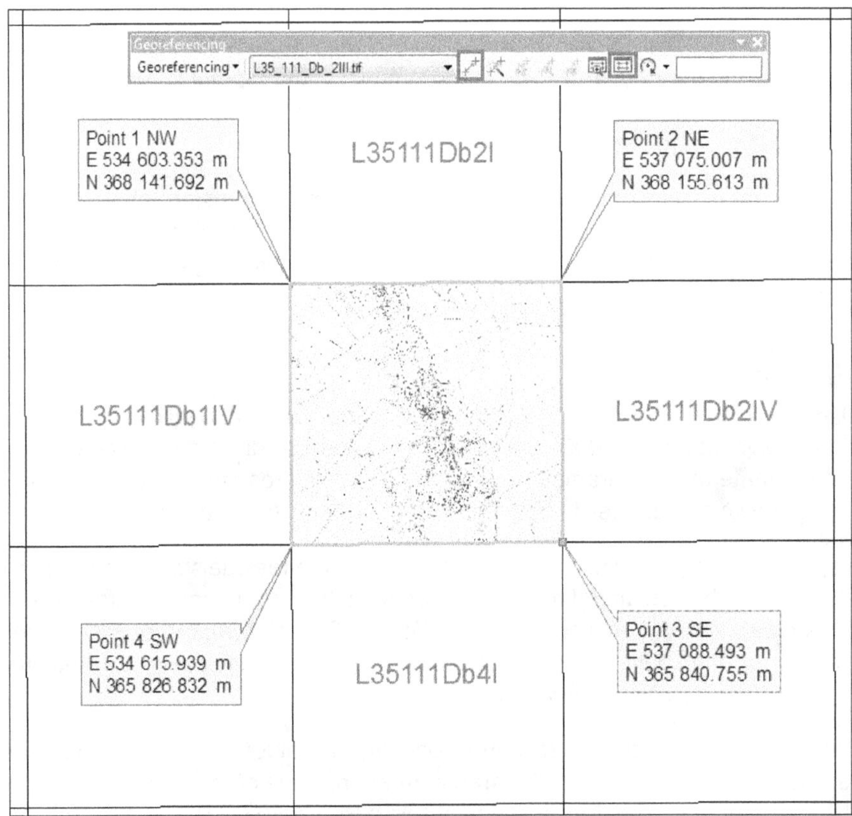

1. Start **ArcMap**, and open an existing map document **Georeferencing.mxd** from `<drive>:\PacktPublishing\Data\CRS`.
2. From the **Bookmarks** menu, select **Raster**. Notice the two **Overview** windows that will help you to see the full extent of raster and the control layer in `.dxf` format. If necessary, rearrange the windows on **Data View**.
3. From the **Customize** menu, navigate to **Toolbars** | **Georeferencing**. On the toolbar, choose **L35_111_Db_2III.tif** as the georeferencing layer.
4. Click on the **View Link Table** button, and uncheck **Auto Adjust** in **Link Table**.
5. In the next steps, you will add four links between the source points and the destination points by selecting the **Add Control Points** button from the **Georeferencing** toolbar. The green pointer indicates where the source point is, and the red pointer indicates where it should be.

6. From the **Bookmarks** menu, select **RasterNW**. In the **Snapping** toolbar, select **Vertex** to snap to the vertices of lines from the control layer.
7. Click on in the first corner of the raster to add the first source point (**Source 1**). Go to the reference layer using the bookmark **Grid 1:5000**. Click on `Point 1 NW` (**Destination 1**) to add the corresponding destination point using the following screenshot as a guide:

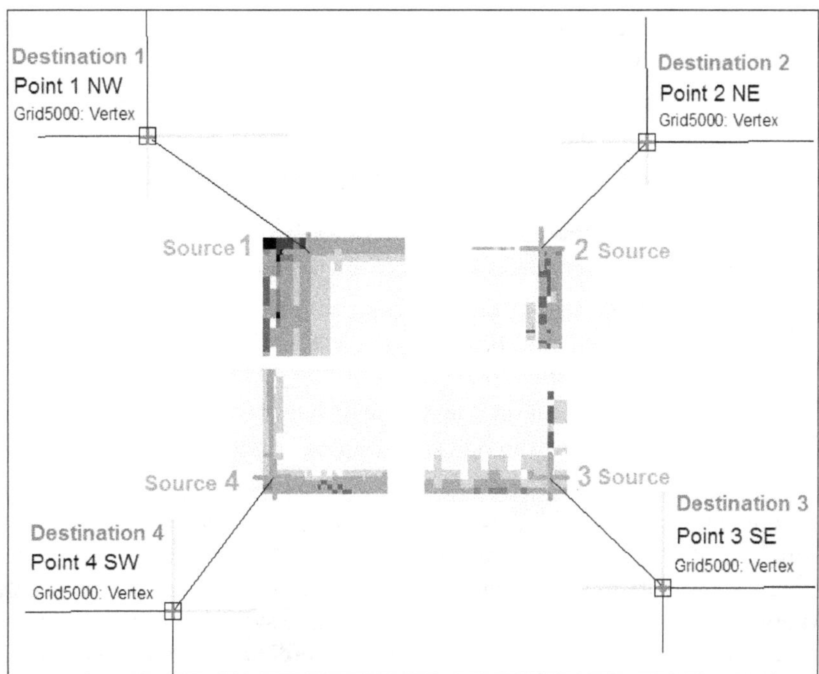

8. Use the bookmarks **RasterNE**, **RasterSE**, and **RasterSW** to repeat step 7 for the next three points. To line up the raster to the reference layer, click **Auto Adjust**, and select the **Transformation: 1st Order Polynomial (Affine)** method in **Link Table**. You should obtain similar values to the ones shown in the following screenshot:

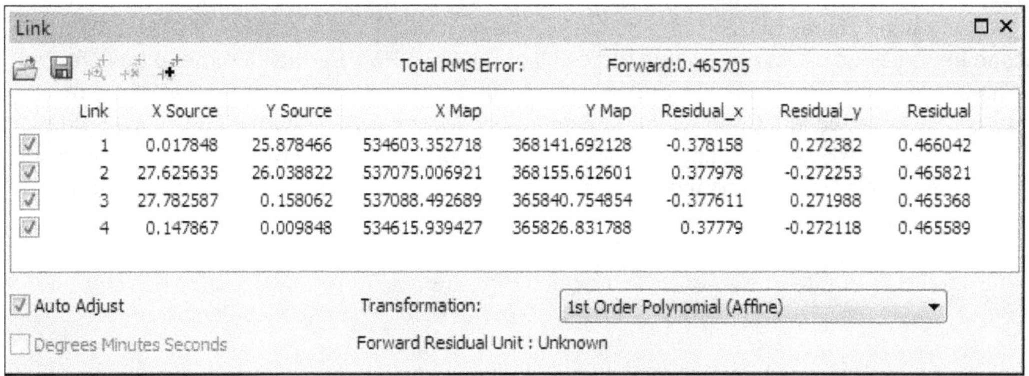

	Link	X Source	Y Source	X Map	Y Map	Residual_x	Residual_y	Residual
☑	1	0.017848	25.878466	534603.352718	368141.692128	-0.378158	0.272382	0.466042
☑	2	27.625635	26.038822	537075.006921	368155.612601	0.377978	-0.272253	0.465821
☑	3	27.782587	0.158062	537088.492689	365840.754854	-0.377611	0.271988	0.465368
☑	4	0.147867	0.009848	534615.939427	365826.831788	0.37779	-0.272118	0.465589

9. Notice the estimated resultant residual values and RMS Error. If you make a mistake while adding links, select the wrong link by clicking on the row and click on the **Delete Link** tool or press the *Delete* key. Add the link again by repeating step 6. If you have not obtained an RMSE equal to or less than 0.5 meter, you should delete the link with larger residual errors and add it again.
10. Save the links to a text file using the **Save** tool from **Link Table**. Save them in `<drive>:\PacktPublishing\Data\CRS\MyGeoref` as `L35_111_DB_2III_EPSG3844.txt`. You can see the file using any text editor (for example, WordPad).
11. You can remove all points from the **Link Table** section with the **Delete Link** tool and restore them with the backup file you just created using the **Load** tool.

    Let's go to the last step of the georeferencing process:

12. On the **Georeferencing** toolbar, select the **Georeferencing** menu, and choose **Rectify** to save the georeferencing adjustments of the raster.
13. Accept the default values for **Cell size**, **Resample Type**, and **Format**. For **Output Location** go to **<drive>:\PacktPublishing\Data\CRS** and select the **MyGeoref** workspace (folder). For **Name**, type `L35_111_Db_2III`. Click on **Save**.
14. Remove the control points from **Link Table**, and close the window. Add the georeferenced raster using the **Add Data** button from the **Standard** toolbar.
15. In the **Table Of Contents** section, right-click on the `L35_111_Db_2III` raster, and select **Properties**. Select the **Source** tab, and examine the **Spatial Reference** section. **Spatial Reference** is **Pulkovo_1942_Adj_58_Stereo_70**. The spatial reference was automatically inherited by the newly created raster from the **Data Frame** coordinate system. To check the coordinate system of the data frame, navigate to **View** | **Data Frame Properties** | **Coordinate System**. Click on **OK**.
16. Save the map document at `<drive>:\PacktPublishing\Data\CRC` as `MyGeoreferencing.mxd`.

How it works...

You have georeferenced a raster using a `.dxf` file as the reference layer. You can add links using the **Insert Links** button from the **Link Table** window. You can also modify the existing coordinate values directly by editing the fields of the links. The **Rectify** command has created a new raster and a world-file in a separate ASCII file with the same name as that of the raster file, with a "w" attached. Open Windows Explorer, go to `<drive>:\PacktPublishing\Data\CRS\MyGeoref`, and open the file `L35_111_DB_2III.tfw` with any text editor to see the content. This file contains the six parameters of the affine transformation associated to the new raster.

For more details regarding the world-file, please refer to `http://resources.arcgis.com/en/help/main/10.2`, and navigate to **Geodata** | **Data types** | **Rasters and images online** in ArcGIS Help (10.2).

At step 10, you saved the links in the `L35_111_DB_2III_EPSG3844.txt` file. If you open this file, you see there is no information about the spatial reference or RMS value. It is recommended that you save the RMS value, because this information should be specified in the Quality and Validity section from Metadata. According to the Infrastructure for Spatial Information in the European Community (INSPIRE) directive, metadata is an important component of an infrastructure for spatial information. **Metadata** means information describing spatial datasets and spatial data services (INSPIRE Directive 2007/2/EC, Article 3).

For further research, please refer to the INSPIRE Directive 2007/2/EC and for elements of metadata, please refer to the adoption/metadata section from: *http://inspire.ec.europa.eu*

Setting a custom coordinate reference system

In this recipe, you will define a custom projected coordinate reference system. Your CRS will have the following parameters:

- **A geodetic datum**:
    - The reference ellipsoid is at Krassowsky 1940 (semi-major axis: 6378245.000 m and inverse flattening: 298.3)
    - The fundamental astronomic point at the Pulkovo observatory with the latitude 59°46'18.550"N and longitude 30°19'42.090"E
    - Prime Meridian at Greenwich.
- **Projection**: A conformal azimuthal perspective projection (stereographic), it is also an oblique stereographic projection on a secant projection plane, with the following parameters:
    - Latitude and longitude of origin (projection pole = center of projection) are φ0 = 460 N and λ0 = 250 E
    - Scale factor (rectangular coordinates from the tangent plane to the secant plane) is 0.999,750,000
    - False easting = 500,000 and false northing = 500,000 (the origin of the rectangular coordinate system is the projected image of the projection pole)

You will consider that ellipsoidal coordinates are transformed first (conformal transformation) into spherical coordinates (considering radius of sphere equal to the mean radius of the ellipsoid at projection origin: λ0 = 250), and then coordinates are projected onto the projection plane.

Getting ready

To double-check all the parameters, please go to `http://epsg.io`, type `3844`, and click on **Search**. You will find the following details:

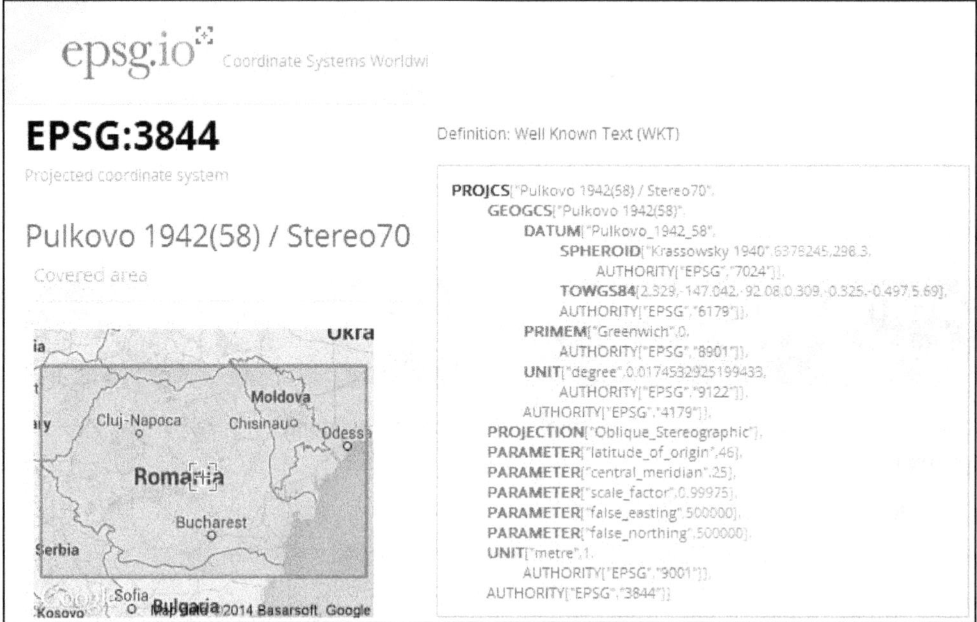

How to do it...

Let's define a custom projected coordinate reference system in ArcCatalog using the parameters from `epsg.io`:

1. Start **ArcCatalog**. In the **Catalog Tree** section, navigate to `<drive>:\PacktPublishing\Data\CRS\Topo5k.gdb`.

2. Right-click on the `Boundaries` feature dataset, and navigate to **Properties** | **XY Coordinate System**. Click on the small, gray globe icon, and navigate to **New** | **Projected Coordinate System**, as shown in the following screenshot:

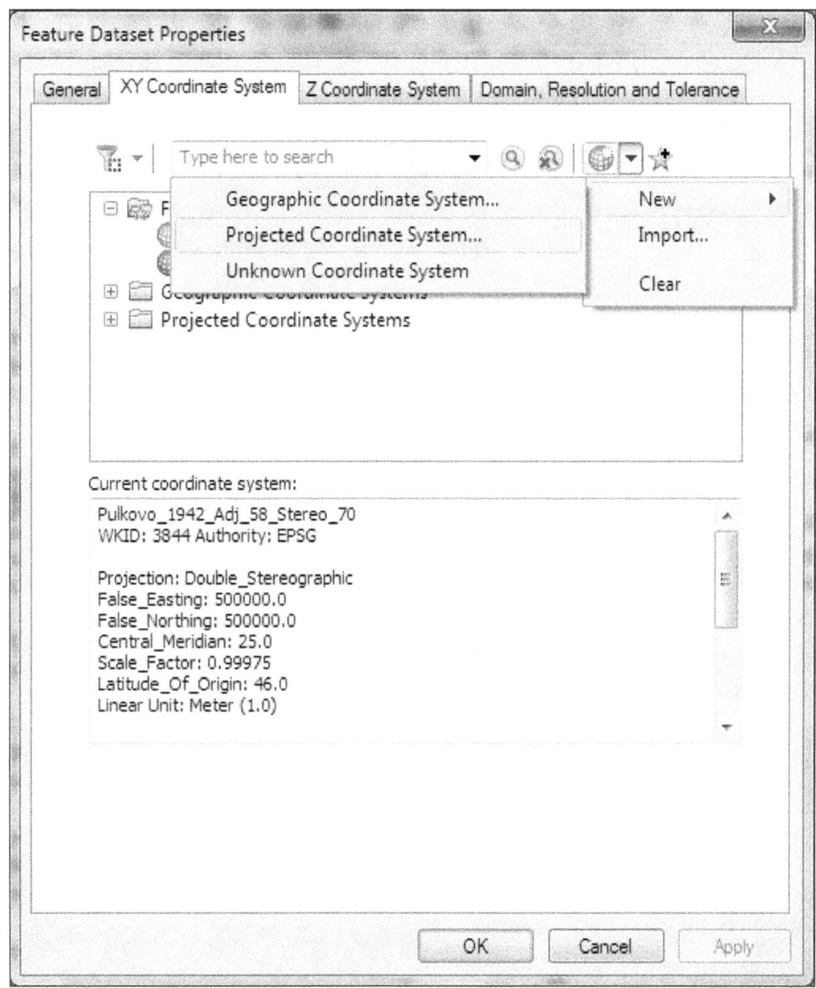

3. In the **New Projected Coordinate System** window (step **1**), for **Name**, type `ObliqueStereographic70`.

4. At step (**2**), you will define the project method. Navigate to **Projection** | **Name**, select **Double_Stereographic**. Type the parameters of projection at step (**3**). For **Linear Unit**, select **Meter** , and the value **1** for **Meters per unit**. The following screenshot guides you through these steps diagrammatically:

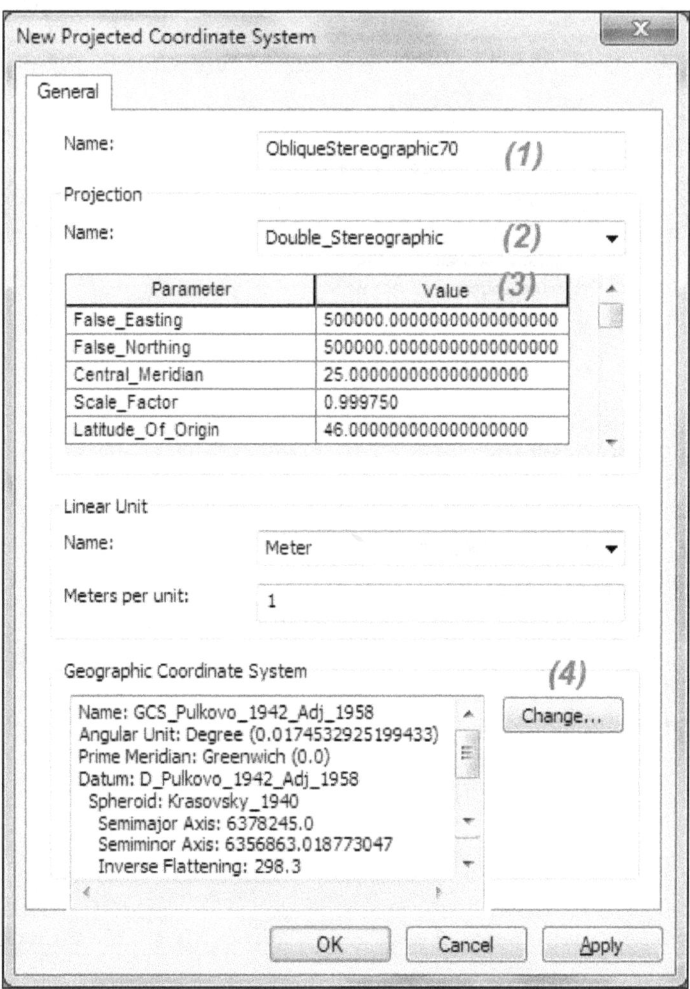

5. At step (**4**) you will choose the datum of projection. Select **Change** to open the **Geographic Coordinate System** window. Navigate to **Geographic Coordinate Systems** | **Europe**, and select **Pulkovo 1942 Adj 1958**. Click on **OK**.

6. In **Feature Dataset Properties**, there is a new folder, **Custom**, that contains your custom projection. Select the custom projection, and right-click to select **Add To Favorites**.
7. Close the window without applying the new projection. Click on **Cancel**.
8. You can see and use the projection from the **Favorites** folder in the **ArcCatalog**, **ArcMap**, or **ArcToolbox** context.

How it works...

When you created the custom projected coordinate reference system, it was created as a virtual **Custom** folder that stores the projection file (`.prj`). You have chosen to export it to the **Favorites** folder in order to have access at the file content. You can find the newly created file at `C:\Users\<user>\AppData\Roaming\ESRI\Desktop10.2\ArcMap\Coordinate Systems`, as shown in the following screenshot:

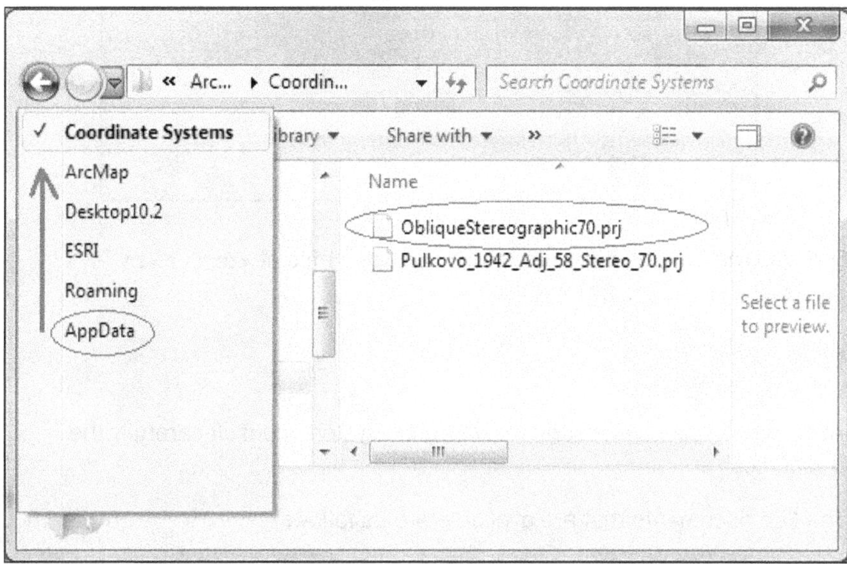

The `.prj` file is a text file that stores the parameters of the geodetic coordinate reference system (for example, datum, prime meridian, and scale factor) and the vertical reference system. To see the content of the file in Windows Explorer, right-click `ObliqueStereographic70`.prj, and choose WordPad to open the file:

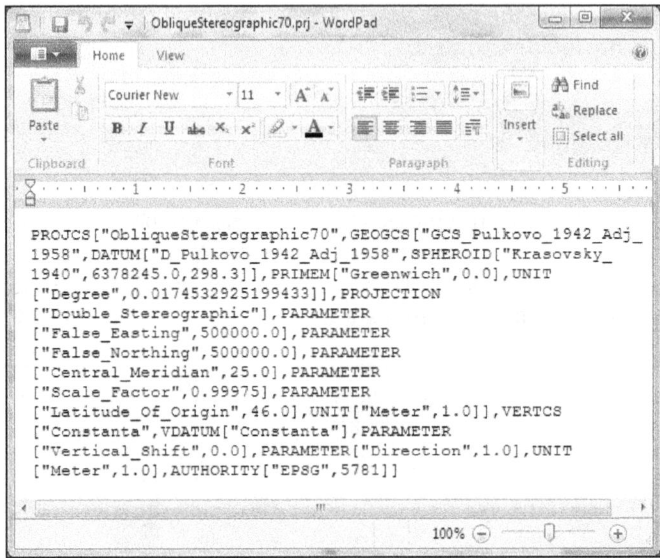

You can find also the `ObliqueStereographic70.prj` file at `<drive>:\PacktPublishing\Data\CRS\MyCRSResults`.

See also

If you want to define your own coordinate reference system, consult carefully the following resources:

- The Esri documents that are available are as follows:
  - `geographic_coordinate_systems.pdf`
  - `projected_coordinate_systems.pdf`
  - `geographic_transformations.pdf`
- INSPIRE specifications on coordinate reference systems are available at `http://inspire.ec.europa.eu`.
- Databases with coordinate reference systems and transformations are available at:
  - `http://www.epsg-registry.org`
  - `http://spatialreference.org`
  - `http://epsg.io`

4
Geoprocessing

In this chapter, we will cover the following topics:

- Spatial joining features
- Using spatial adjustment
- Using common tools
- Creating models

Introduction

According to the online **ESRI GIS Dictionary**, a **geoprocess** is a GIS operation used to manipulate data. A geoprocess implies creating new data based on existing data. In general, this happens in the following circumstances:

- When data is copied, imported, or added from one source to another source (for example, a geodatabase)
- During data conversion (for example, loading data in a feature class from a CAD file)
- When analyzing data, that is, combining features or/and attributes from one or more dataset to create a new dataset

This chapter has the following scenarios:

- Counseling the local public administration
- Promoting alternative methods of transportation cycling
- Initiating a pilot project named **VeloGIS**

The goal is to analyze the possibility of creating a cycling infrastructure in our city. We already have a file geodatabase model named `VeloGIS.gdb`, and we received five CAD datasets from two departments of municipality.

Spatial joining features

The **Spatial Join** tool is an overlay geoprocessing tool. This tool will join the attributes coming from a feature class to another feature class, based on their spatial relationship. The new feature class will have the geometry of the target feature class and the attribute fields of both the input feature classes. The spatial join is a classical solution when you want to extract data from CAD files.

Getting ready

In this recipe, you will add data from the `.dxf` source datasets to an existing geodatabase named `VeloGIS.gdb`. Your geodatabase contains one feature dataset named `Planimetry` and five empty feature classes: `Buildings1`, `Parcels1`, `RoadCenterLine`, `SideWalk`, and `Streets1`.

There are five CAD files: `Buildings.dxf`, `Parcels.dxf`, `RoadCenterLine.dxf`, `SideWalk.dxf`, and `Streets.dxf`.

How to do it...

Follow these steps to populate your `VeloGIS` geodatabase with features from the `.dxf` files, using ArcCatalog:

1. Start ArcCatalog. Navigate to `<drive>:\PacktPublishing\Data\Geoprocessing`, and select the `DXF` folder to see the five `.dxf` files: `Buildings`, `Parcels`, `RoadCenterLine`, `SideWalk`, and `Streets`.
2. Expand the `Buildings` CAD feature dataset, and preview the features from the CAD `Polygon` feature class. Right-click on the feature dataset, and navigate to **Properties** | **General**. The **Coordinate Systems** field is unspecified. Click on the **Edit** button, navigate to **Add Coordinate System** | **Import**, and then go to `<drive>:\PacktPublishing\Data\Geoprocessing\VeloGIS.gdb`. Select the `Planimetry` feature dataset, and navigate to the **Add** | **OK** buttons to return to **CAD Feature Dataset Properties**. Explore by yourself the contents of the **Coordinates** and the **Details** tab. Click on **OK** to close the dialog. Repeat the step for `Parcels`, `SideWalk`, and `Streets`.

3. Go to <drive>:\PacktPublishing\Data\Geoprocessing\ VeloGIS.gdb\Planimetry. Right-click on the Buildings1 feature class and navigate to **Load** | **Load Data**. Set the following parameters from the panels:
   - Set the **Input data** parameter as the ...\Data\Geoprocessing\DXF\Buildings.dxf: Polygon feature class; click on **Add**
   - Check **I do not want to load all features into a subtype**
   - **Target Field** (destination) and **Matching Source Field** (source):

     ```
     HCDC [int] = <None>
     Structure [string] = <None>
     State [string] = <None>
     ```

   - Check **Load all of the source data**
4. Click on **Finish** to load the features. Preview the Buildings1 feature class in ArcCatalog to check if the polygon streets were loaded.
5. Repeat step 3 to populate the Parcels1, SideWalk1, and Streets1 feature classes with features.

   Follow these steps to export the specific attributes as a standalone point feature class from the Annotation CAD feature class in ArcCatalog.

6. Let's return to ...\Data\Geoprocessing\DXF\Buildings.dxf. Expand the Buildings CAD feature dataset. Right-click on the Annotation feature class, and navigate to **Export** | **To Geodatabase (single)**.
7. For **Output Location**, select VeloGIS.gdb, and click on **Add**. For **Output Feature Class**, type DestinationBuilding. For **Expression (optional)**, select the **SQL** icon from the right-hand side. In the **Query Builder** dialog, build the following expression: "Layer" = '0'. How do you do that? Double-click on "Layer" and select the **Equal** button. Click on the **Get Unique Values** button, and double-click on the 0 value to add it into the expression. Click on **OK** to return to the **Feature Class to Feature Class** dialog.
8. In **Field Map (optional)**, delete all the fields except **Text (Text)**. The **Text** field contains the values for the **Destination** field in the Building feature class. Click on **OK**.
9. Right-click on the Annotation feature class once again, and navigate to **Export** | **To Geodatabase (single)**.
10. For **Output Location**, select VeloGIS.gdb, and click on **Add**. For **Output Feature Class**, type StructureBuilding. For **Expression (optional)**, select the **SQL** icon from the right-hand side. In the **Query Builder** dialog, build the following expression: "Layer" = 'Material and State' AND "Height" = 1.5. Click on **OK** to return to the **Feature Class to Feature Class** dialog. In **Field Map (optional)**, delete all the fields except **Text (Text)**. Click on **OK**.

11. Right-click on the `Annotation` feature class, and navigate to **Export** | **To Geodatabase (single)**. For **Output Location**, select `VeloGIS.gdb`, and click on **Add**. For **Output Feature Class**, type `StateBuilding`. For **Expression (optional)**, build the following expression: `"Layer" = 'Material and State' AND "Height" = 1`. Click on **OK** to return to the **Feature Class to Feature Class** dialog. In **Field Map (optional)**, delete all the fields except **Text (Text)**. Click on **OK**.

    Follow these steps to add the attributes to the `Buildings` feature class using the **Spatial Join** tool in ArcCatalog.

12. Open **ArcToolbox**, and navigate to **Analysis Tools** | **Overlay**. Double-click on the **Spatial Join** tool, and follow the steps shown in the following screenshot:

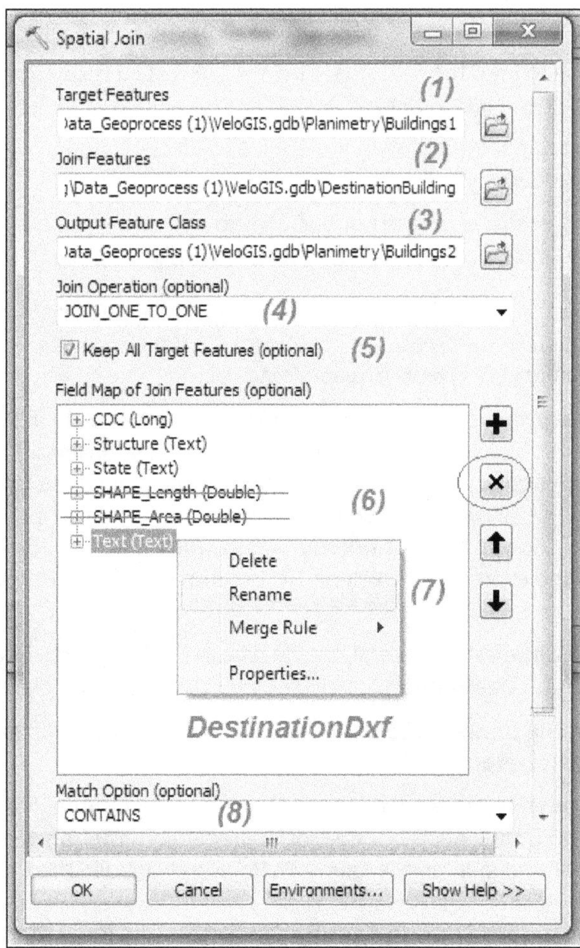

13. For **Target Features**, select `VeloGIS.gdb\Planimetry\Building1`, and click on **Add** (step **1**). For **Join Features**, select `VeloGIS.gdb\DestinationBuilding`, and click on **Add** (step **2**). For **Output Feature Class**, type `Building2`, and click on **Save** (step **3**). For **Join Operation (optional)**, select **JOIN_ONE_TO_ONE** (step **4**). Check the **Keep All Target Features (optional)** option (step **5**).
14. In the **Field Map of Join Features (optional)** section, erase the **SHAPE_Length** and **SHAPE_Area** fields (step **6**). Rename the **Text (Text)** field to `DestinationDxf` (step **7**). For **Match Option (optional)**, select `CONTAINS`, and click on **OK** (step **8**).
15. Again use the **Spatial Join** tool to add two more fields: the **StructureDxf** field from the `StructureBuilding` standalone feature class and the **StateDxffield** field from the `StateBuilding` standalone feature class, as shown in the following two screenshots:

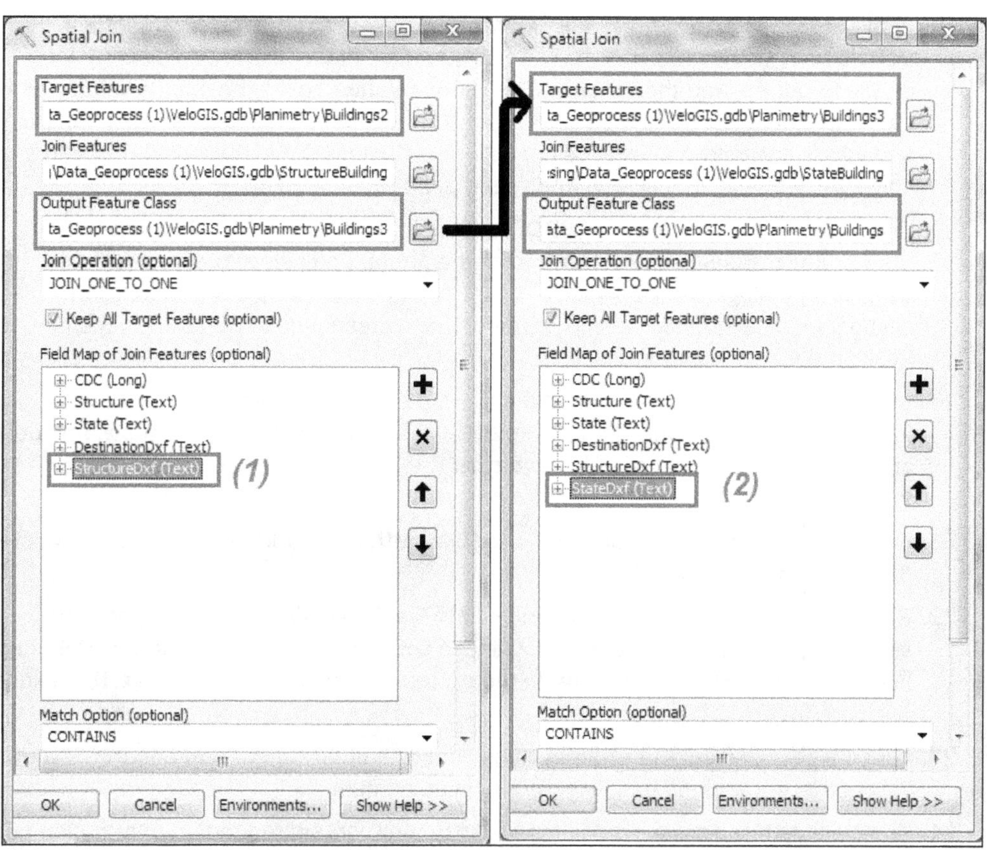

16. We have obtained the `Building` feature class with the field attributes: **DestinationDxf**, **StructureBuilding.dxf**, and **StateBuilding.dxf**. We should populate the **Destination** fields (the **CDC** subtype field), **Structure** (domain field), and **State** (domain field) with the corresponding fields from the preceding screenshot. Inspect by yourself the codes of the subtype field and the valid values for the domains.

    Follow these steps to add subtypes and domain values in the `Buildings` feature class using the ArcMap context menu.

17. Close the ArcCatalog application. Open ArcMap, and add the `Building` feature class in the **Table Of Contents** section. Right-click on the `Building` layer, and select **Open Attribute Table**. Erase the **Join_Count** and **TARGET_FID** fields by right-clicking on the field and selecting **Delete Field**. Click on the drop-down arrow next to the **Table Options** button, and select **Select by Attributes**.
18. In the **Select by Attributes** dialog, for **Method**, select **Create a new Selection**. Double-click on `DestinationDxf` to add the field in the expression section. Click on the **IS** button, and then click on **Get Unique Values** to list the unique values for `DestinationDxf`. Double-click on the `NULL` value. In the expression section, you should see the following expression: `DestinationDxf IS NULL`. Click on **Verify** to check the expression. Click on **Apply**, and don't close the dialog.
19. Right-click on the **Destination (CDC)** name field, and select **Field Calculator**. In the warning message, check **Don't warn me again**, and select **Yes**. In the **Field Calculator** dialog, check if you see the `CDC=` expression above the expression section. In the expression section, type `0`; this corresponds to the `Unknown` subtype. Click on **OK**.
20. Return to the **Select by Attributes** dialog. In the expression section, build the following expression: `DestinationDxf = 'dwelling '`. Click on **Apply**, and don't close the dialog. Repeat step 18, and type `1`; this corresponds to the `Dwelling` subtype. Click on **OK**.
21. Repeat the steps for all values from the **DestinationDxf** field to populate all rows of the **Destination** column.
22. Right-click on the **Structure** field, and select **Field Calculator**. In the expression section, add the **StructureDxf** field. Click on **OK**. Right-click on the **State of Building** field, and select **Field Calculator**. In the expression section, add the **StateDxf** field. Click on **OK**.
23. For both fields, select the **<Null>** values, and with **Field Calculator**, add the `"Unk"` values that correspond to the `Unknown` value domain.
24. Erase all the three source **dxf** fields because they contain redundant information now.

25. In the **Catalog** window, erase all redundant feature classes, such as `Buildings1/2/3` and `Destination/ Structure/StateBuilding`, as shown in the following screenshot:

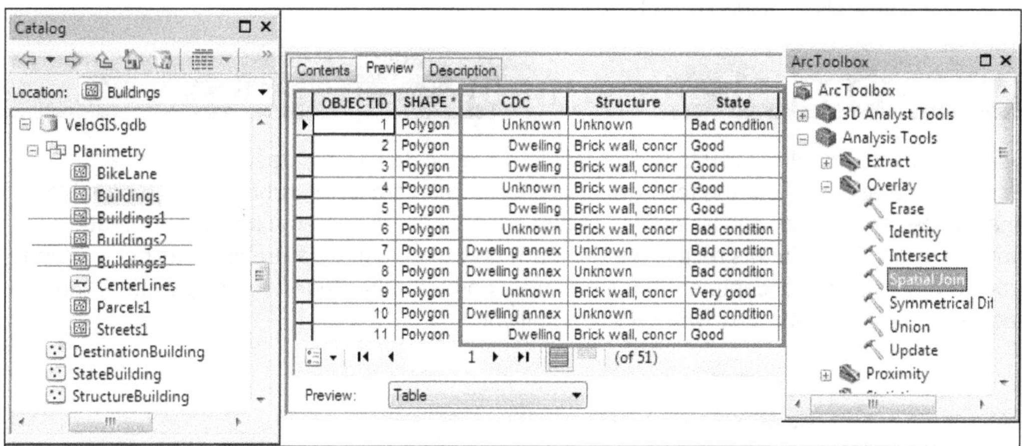

26. Add features and attributes to the `Parcels`, `SideWalk`, and `Streets` feature classes from the corresponding `.dxf` files.

You can find the `VeloGIS` file geodatabase and the expressions files (steps 10 and 11) in `...\Data\Geoprocessing\MySpatialJoinResults`.

How it works...

At step 2, we defined the coordinate reference system for every `.dxf` dataset. ArcCatalog has created a projection file (`prj`) for every CAD feature dataset. Before importing a CAD file into the `VeloGIS.gdb` geodatabase, it is necessary to examine carefully the table of the CAD annotation, polygon, or polyline feature class. It's important to identify the relevant attributes for our feature class before loading the features and attributes in our geodatabase.

There's more...

Open **Windows Explorer**, and notice the size of our `VeloGIS.gdb` file geodatabase—it's around 1.5 MB. In ArcCatalog, right-click on `VeloGIS.gdb`, and navigate to the **Administration | Compact Database** option. Read carefully what the compact process means. Return to Windows Explorer, and notice the size of our `VeloGIS.gdb`—it's around 500 KB.

See also

- If you want to see all your attribute values for the `Building` layer as complex labels in **Data View**, please refer to *Chapter 6, Building Better Maps*

Using spatial adjustment

Spatial adjustment allows you to spatially adjust the location of a dataset based on a reference dataset through the following processes: transformation, rubbersheet, and edge snap. In general, these processes are used to improve the planimetric accuracy of features from a feature class. In this recipe, we will learn how to adjust a feature class using an affine transformation. The concepts and recommendations from the *Georeferencing raster data* recipe of *Chapter 3, Working with CRS*, are still valid.

Getting ready

In the previous recipe, we added data from the `.dxf` source datasets to the `VeloGIS` geodatabase as feature classes. The vectors that came from the `RoadCenterLine.dxf` dataset have a local *cartesian coordinate system*. We will consider that the derivate `CenterLines` feature class has the *source coordinate system*.

All other `dxf` datasets are in the right coordinate system, that is, Pulkovo 1942 Adj 58 Stereo 70 (EPSG:3844). We will consider that the derivate `Streets` and `Parcels` feature classes have the *destination or target coordinate system*.

The displacement links are connecting the source location to the destination location. The affine transformation needs at least three links in order to perform the transformation. We will create those links manually using a control point file named `ControlPoints.txt`. This text file contains a list of five destination coordinates (*X*, *Y*).

How to do it...

Follow these steps to perform a spatial adjustment in ArcMap:

1. Start ArcCatalog. Go to `<drive>:\PacktPublishing\Data\Geoprocessing`, and select the `DXF` folder to see the `RoadCenterLine.dxf` dataset. Expand the `RoadCenterLine` CAD feature dataset and preview the features from the CAD `Polyline` feature class. Right-click on the feature dataset, and navigate to **Properties** | **General**. The coordinate system is not specified. Click on the **Coordinates** tab to see **CAD Dataset Extends**. The coordinate values are very small, and this indicates that vectors were created using an ungeoreferenced scanned map as background source.

2. Go to `..Geoprocessing\ VeloGIS.gdb\Planimetry`. Right-click on the `CenterLines` feature class, and navigate to **Load** | **Load Data**. Set the following parameters from the panels:
    - Set the **Input data** parameter as `...\Geoprocessing\DXF\RoadCenterLine.dxf:Polyline`
    - Check **I do not want to load all features into a subtype**
    - **Target Field** (destination) and **Matching Source Field** (source):

      ```
      Type [short int] = <None>
      Name [string] = <None>
      ```

    - Check **Load all of the source data**
3. Click on **Finish** to load the features. Preview the `Buildings1` feature class in ArcCatalog to check if the polygon streets were loaded.
4. Preview the `CenterLines` feature class in ArcCatalog to check whether the polyline streets were loaded.
5. Close ArcCatalog. Open ArcMap, and add the `Streets`, `Parcels`, and `CenterLines` feature classes. In the **Table Of Contents** section, right-click on the `CenterLines` layer, and select **Zoom To Layer**. Inspect the data.
6. Let's add the **Spatial Adjustment** toolbar by navigating to **Customize** | **Toolbars**. On the **Editor** toolbar, navigate to **Editor** | **Start Editing**.
7. We will declare the layer we want to adjust from the **Spatial Adjustment** toolbar. Navigate to **Spatial Adjustment** | **Set Adjust Data**. Click on the **All features in these layers** option, and check only `CenterLines`. Click on **OK**.
8. We will choose the adjustment method by navigating to **Spatial Adjustment** | **Adjustment Methods** | **Transformation-Affine**.
9. On the **Editor** toolbar, go to **Editor** | **Snapping** | **Snapping Toolbar** and check only **End Snapping**.
10. Open **View Link Table** from the **Spatial Adjustment** toolbar. The table is empty. You will add the control points from a text file. Navigate to **Spatial Adjustment** | **Links** | **Open Control Points File**. Select the `ControlPoints.txt` file from `<drive>:\PacktPublishing\Data\Geoprocessing`. Click on **Open**.

11. There are five *destination points*, as shown in the following screenshot:

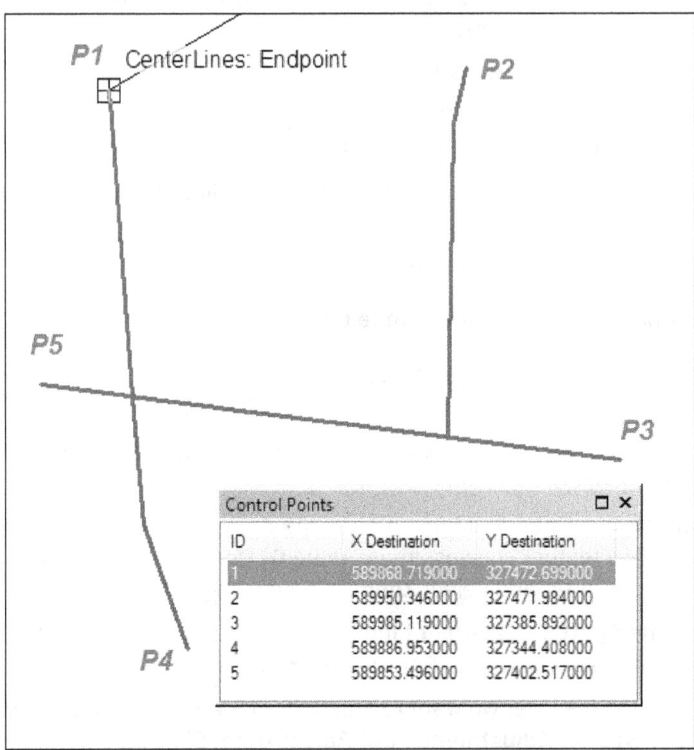

ID	X Destination	Y Destination
1	589868.719000	327472.699000
2	589950.346000	327471.984000
3	589985.119000	327385.892000
4	589886.953000	327344.408000
5	589853.496000	327402.517000

12. Let's add the *source points*. Right-click on the first point from the **Control Points** window, and select **Add Link**. Click on the first point (**P1**) from the ArcMap display area to connect the link with the source point. Use the *F5* key from time to time to refresh the view and see the links.

13. Repeat step 10 for all points. Close the **Control Points** window, and open **View Link Table** from the **Spatial Adjustment** toolbar, as shown the following screenshot:

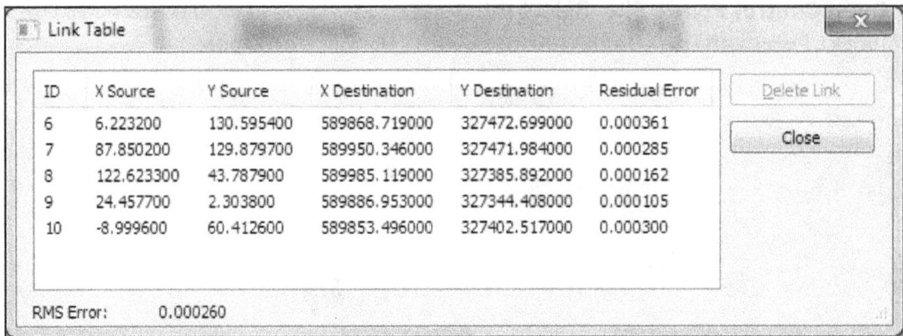

ID	X Source	Y Source	X Destination	Y Destination	Residual Error
6	6.223200	130.595400	589868.719000	327472.699000	0.000361
7	87.850200	129.879700	589950.346000	327471.984000	0.000285
8	122.623300	43.787900	589985.119000	327385.892000	0.000162
9	24.457700	2.303800	589886.953000	327344.408000	0.000105
10	-8.999600	60.412600	589853.496000	327402.517000	0.000300

RMS Error: 0.000260

14. Notice the estimated individual residual error and the RMS error. Save the links to a text file by navigating to **Spatial Adjustment** | **Links** | **Save Links File**. Save it in `<drive>:\PacktPublishing\ Data\Geoprocessing` as `AdjustmentLinks.txt`. You can see the file using any text editor (for example, WordPad).
15. Close **View Link Table**. Go to **Spatial Adjustment**, and select **Adjust**. Your features from the `CenterLines` layer were transformed to destination coordinate system.
16. In the **Table Of Contents** section, right-click on the `Streets` layer, and select **Zoom To Layer**. Inspect the results. Stop the edit session by navigating to **Editor** | **Stop Editing**.
17. Save your map document to `<drive>:\PacktPublishing\ Data\Geoprocessing as MySpatialAdjustment.mxd`.

How it works...

It is a useful practice to apply topological rules before and after the spatial adjustment for all new feature classes derived from the `.dxf` datasets. Also, don't forget to save the RMS value; it could be mentioned in the history of your dataset (*Lineage* section) from *Metadata*.

There's more...

Using the **Attribute Transfer** tool, we will update the name of the street from the `CenterLines` and `Parcels` layers based on the attribute values of the `Streets` layer. Follow these steps to add attribute values using the **Attribute Transfer** tool from the **Spatial Adjustment** toolbar:

1. Open `MySpatialAdjustment.mxd` with ArcMap. In the **Table Of Contents** section, we have `Streets`, `Parcels`, and `CenterLines` layers. Let's add the `SideWalk` features from `VeloGIS.gdb\Planimetry`.
2. Start the edit session by navigating to **Editor** | **Start Editing**. On the **Snapping** toolbar, check only **Edge Snapping**.

3. On the **Spatial Adjustment** toolbar, navigate to **Spatial Adjustment** | **Attribute Transfer Mapping**, and go through all the steps shown the following screenshot:

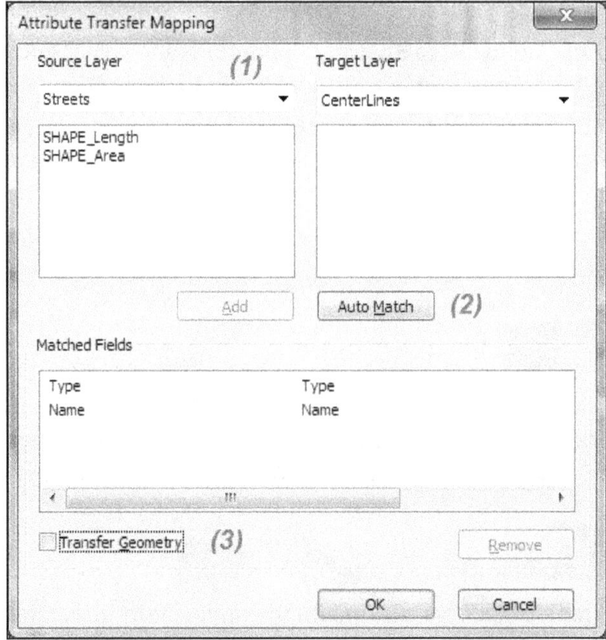

4. Uncheck the **Transfer Geometry** option (step **3**), and click on **OK**.
5. Open attribute tables for the `Streets` and `CenterLines` layers, as shown in the following screenshot:

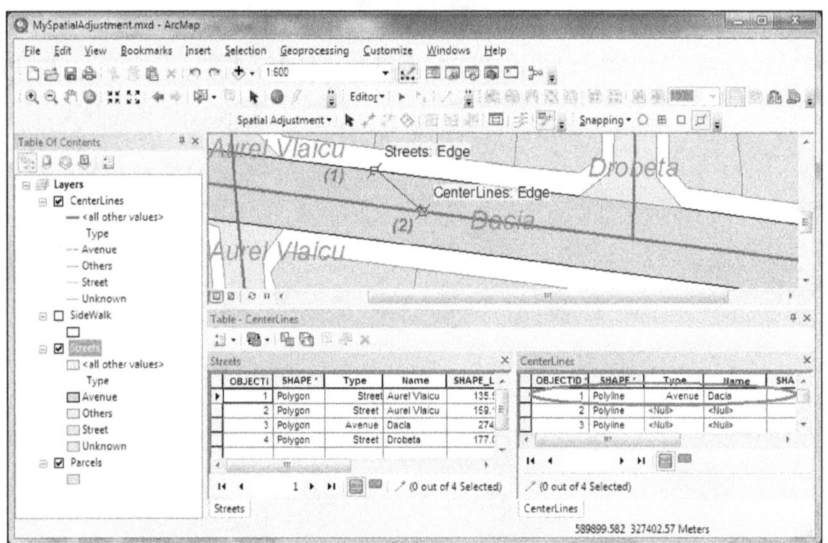

6. Select the **Attribute Transfer** tool from the **Spatial Adjustment** toolbar.
7. Click on the edge of the street in the `Streets` layer (step **1**). Click on the feature in the `CenterLines` layer (step **2**). Your features will be flashed, and the **Type** and **Name** columns of the `CenterLines` layer will be updated.
8. Repeat step 7 for every feature from the `CenterLines` layer. Save the edit by navigating to **Editor** | **Save Edits**. To update the **StreetName** field for the `Parcels` layer, please repeat steps 3 to 7.

In the **Attribute Transfer Mapping** window, we should select the `Streets` layer as source and the `Parcels` layer as target. To add the source (**Name**) and target (**StreetName**) fields in the **Matched Fields** section, use the **Add** button

9. Save and stop the edit session; save your map document.

You can find the results of this section at `...\Data\Geoprocessing\MySpatialAdjustmentResults`.

Using common tools

There are three main classes of geoprocessing tools:

- Data extraction (for example, clip and select)
- Overlay (for example, intersect and spatial join)
- Proximity (for example, buffer and point distance)

With geoprocessing tools, we can perform geoprocessing operations. The geoprocessing operations are used to obtain new feature classes or nonspatial tables for a specific analysis.

Getting ready

In this exercise, we will perform an analysis to identify the possible consequences of creating a cycling infrastructure on Dacia Avenue, taking into account the existing road network, as shown in the following image:

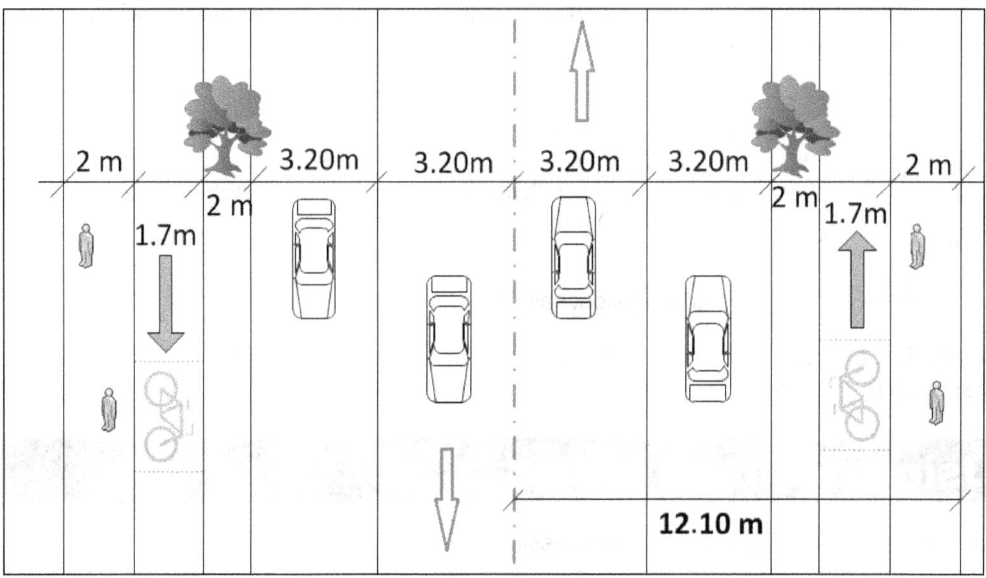

Both sides of the street contain the following elements:

- One 2-meter-wide footway for pedestrians' movements
- Two 3.2-meter-wide lanes for moving and stationary vehicles
- One 2-meter median used to separate the roadway from the cycling infrastructure
- One 1.7 meter cycle lane

Suppose that we add two 1.7-meter-wide cycle lanes to the existing road network. We should answer the following questions:

- What is the total area of the expropriated parcels?
- How many buildings should be demolished?

We will continue to use the feature classes from the previous recipe *Using spatial adjustment*.

How to do it...

Follow these steps to analyze your data using different geoprocessing operations:

1. Start ArcMap. Open the existing map document `MySpatialAnalysis.mxd` from `<drive>:\PacktPublishing\Data\ Geoprocessing\MySpatialAdjustmentResults`. At the end of this exercise, we will obtain the following results:

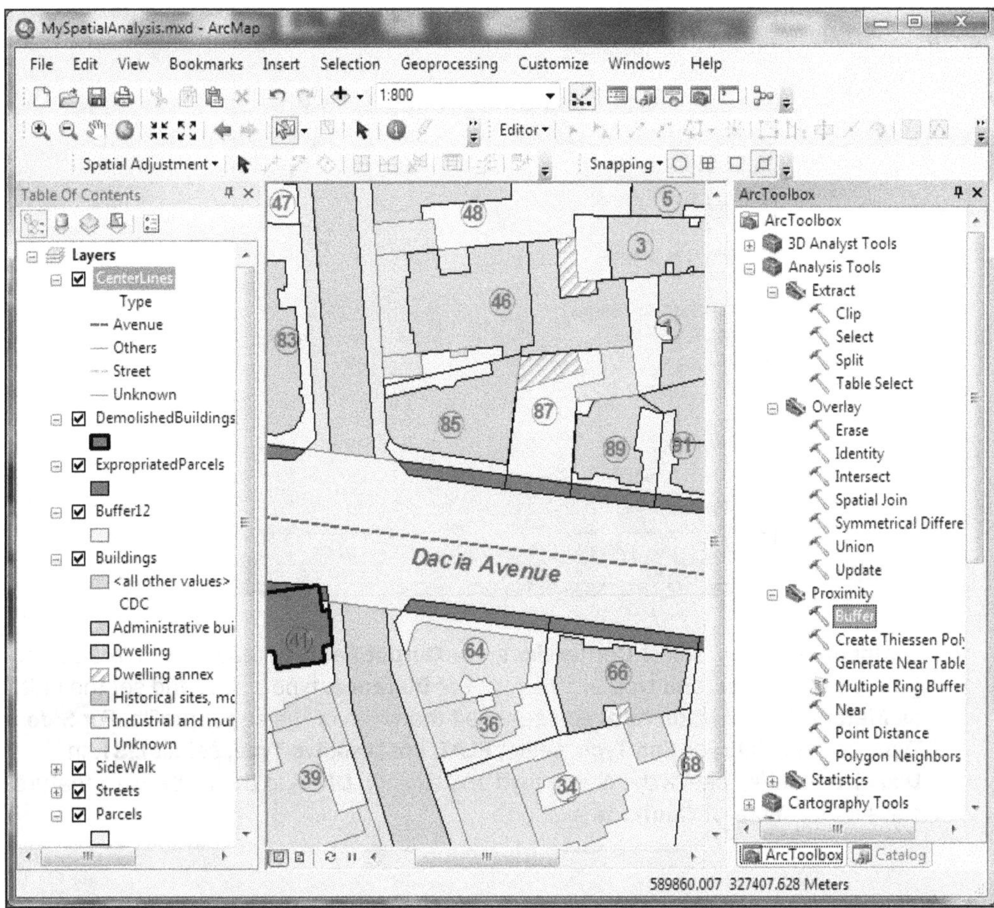

2. In the **Table Of Contents** section, right-click on the `CenterLines` layer, and navigate to **Properties** | **Definition Query** | **Query Builder**. Build the following expression: `Name = 'Dacia'`, and click on **OK**. Now, you see only the `Dacia Avenue` feature.

3. Open **Catalog** and **ArcToolbox**. In **ArcToolbox**, expand **Analysis Tools**. Go to the **Proximity** toolset, and double-click on the **Buffer** tool to open the dialog, as shown in the following screenshot:

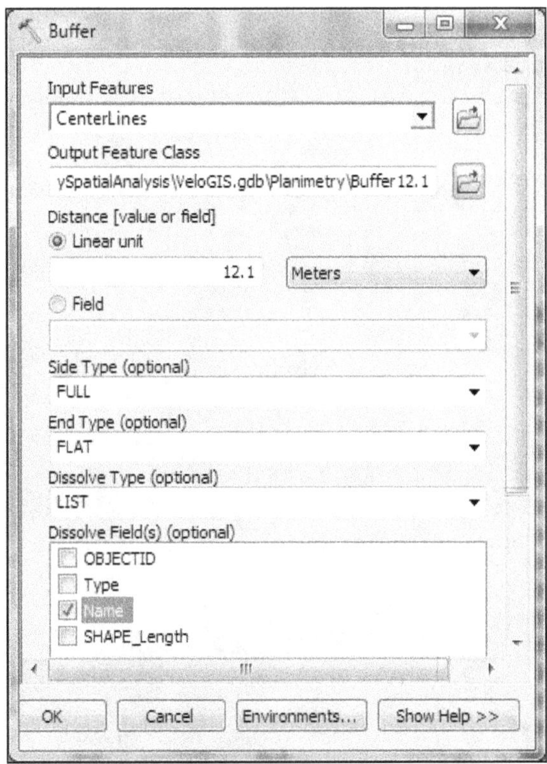

4. For **Input Features**, select **CenterLines**. For **Output Feature Class**, go to `VeloGIS.gdb\Planimetry`, and type `Buffer12`. For **Distance**, type `12.1`, and set the unit as **Meters**. The `12.1` meters represent half the width of the street profile. For **Side Type**, select **FULL**. For **End Type**, select **FLAT**. For **Dissolve Type**, select **LIST**. In **Dissolve Field(s)**, check the **Name** field and click on **OK**. Add the `Buffer12` feature class in the **Table Of Contents** section.

5. Let's extract the parcels that should be expropriated from the `Parcels` feature class:

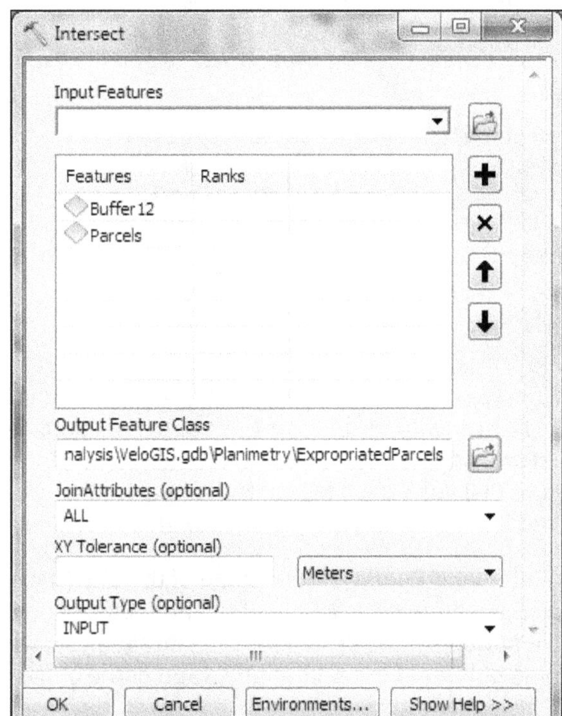

6. In **ArcToolbox**, go to the **Analysis Tools** | **Proximity** toolset, and double-click on the **Intersect** tool. As **Input Features**, select the `Buffer12` and `Parcels` layers. For **Output Feature Class**, go to `VeloGIS.gdb\Planimetry`, and type `ExpropriatedParcels`. Click on **Save**. For **Join Attributes (optional)**, select **ALL**. For **Output Type**, select **INPUT**. Click on **OK**. `ExpropriatedParcels` is added to the **Table Of Contents** section. Right-click on `ExpropriatedParcels`, and select **Open Attribute Table**. Right-click on the **SHAPE_Area** field, and select **Statistics**. Inspect the statistics.

   Next, you will check if there are buildings that should be demolished.

7. From the **Selection** menu, choose **Selection By Location**. For the **Selection** method, choose **Select features from**. As **Target layer(s)**, check **Buildings**. As **Source layer**, choose **Buffer12**. For **Spatial selection method for target layer feature(s)**, choose **Intersect the source layer feature**. Check **Apply a search distance** and type `0.5` with the unit as **Meters**. Click on **OK**.

8. Save the selected buildings in a new feature class by right-clicking on the `Buildings` layer and selecting **Data Export Data**. For **Export**, choose **Selected features**. For **Output feature class**, go to `VeloGIS.gdb\Planimetry`, and type `DemolishedBuildings`. Click on **OK** and **YES** to add the exported data as a layer in your map.
9. Save the map document as `MySpatialAnalysis.mxd` to `...\Data\Geoprocessing\MySpatialAdjustmentResults`.

You can find the results of this section at `...\Data\Geoprocessing\MySpatialAnalysisResults`.

How it works...

Because we decided to analyze only Dacia Avenue, at step 2, we used the **Definition Query** option to display only the `Dacia` features from the `CenterLines` layer. The **Buffer** tool will create polygon zones just around the drawn features. By choosing **LIST** for **Dissolve Type**, and the **Name** field for **Dissolve Field(s)**, we will tell the **Buffer** tool to create a single polygon zone around all features named Dacia. In this way, we worked with the minimum necessary data.

At step 6, we chose the **Intersect** (**Overlay**) tool instead of the **Clip** (**Data extraction**) tool because we want the output feature class to inherit the attributes of both the input feature classes. If we explore the attribute table of `ExpropriatedParcels`, we will notice some difference in `StreetName` of the `Parcels` layer and `Name` of the buffer (Dacia). In this way, we can make the following identifications:

- The errors in attribute values for the `Parcels` feature class
- The parcels that belong to the neighboring streets and that will be affected by the new cycling infrastructure

Creating models

According to the online ESRI GIS Dictionary, a **model** is an abstraction of reality used to represent objects, processes, and events. In ArcGIS, a geoprocessing model is created with the **Model Builder** built-in application, as shown in the following screenshot:

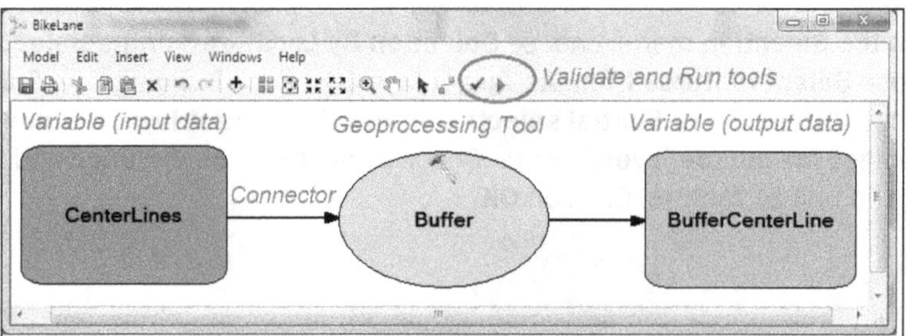

A model contains a process or more connected processes that are automatically executed when the model is run. A process consists of three elements: an input dataset (blue element), a tool (orange element), and an output dataset (green element). A geoprocessing model helps us to use various analysis operations and to create new data in a continuous workflow. In this way, the geoprocessing workflow can be reused and shared with other users. In `ModelBuilder`, you can use three basic types of elements:

- Variables (for example, input or derived data, buffer distance value, and SQL expression)
- Tools (for example, system tools such as buffer and intersect)
- Connectors (for example, data connector)

Another important aspect of a model is **Environment settings**. **Environment settings** refers to additional parameters, such as output geodatabase for the results, output coordinate system, or if the field domain descriptions should be transferred to the derived data. **Environment settings** has a hierarchical approach and affects the result of the workflow.

For more details about **Environment settings**, please refer the online *ArcGIS Help (10.2)* by navigating to **Desktop** | **Geoprocessing** | **Environment settings**.

Getting ready

In this exercise, we will build a model that will repeat the analysis steps from the previous recipe *Using common tools*.

We will use the following feature classes and tools in the **BikeLane** model:

Feature classes	**Tools (ArcToolbox)**
`CenterLines`	**Buffer** (`Distance = 12.1` as model parameter)
(Model parameter)	**Select** (`"Name" = 'Dacia'`)
	Intersect
`Parcels`	**Summary Statistics** (**SUM** for `SHAPE_Area`)
(Model parameter)	
	Select Layer By Location
`Buildings`	**Feature Class to Feature**
(Model parameter)	

The following workflow model provides an overall view of the `VeloGIS` project:

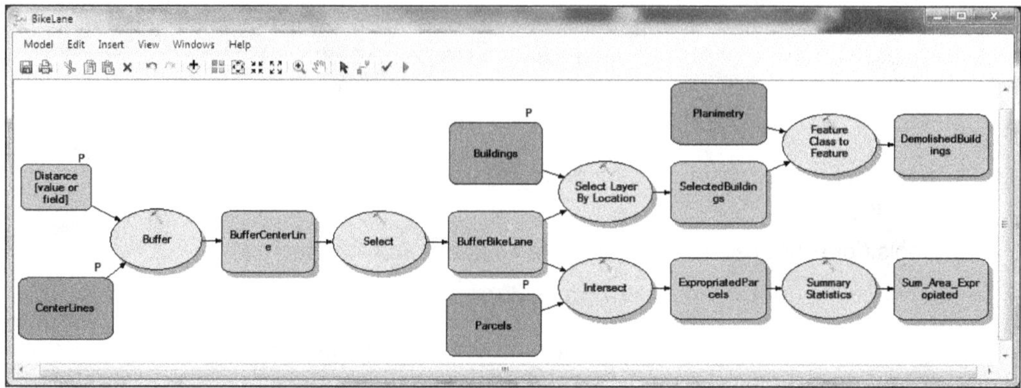

The preceding diagram shows five geoprocessing tasks:

- Create a buffer of 12.1 meters around the Dacia Avenue center line. The output is intermediate data because it is used as an input for the intersect process (the **Intersect** tool).
- Intersect the buffer with the `Parcels` layer to obtain the parcels that should be expropriated. The `ExpropriatedParcels` output is considered as an intermediate result because it is used as an input for the statistic process (the **Summary Statistics** tool).
- Create a file geodatabase table that contains the total area of parcels. The `Sum_Area_Expropiated` output table is the final data.
- Select the buildings that intersect with the 12.1-meter buffer. The output is an intermediate result, and it is used as an input for the process of creating features (the **Feature class to Feature** tool).
- Create a new feature class based on the selected buildings that should be demolished. The `DemolishedBuildings` output is the final data.

How to do it...

Follow these steps to build and use a simple model in the `ModelBuilder` environment:

1. Start ArcMap. Open the existing map document `MyModel.mxd` from `<drive>:\PacktPublishing\Data\Geoprocessing\MyModel`. Our map contains the same layers from the previous recipe. Open **Catalog** and **ArcToolbox**.
2. In **Catalog**, set the default geodatabase to `VeloGIS.gdb`. Right-click on the geodatabase, and choose **Make Default Geodatabase**.

3. Let's create the model. Right-click on `VeloGIS.gdb`, and navigate to **New** | **Toolbox**. Right-click on the new **Toolbox**, and navigate to **New** | **Model**.
4. From the **Model** menu, navigate to **Model** | **Properties** | **General**. For **Name** and **Label**, type `BikeLane`. For **Description**, type the project description from the *Getting ready* section.
5. Check the **Store relative path names** box. Click on **Apply**. Select the **Environments** tab, and expand **Workspace**. Check **Current Workspace**, and select **Values** to confirm that `VeloGIS.gdb` is your current workspace. Click on **OK**, **Apply**, and again on **OK** to return to the blank `BikeLane` model.
6. From the **Model** menu, navigate to **Model** | **Diagram Properties** | **Symbology**. Select **Style 2**, and click on **OK**.
7. Let's add some processes to the `BikeLane` model. Please follow the workflow diagram from the *Getting ready* section. Navigate to **ArcToolbox** | **Analysis Tools** | **Proximity**, select and drag-and-drop the **Buffer** tool to the `BikeLane` model window.

   The white model elements are in the first state, that is, *they are not ready to run*. This is because it is missing the input. You will add the input, as shown in the following steps.

8. From the **Insert** menu, select **Add Data** or **Tool**. From `<drive>:\PacktPublishing\Data\Geoprocessing\MyModel\ VeloGIS.gdb\Planimetry`, select the `CenterLine` feature class. Click on **Add**.
9. The `CenterLine` feature class will become the input element for the **Buffer** tool. To add a connection, select the **Connect** tool from the **Standard** toolbar. First click on `CenterLine`, and then click on the **Buffer** tool and choose **Input Features**.
10. With the **Select** tool (black arrow), double-click on the **Buffer** tool, and set the following parameters:
    - **Input Feature**: For this, the `CenterLines` feature class is already mentioned
    - **Output Feature Class**: For this, the default path is already mentioned; type the name of the new feature class—`BufferCenterLine`
    - **Distance**: This should be `12.1` with the unit as **Meters**
    - Set the last parameters from steps 3 and 4 of the *Using common tools* recipe; click on **OK**

    Now, your colored model elements are in the second state, that is, *they are ready to run*. You will continue to add tools, as shown in the following steps.

11. Navigate to **ArcToolbox** | **Analysis Tools** | **Extract**, and select and drag-and-drop the **Select** tool to the `BikeLane` model window.

12. The `BufferCenterLine` output will become the input element for the **Select Data** tool. To add a connection, select the **Connect** tool, click on `BufferCenterLine`, click on the **Select Data** tool, and choose **Input Data Element**. All our model elements are colored. With the **Select** tool (black arrow), double-click on the **Select** tool. Set the following parameters:
    - **Input Feature**: This already has the `BufferCenterLine` layer
    - **Output Feature Class**: For this, the default path is already mentioned; type only the name of the feature dataset and new feature class: `Planimetry\BufferBikeLane`
    - **Expression (optional)**: For this, type `"Name" = 'Dacia'`; click on **OK**.
13. Navigate to **ArcToolbox** | **Analysis Tools** | **Overlay**, and select and drag-and-drop the **Intersect** tool to the `BikeLane` model window.
14. The `BufferBikeLane` output will become the input element for the **Intersect** tool. To add a connection, select the **Connect** tool, click on `BufferBikeLane`, click on the **Intersect** tool, and choose **Input Features**. You will add a second input layer. From the **Insert** menu, select **Add Data or Tool**, and select the `Parcels` feature class. Click on **Add** to close and return to the `BikeLane` model window. Connect `Parcels` to the **Intersect** tool, and choose **Input Features**.
15. With the **Select** tool (black arrow), double-click on the **Intersect** tool. For **Output Feature Class**, the default path is already mentioned; type only the name of the feature dataset and new feature class: `Planimetry\ExpropriatedParcels`. Leave the other parameters unchanged, and click on **OK**.
16. Click on the **Auto Layout** and **Full Extent** buttons to rearrange the elements and better see the whole ready-to-run model. Navigate to **Model** | **Save** to save the changes.
17. Navigate to **ArcToolbox** | **Analysis Tools** | **Statistics**, and select and drag-and-drop the **Summary Statistics** tool to the `BikeLane` model window. Connect `ExpropriatedParcels` to the **Summary Statistics** tool, and choose **Input Table**. The model elements remain white, that is, they are not ready to run. Double-click on the tool. For **Output Table**, type `SUM_Area_Expropriated` at the `VeloGIS.gdb` level. For **Statistics Field(s)**, select **SHAPE_Area**. For **Statistic Type**, choose **SUM**. Click on **OK**.
18. To validate the model, select the **Validate Entire Model** tool. To run the model, navigate to **Model** | **Run Entire Model**. Click on **Close** after the process is completed.

    Now, your colored model has a drop shadow behind the tools and derived data elements. The model is in the third state, that is, *it has already been run*.

19. Close the model. In the **Catalog** window, check if the new feature classes were created.
20. We can use our **BikeLane** tool by right-clicking on it and selecting **Open**. Into the **BikeLane** dialog window, we can control only the **Environments** option and the **OK** button. We could also see the project description if we click on **Show Help**. If we click on **OK**, we can again run our tool, and our results will be in superscript in the geodatabase. Click on **OK** and on **Close**.
21. You can find the intermediated model at `...\Data\Geoprocessing\MyIntermediatedModel`.
22. As we could notice, there is some inconvenience when we use the **BikeLane** tool; we don't have control over the parameters of `BikeLane`, and the feature classes were not added in the **Table Of Contents** section.

    Let's declare the model parameters for `BikeLane` in order to have more control over the **BikeLane** tool results.

23. In **Catalog**, right-click on the **BikeLane** tool, and select **Edit**. Right-click on the `CenterLines` input, and select **Model Parameter**. Our blue input will have a letter **P**. Right-click on the **Buffer** tool, and navigate to **Make Variable | From Parameter | Distance [value or field]**. Click on **Auto Layout** to see all elements. To declare this value variable as a model parameter, right-click on the **Distance [value of field]** input value, and select **Model Parameter**. Right-click on the `Parcels` input, and select **Model Parameter**.
24. As a final improvement, we will specify the layers to be added in the **Table Of Contents** section. Right-click on the `BufferBikeLake` derived data, and select **Add To Display**. Repeat this step for the `ExpropriatedParcels` and `SUM_Area_Expropriated` derived data.

25. Save and close the model. In **Catalog**, double-click on the **BikeLane** tool. We should see the following screenshot:

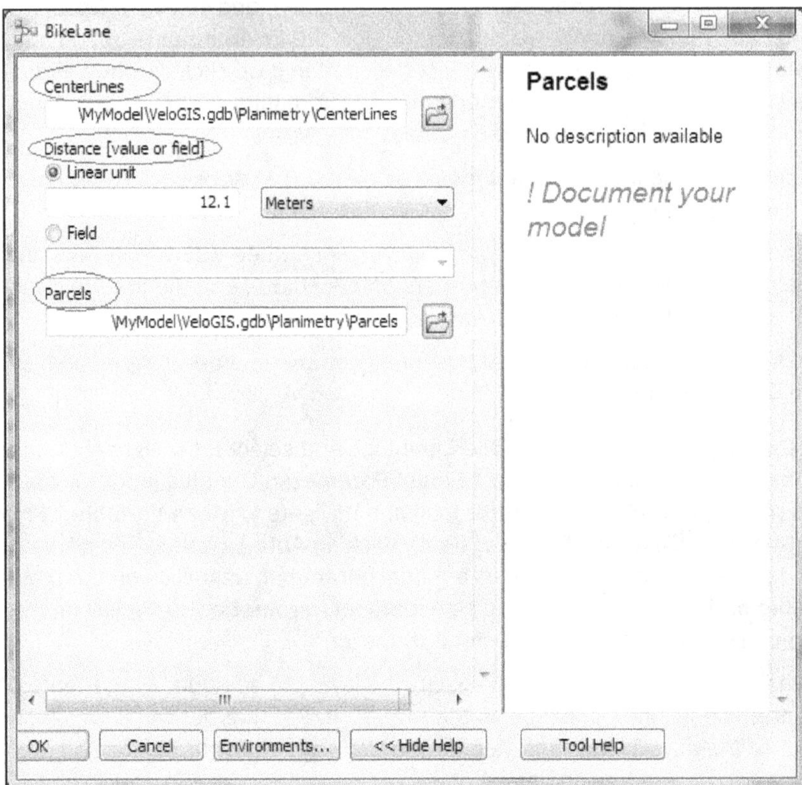

26. You can use feature classes with different names and change the buffer distance. Close the tool window.

27. Try to complete the model with more processes using the following three segments:
    - The workflow diagram from the *Getting ready* section
    - Steps 7 to 8 from the *Using common tools* recipe
    - The final model from the `MyModelResults` folder

How it works...

From the beginning, we set one of the most important parameters: `Workspace`. We defined the current workspace at the application-level environment (for all tools). The `BikeLane` model will inherit this environment setting parameter, and all intermediate and final data results will be stored by default in `VeloGIS.gdb`. If you want to check and modify the current and scratch workspace at the application level, please navigate to **Geoprocessing** | **Environments** | **Workspace**.

At step 5, we used relative paths for our model. If we change the folder where we keep the `VeloGIS.gdb` geodatabase, the model will still run properly.

At step 17, we mentioned that the `SUM_Area_Expropriated` table will be saved at the geodatabase level. A nonspatial table cannot be created in a feature dataset. Please recall from *Chapter 1, Designing Geodatabase*, that a feature dataset is a collection of feature classes that have the same coordinate system. A nonspatial table does not have a spatial component.

If we try to run the **BikeLane** tool from `...\Data\Geoprocessing\MyModelResults\VeloGIS.gdb`, using the ArcCatalog application, an error will occur: **ERROR 000732 Buildings: Dataset Buildings does not exist or is not supported**. This is because we have used the **Select Layer By Location** tool to select the buildings from the `Buildings` layer in ArcMap. This tool selects the feature from a layer (input data). A layer is specific to an ArcMap environment. In conclusion, our final model will run only in ArcMap. Open `MyModelResults.mxd`, and use the **Catalog** window to run the **BikeLane** tool.

See also

- If you want to change the symbols for your layers in ArcMap, please refer to *Chapter 5, Working with Symbology*

5

Working with Symbology

In this chapter, we will cover the following topics:

- Managing styles
- Modifying symbols
- Creating custom symbology
- Using proportional symbology
- Using the symbol levels
- Using representation

Introduction

An important aspect of map design is data **symbology**. For a final cartographic product, cartographers must classify, generalize, and symbolize different spatial data. A map reader should be able to recognize the symbols used by cartographers, and analyze and understand the reality represented in the final map.

Symbols can vary in graphic characteristics in order to express differences, hierarchy, and importance of the represented features from a qualitative or quantitative point of view. Choosing the right size, color, shape, or position (center or offset) for a symbol is crucial for the final message of the map.

In this chapter, you will learn how to create and store a set of customized symbols, and draw rules in your maps using **Style Manager** and **Representations**. You will also apply the proportional symbology according to quantitative values of the attribute value.

Managing styles

A style is a collection of symbols, colors, and map elements (such as legend, scale, north arrow, border, and reference system) used to create a map. A style set organizes symbols and map elements in a predefined structure. You can use the system styles or personal styles that contain certain elements for specific types of maps. In a system style, the gray folder icons indicate that the style set is read-only. In a new style set, the white folder icons indicate that they are empty, and the yellow folder icons indicate that they are populated with customized elements, as shown in the following screenshot:

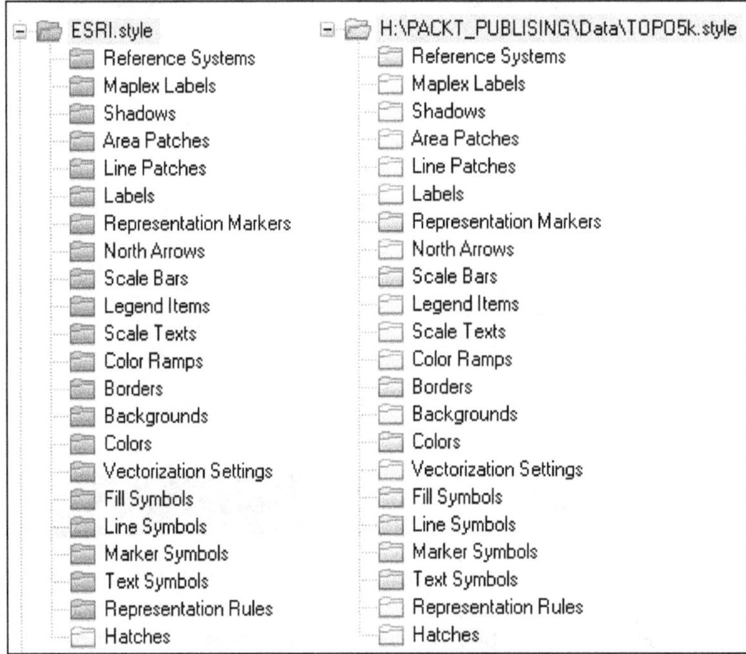

A style set is actually a single file with a .style extension that can be read in ArcMap. The style file can be accessed with Windows Explorer and saved as a backup or sent to other users by e-mail. A personal style is not dependent on the ArcMap document (*.mxd) in which it was created and can be used by different map documents.

The system style files are installed at: <ArcGIS_HOME>\Desktop10.2\Styles.

The default personal style is saved at (Windows operating system): <install_drive>:\Users\<username>\AppData\Roaming\ ESRI\Desktop10.2\ArcMap.

Getting ready

You can work with a style in ArcMap using the **Style Manager** dialog box. The **Style Manager** dialog box lets you create and organize styles and their contents. You can create, cut, copy, paste, rename, and modify the style contents in personal styles. You can also copy symbols, colors, or other map elements from system styles (read-only) and paste to your personal styles, where you can modify them. Before you modify default ArcGIS symbols or create new symbol libraries, you should create a new style set in which to keep all of your changes. In the next step, you will create a new style set for the purpose of applying standard symbols for a topographic map at scale 1:5,000.

How to do it...

Follow these steps to create a new style set using the **Style Manager** dialog box:

1. Start ArcMap, and open a new blank map document.
2. Navigate to **Customize menu** | **Style Manager**. The **Style Manager** dialog box shows two style sets. The first one is a default personal style and the second one, named ESRI.style, is a default system style.
3. Navigate to **Styles** | **Create New Style**, as shown in the following screenshot:

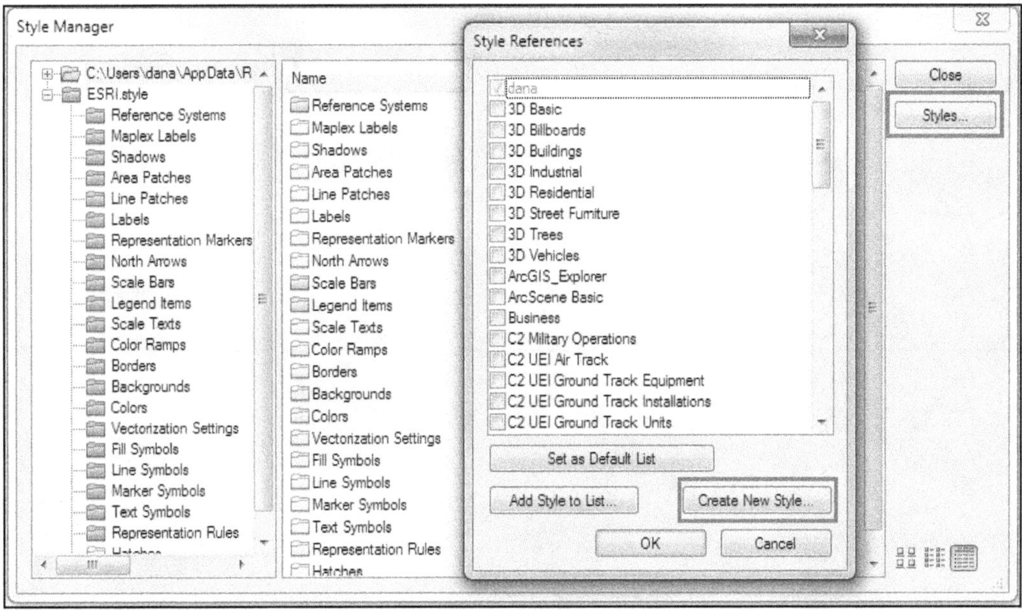

4. In the **Save As** dialog box, navigate to <drive>:\PacktPublishing\Data, type TOPO5k, and click on **Save**. Click on **OK**. The structure of TOPO5k.style is empty.

Now, copy a linear symbol from `ESRI.style` and paste it into your `TOPO5k.style`, where you can rename and modify the symbol:

1. In the **Style Manager** dialog box on the left-hand side, navigate to **ESRI.style** | **Line symbols**. In **Style Manager**, right-click on the symbol named `River, Navigable` and select **Copy**.
2. In **Style Manager** on the left-hand side, navigate to **TOPO5k.style** | **Line symbols**. In **Style Manager** on the right-hand side, in the empty space, right-click and select **Paste**. You should see something similar to the following screenshot:

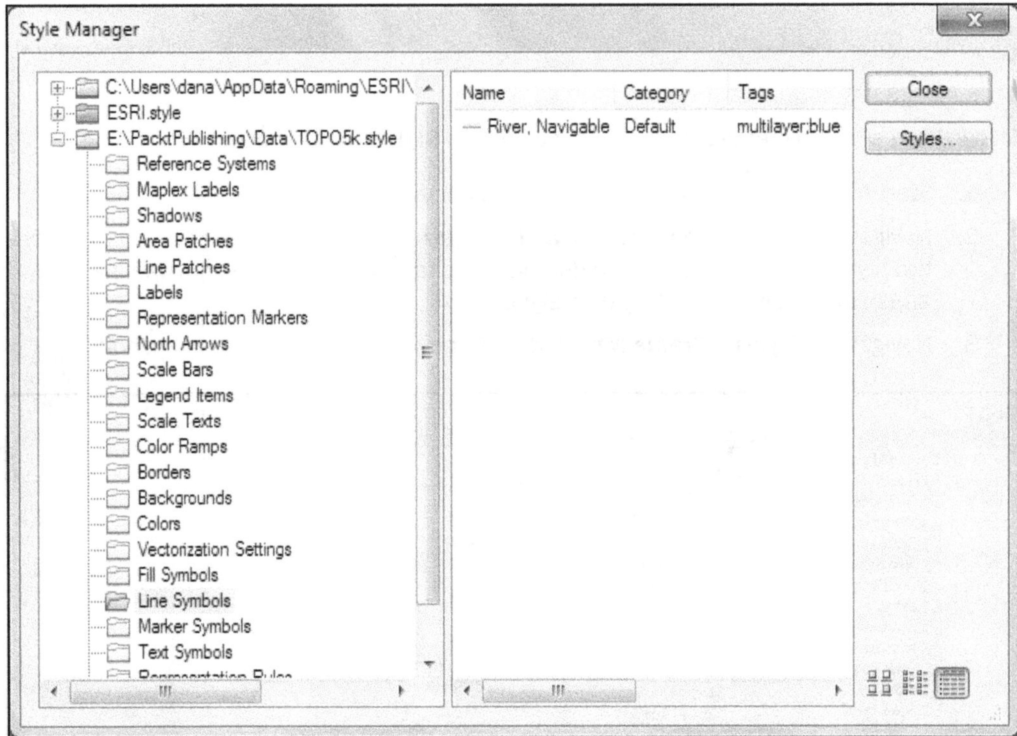

3. The symbol was saved in your personal style set. Notice that the white symbol folder icon became yellow. This indicates that the folder is populated with customized elements.
4. Close the **Style Manager** dialog box. The `TOPO5k.style` folder you created is saved even if your map document is not. Close ArcMap.

How it works...

The **Style References** dialog box lists all the system style files and the default personal style. The checked styles will be referenced into your map document. Referencing one or more styles means that its contents will be displayed in the **Symbol Selector** window. The **Style Manager** dialog box lets you see the styles you created before. You can add more styles to **Style Manager** by navigating to **Styles** | **Add Style to List**, and if you don't want to see one of them anymore, uncheck it in **Style References**. The unchecked style files are still in **Style References** and in your directory too. To erase them from your directory, you should use Windows Explorer or ArcCatalog applications.

In ArcCatalog, navigate to **Customize** | **ArcCatalog Options**. Select the **File Types** tab. Define the `.style` type by selecting **New Type**. Fill the **File extension** and **Description of type** sections. In ArcCatalog navigate to **View** | **Refresh**. With those settings, ArcCatalog will be able to see the style files.

There's more...

You can change the list of styles displayed in the **Style References** dialog box by changing the style path setting with the **ArcMap Advanced Settings Utility** parameter located at `<ArcGIS_HOME>\Desktop10.2\Utilities\AdvancedArcMapSettings.exe`.

Double-click `AdvancedArcMapSettings.exe` to open **ArcMap Advanced Settings Utility** and select the **System Paths** tab. In the **All users** section, go to **Styles path**. You can change the location of `.style` files by navigating to `<drive>:\PacktPublishing\Data`, as shown in the following screenshot:

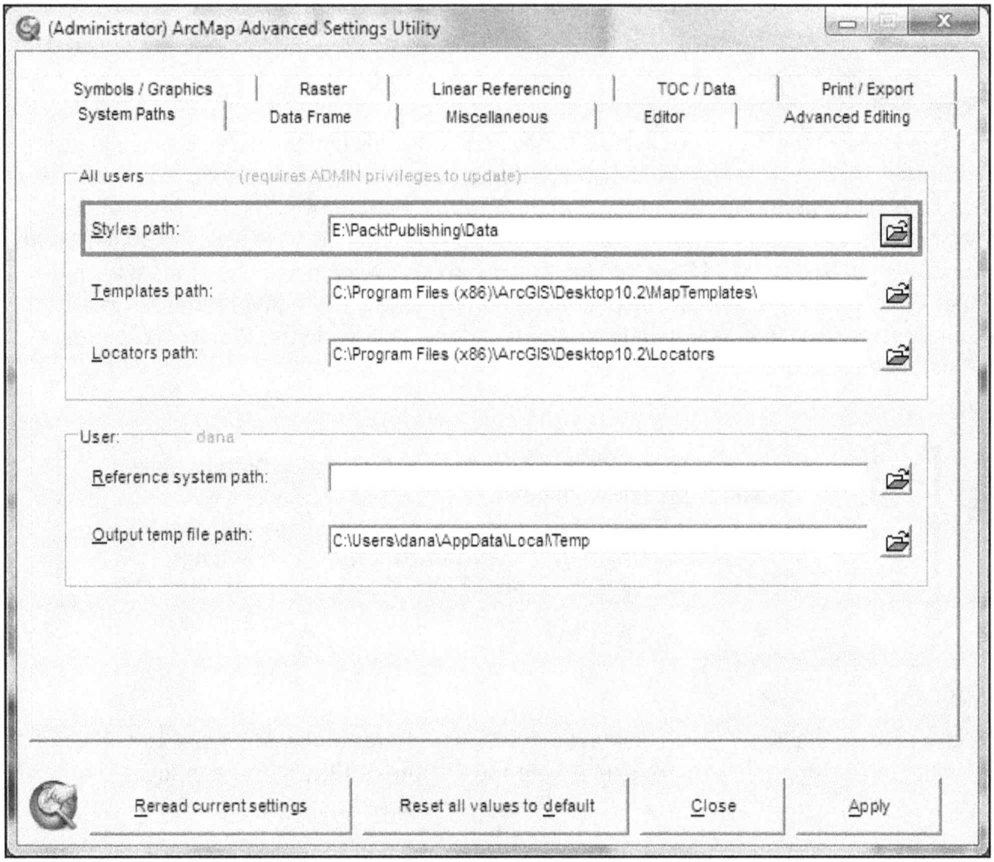

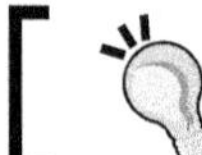

> If ArcMap is open, you must close the session and restart the application for your changes to take effect.

When you restart ArcMap, the **Style References** dialog box will show all your style files created in the previous steps. You can also **Reset all values to default** in **ArcMap Advanced Settings Utility**.

See also

- For information about modifying and creating new symbols, please refer to the upcoming *Modifying symbols* and *Creating custom symbology* recipes

Modifying symbols

The geographic features are graphically represented by symbols. Symbols have characteristics that can be varied in order to emphasize the differences, hierarchy, or importance of the features on a map.

The graphic characteristics or visual variables are referring to symbol size, color value (gray tone), hue (color), texture (pattern), orientation, and shape. These variables are applied individually or in combination to point, line, and area symbols depending on the type of data to be represented on a map.

For qualitative data, you can vary the following symbol characteristics: hue (color), texture, orientation, and shape.

For quantitative data, you can vary the following two symbol characteristics: size and color value (gray tone).

Getting ready

In the next step, we will apply standard topographic symbols for a topographic map at scale `1:5,000`.

Symbolize point and line features and try by yourself to symbolize polygon features. Firstly, symbolize one point feature class named `Elevation` and one polyline feature class named `WatercourseL`, as shown in the following screenshot:

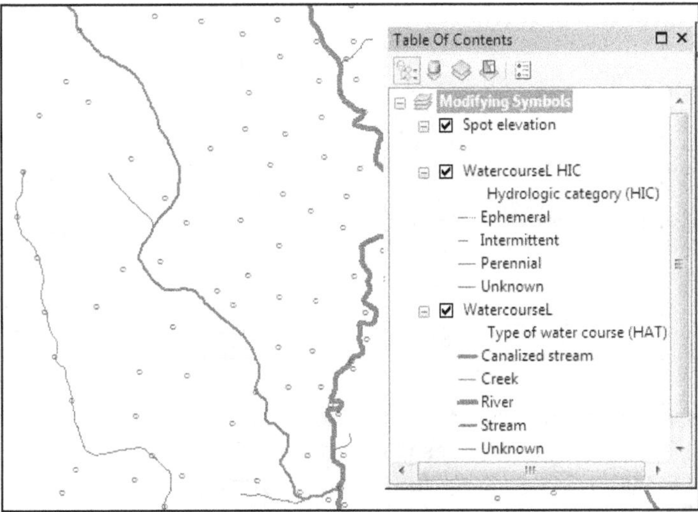

Secondly, add again `WatercourseL` in your map document, and symbolize it based on the field named `HIC`, which defines the hydrologic category.

How to do it...

Follow these steps to symbolize the `Elevation` and `TriangulationPoint` feature classes:

1. Start ArcMap and open a new map document. Click on **Add Data** button. Load the `Elevation` feature class from `<drive>:\PacktPublishing\Data\TOPO5000.gdb\Relief`.
2. In the **Table Of Contents** section, right-click on the `Elevation` layer and navigate to **Properties** | **Symbology** | **Features** | **Single symbol**. In the **Symbol** section, click on the symbol to open the **Symbol Selector** dialog box, as shown in the following screenshot:

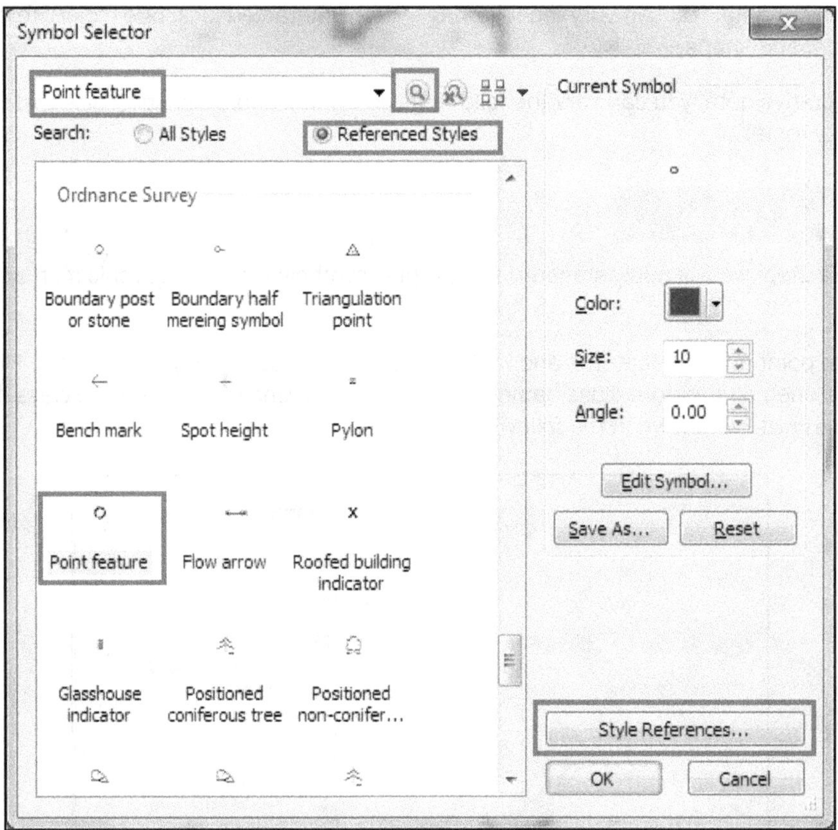

3. In the **Symbol Selector** dialog, navigate to **Style References** | **Ordnance Survey**. Click on **OK** in **Style References** to return to the **Symbol Selector** dialog . In the **Symbol Selector** dialog, check **Referenced Styles**, type `point feature` and click on the **Search** button on the right. This will find the symbol `Point feature` from `Ordnance Survey.style`.
4. Select the `Point feature` symbol. Let's change the `Point feature` symbol.

5. To change color, navigate to **Color** | **More Colors**. In the **Color Selector** dialog, select the **Color** tab and choose the **RGB (Red Green Blue)** color model. To set the color to sepia, type the values `112`, `66`, and `20` as shown in the following screenshot:

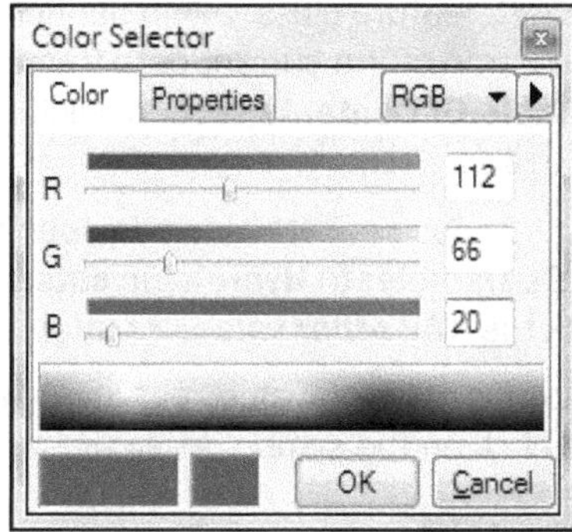

6. Click on **OK**. For **Size**, type `10` points. Select the **Edit Symbol** button to notice that your symbol is a **Character Marker Symbol** that has a single layer. Click on **OK** three times to close all dialog windows.
7. In the **Table Of Contents** section, right-click on the `Elevation` layer and select **Save As Layer File**. Save your layer as `Spot Elevation.lyr` to `<drive>:\PacktPublishing\Data`.
8. Navigate to **File** | **Save as**, and save your map document as `ModifySymbol.mxd`.

   Follow these steps to symbolize the `WatercourseL` feature class:

9. Click on the **Add Data** button, and load the `WatercourseL` feature class from `...\Data\TOPO5000.gdb\Hidrography`.
10. In the **Table Of Contents** section, the `WatercourseL` layer is symbolized based on its subtypes by default (`HAT` field). Double-click on the `WatercourseL` layer to open the **Layer Properties** window.
11. Uncheck the **<All other values>** parameter. Double-click on the line symbol of subtype with the value `4 Canalized stream`. In the **Symbol Selector** dialog, select **Color**: `Cretan Blue` (10th column, 3rd row). For **Width**, select **2**, and click on **OK**.
12. Symbolize the rest of the subtypes as follows:
    - `3 Creek`: color `Cretan Blue`, width `1`
    - `2 River`: color `Cretan Blue`, width `3`
    - `1 Stream`: color `Cretan Blue`, width `1.7`
    - `0 Unknown`: color `Mars Red`, width `0.5`
13. Select **Label**: `HAT`, type `Type of water course`, and click on **OK**.

14. Save your WatercourseL layer as WatercourseL_HAT.lyr to <drive>:\PacktPublishing\Data.
15. In the **Table Of Contents** section, uncheck the Watercourse layer.
16. To add the WatercourseL feature class, again right-click on the layer and select **Copy**. Right-click on the name of the current data frame called **Layers** and select **Paste Layer(s)**. In the **Table Of Contents** section, click on the name of the layer to rename it as WatercourseL HIC.
17. Double-click on the WatercourseL HIC layer to open the **Layer Properties** window. Change the **Value Field** parameter to **Hydrologic category** and click on the **Add All Values** button. Uncheck the **<All other values>** button.
18. Double-click on the following subtype value: 3 Ephemeral. In the **Symbol Selector** dialog, scroll down and click on the symbol Stream, Intermittent from the **ESRI** default section.
19. For **Color**, select Cretan Blue (10th column, 3rd row). For **Width**, select **1**, and click on **OK**.
20. Symbolize the remaining subtypes as follows:
    - 2 Intermittent: Dashed 6:6, color Cretan Blue, width 1
    - 1 Perennial: River, color Cretan Blue, width 1
    - 0 Unknown: River, color Mars Red, width 1
21. Save the WatercourseL layer as WatercourseL_HIC.lyr to <drive>:\PacktPublishing\Data.
22. Navigate to **File** | **Save as** and save your map as MyModifySymbols.mxd at <drive>:\PacktPublishing\Data\MapDocuments.

You can find the results at ...\Data\MapDocuments\Chapter5\ModifySymbol.mxd.

How it works...

Firstly, you have symbolized a one-point feature class named Elevation. The symbols represent qualitative (nominal) data and have the same size with the same hue (color).

Secondly, you symbolized one polyline feature class, WatercourseL based on the field named HAT, which classifies the rank of rivers. The symbols represent quantitative (ordinal) data and have different sizes with the same hue (color). You created a hierarchy by varying the size of symbol. When you use categories to symbolize the data, the **<All other values>** value refers to all other possible field values that are not defined in the list of the unique values (unmatched values).

You unchecked the **<All other values>** value because your layer is symbolized by default based on the subtypes and you assume that your data does not contain errors. You can use the **<All other values>** value if you want to visually validate the attribute values of a subtype or domain field.

You symbolized once again `WatercourseL` based on the field named `HIC`, which defined the hydrologic category. The symbols represent qualitative (nominal) data and will have a different shape with the same hue (color). You showed differences by varying the shape of symbols. You have saved symbolized layers in a separate layer file (`.lyr`). This file saves the path to the physical store of the source data and the way you choose to symbolize the features. Now, try by yourself to symbolize by navigating to the **TOPO5000.gdb** | **LandUse** | **LandUse** polygon feature class base on the field named `CAT`, which defines land use categories. The symbols will represent qualitative (nominal) data and will have a different texture with the same hue (color). You will show the differences by changing the texture of the symbol.

There's more...

Hue refers to the colors from the visible portion of the electromagnetic spectrum (wavelengths of the electromagnetic radiation from 0.4μm to 0.7μm). Hue is defined by mixing the colors Red, Green and Blue (RGB) or by mixing the colors Cyan, Magenta and Yellow (CMY). RGB is the main color system used for televisions and computer screens. CMYK is the main color system used for desktop printing (inkjet or laser) and commercial offset printing. CMYK refers to four basic ink colors: cyan, magenta, yellow and black (the K). HSV is a three-dimensional color space that link hue with value (lightness or darkness) and saturation (colorfulness or purity). When you select color for a printed map, you should use the **Color Calibration Chart** from: `<ArcGIS_HOME>\Desktop10.2\plotters\calibrate.mxd`.

Creating custom symbology

A symbol is composed of one or more graphic layers. Every symbol layer uses a point, line or area symbol, text, and color to represent a geographic feature. A symbol layer has its own graphic characteristics, such as size, color value, hue, texture, orientation, and shape. Depending on the number, order of drawing, size, form, color, offset, or other graphic characteristics, symbol layers overlap to form a single and distinct symbol.

Getting ready

In the next step, you will create standard topographic symbols for a topographic map at the scale `1:5,000`. Firstly, you will create a point symbol for the thorny shrub feature, as shown in the following screenshot:

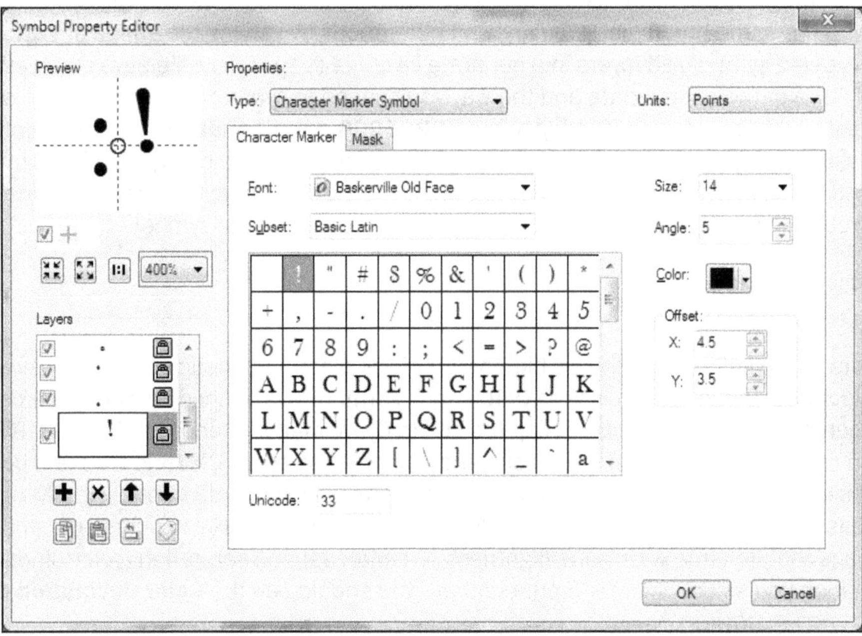

Secondly, you will create a pattern for polygon features from `LandUse` in accordance with the subtype classification field (`CAT`).

How to do it...

Follow these steps to create a new point symbol and an area symbol (pattern):

1. Start ArcMap and open a new map document.
2. Navigate to **Customize** | **Style Manager** | **Styles** | **Add Style to List**.
3. Load the custom style file from `<drive>:\PacktPublishing\Data\TOPO5k.style`. Click on **OK**.
4. In **Style Manager** on the left-hand side, select the `TOPO5k.style/ Marker Symbols` folder.
5. On the right-hand side, in the empty space, right-click and navigate to **New** | **Marker Symbol** to open the **Symbol Property Editor** dialog box.
6. In the **Layers** section, use the **Plus Sign** button to add four symbol layers.

7. Select the first layer. In the **Properties** section, for **Type**, select **Character Marker Symbol**, and for **Units**, select **Points**.
8. In the **Character Marker** tab, choose **Font**: `ESRI Default Marker`, **Subset**: `Basic Latin`, **Unicode**: `40`, and **Size**: `2.83`.
9. Select symbol layer number two, and set **Type**: **Character Marker Symbol** and **Units**: `Points`. In the **Character Marker** tab, choose **Font**: `ESRI Default Marker`, **Subset**: `Basic Latin`, **Unicode**: `33`, **Size**: `2.83`, **OffsetX**: `-2.55`, and **OffsetY**: `-3.40`.
10. Select symbol layer number three, and set **Type**: **Character Marker Symbol** and **Units**: `Points`. In the **Character Marker** tab, choose **Font**: `ESRI Default Marker`, **Subset**: `Basic Latin`, **Unicode**: `33`, **Size**: `2.83`, **OffsetX**: `-2.55`, and **OffsetY**: `3.11`.
11. Select layer number four, and set **Type**: **Character Marker Symbol** and **Units**: `Points`. In the **Character Marker** tab, choose **Font**: `Baskerville Old Face`, **Subset**: `Basic Latin`, **Unicode**: `33`, **Size**: `14`, **Angle**: `5`, **OffsetX**: `4.5`, and **OffsetY**: `3.5`.
12. Click on **OK**. Rename the symbol `MyThornyShrub`, and **Category**: `LAND USE`.

Let's create a pattern for the `LandUse` polygon features, as shown in the following screenshot:

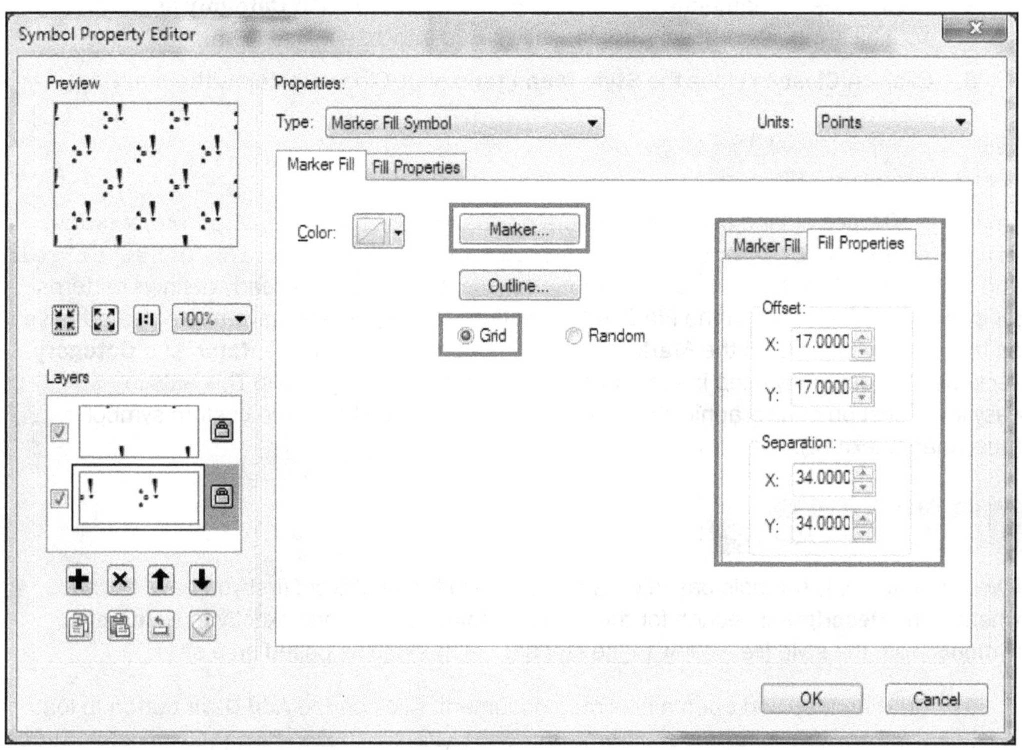

The following steps will show you how to create `LandUse` polygon features:

1. In **Style Manager** on the left-hand side, select the `TOPO5k.style/Fill Symbols` folder.
2. In **Style Manager** on the right-hand side, in the empty space, right-click and navigate to **New** | **Fill Symbol**. In the **Symbol Property Editor** section, go to the **Layers** section, and use the **Plus Sign** button to add one more symbol layer.
3. For the first symbol layer, set **Type**: `Marker Fill Symbol` and **Units**: `Points`. In the **Marker Fill** tab, select the **Marker** button, and choose `MyThornyShrub` from `TOPO5k.style`. Click on **OK**. Select the **Grid** check box.
4. In the **Fill Properties** tab, set **SeparationX**: `34` and **SeparationY**: `34`.
5. Select the second layer, and set **Type**: `Marker Fill Symbol` and **Units**: `Points`. In the **Marker Fill** tab, select the **Marker** button, and choose `MyThornyShrub` from `TOPO5k.style`. Click on **OK**. Select the **Grid** check box.
6. In the **Fill Properties** tab, set **OffsetX**: `17`, **OffsetY**: `17`, **SeparationX**: `34`, and **SeparationY**: `34`.
7. Click on **OK**. Rename the symbol as `MyThornyShrub`, and **Category** as `LAND USE`. Your fill symbol has been automatically saved into `TOPO5k.style`.
8. Click on **Close** to close the **Style Manager** dialog. Close ArcMap without saving.

How it works...

There are 10 types of land use in the feature class `LandUse`: `Arable`, `Pasture`, `Meadow`, `Vineyard`, `Fruit orchard`, `Forest`, `Hydrography`, `Transportation`, `Other terrain` and `Unproductive`. In `...\Data\TOPO5000.style`, there are already defined patterns for every type of land use in the **Fill Symbol** section. These patterns have been created based on the point symbols from the **Marker Symbol** section. Notice that the **Name** and **Category** sections are constantly used in the same way for point and fill symbols. This aids in easy identification of all graphic characteristics that describe the same custom symbol (hue, shape, texture).

There's more...

If you want to apply symbols based on a custom style for a subtype, first you have to check whether the **Description** section for the subtype **Code** corresponds exactly to **Name of symbols** from the style file. Follow these steps to apply symbols based on a style:

1. Start ArcMap and open a new map document. Click on the **Add Data** button to load the `LandUse` feature class from `<drive>:\PacktPublishing\Data\TOPO5000.gdb\LandUse`.
2. In the **Table Of Contents** section, right-click on the `LandUse` layer, and navigate to **Properties** | **Symbology** | **Categories** | **Match to symbols in a style**.

3. Select the style file named `TOPO5000.style` from `<drive>:\PacktPublishing\Data` folder, and press the **Match Symbols** button. Click on **OK**.
4. Inspect the results, and save your map document as `MyCustomSymbology.mxd` to `...\Data\MapDocuments`.

Using proportional symbology

Point symbols represent discrete physical features (for example, triangulation points, address locations, fire hydrants, or street signs) or conceptual features (for example, cities by population, population by counties, birth rate, or unemployment rate by country, or disease outbreaks between 1990 and 2014 by region).

ArcMap offers four tools to represent quantitative data:

- **Graduated colors**: This varies the hue (color) to symbolize the attribute values grouped into a given number of classes.
- **Graduated symbols**: This varies the size and keeps the hue (color) constant to symbolize the attribute values grouped into a given number of classes.
- **Dot density**: This keeps the size and hue (color) constant to symbolize the data that is not classified; the number of dots is tied to a certain attribute value.
- **Proportional symbols**: This keeps the hue (color) and shape constant and varies the size that is tied to the highest and lowest attribute values and a given number of graduated symbols.

Getting ready

In this recipe, you will use point symbols to represent the quantitative attributes of the `Counties` polygon feature, as shown in the following screenshot:

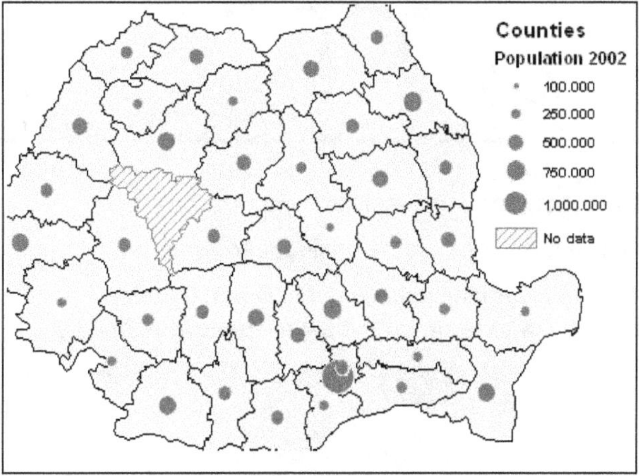

All the point symbols will have the same color, same shape (circle for instance) and will vary in size to show the exact attribute values.

How to do it...

Follow these steps to visualize quantitative data using the proportional symbols method:

1. Start ArcMap and open a new map document. Load the `Counties` feature class from: `<drive>:\PacktPublishing\Data\TOPO5000.gdb\Boundaries`.
2. In the **Table Of Contents** section, double-click on the `Counties` layer to open **Symbology** in the **Layer Properties** dialog.
3. Navigate to **Quantities** | **Proportional symbols**. This will display the **Layer Properties** dialog, as shown in the following screenshot:

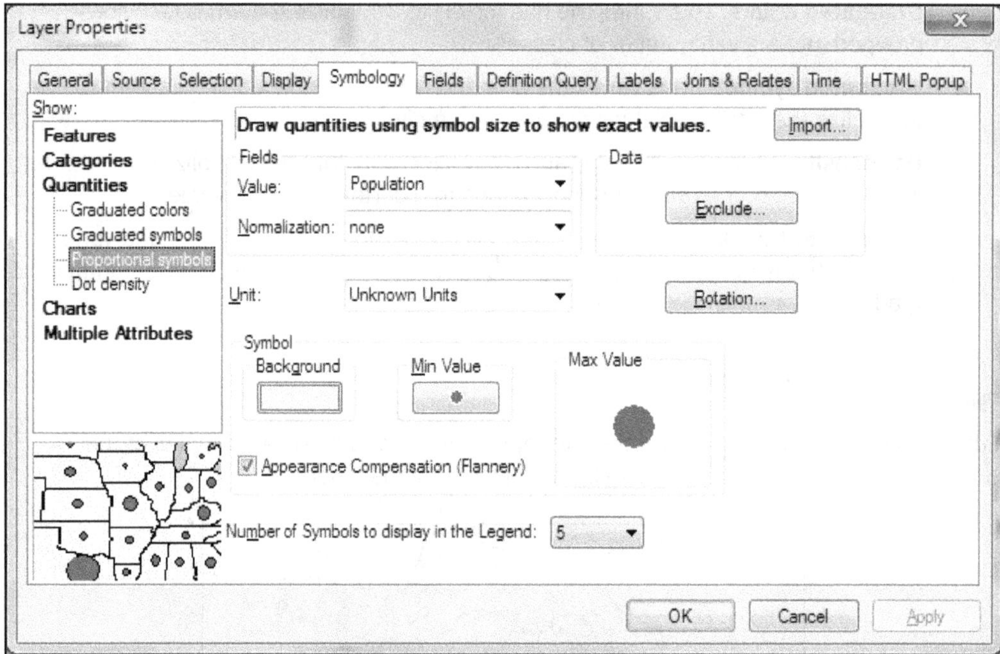

4. In the **Fields** section, for **Value**, select `Population`.
5. In the **Symbol** section, click on the background symbol.
6. In the **Symbol Selector** dialog, choose **Fill Color**: `Olivine Yellow` (6th column, 1st row), and click on **OK** to return to the **Layer Properties** dialog.

7. Click on the **Appearance Compensation (Flannery)** check box. This option will slightly increase the size of circular symbols in order to avoid the wrong perception of symbol sizes on your map.
8. Navigate to **Min Value** | **Edit Symbol**. In the **Symbol Property Editor** dialog, for **Color**, choose the **Fir Green** option (7th column, 7th row).
9. For **Size** choose `6`. Select **Edit Symbol** and uncheck the **Use Outline** box. Click on the **Mask** tab and select **Style Halo** with **Size**: `1`.
10. Click on the **OK** button twice to return to the **Layer Properties** dialog.
11. For **Number of Symbols to display in the Legend**, select `5`.
12. Click on the **Exclude** button to open the **Data exclusion Properties** dialog. Select the **Query** tab, and build the following expression: `Population = 0`.
13. Select the **Legend** tab and the check box **Show symbol for excluded data**. Navigate to **Symbol** | **Edit Symbol**. In the **Layers** section, use the **Plus Sign** button to add a new symbol layer.
14. Select the new symbol layer. Set **Type**: **Line Fill Symbol** and **Units**: `Points`. In the **Line Fill** section, choose **Color**: `Mars Red` (2nd column, 3rd row), and **Angle**: `45`.
15. Click on **OK** twice to return to the **Data Exclusion Properties** dialog. In the **Data Exclusion Properties** dialog, for **Label**, type `No data`.
16. Click on **OK**. Inspect the results and save your `Counties` layer as `Counties.lyr` to `<drive>:\PacktPublishing\Data`.
17. Save your map as `MyProportionalSymbology.mxd` at `...\Data\MapDocuments`.

You can find the results at `...\Data\MapDocuments\Chapter5\ProportionalSymbology.mxd`.

How it works...

You created a correlation between symbol size and ratio data value (quantitative data) by using proportional symbols. After you have gone through steps 7 to 9, you will notice that the **Max Value** symbol has changed proportionally. Since the polygon feature class contains an element with the value `0` for population, you have excluded this data from steps 12 to 16.

There's more...

The position of the circles can fall at the edge or outside of the polygon. This is because the circles are positioned at the centroid of their polygons. The centroid of the irregular polygons might lie outside the polygon, as shown in the following screenshot:

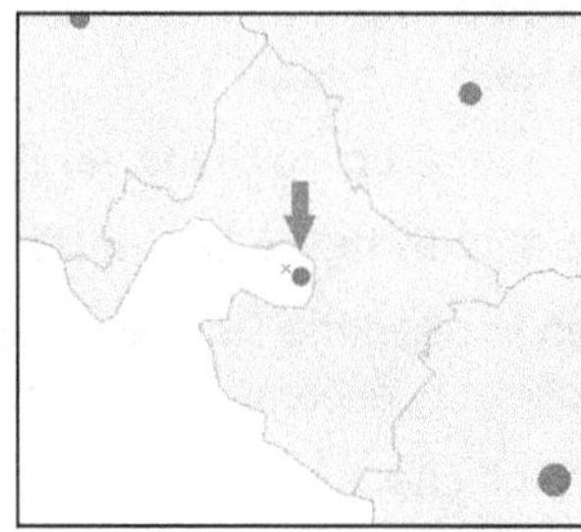

If you want to edit one of the circles' position, you should convert them into graphics. Right-click on the `Counties` layer, and navigate to **Convert Features to Graphics** | **Only draw the converted graphics**. The **Draw** toolbar offers the tools to edit graphics. To see the polygons from `Counties`, right-click on the layer and navigate to **Properties** | **Display** | **Feature Exclusion** | **Restore All**. By converting the symbols to graphics, you will remove the link between the symbol and the associated attribute values. This means they will no longer adjust to changes to the attribute values.

Using the symbol levels

As you saw in the *Creating custom symbology* recipe, your symbols have one or many layers. By default, ArcMap controls the drawing order for symbols. The features are drawn based on the order in the **Table Of Contents** section and based on the order from the attribute table of every feature class. The symbol levels allow you to control the order of the feature symbols in **Data View**.

Getting ready

First, you will prepare data for test symbol layers by applying custom symbols to the `RoadL` feature class subtype. The symbols are already created in `TOPO5000.style`. When you visualize the data at scale `1:5,000`, you will see a lot of graphic errors. To see the errors, use an existing map document named `SymbolLevelRoad.mxd` from `<drive>:\PacktPublishing\Data`. Use the **Default View** and **Advanced View** options to control the drawing order of the symbol layers.

How to do it...

Follow these steps to use layer symbols to correct all graphic errors:

1. Start ArcMap and open an existing map document SymbolLevelRoad.mxd from <drive>:\PacktPublishing\Data.
2. In the **Table Of Contents** section, right-click on the RoadL layer and navigate to **Properties** | **Symbology** | **Categories** | **Match to symbols in a style**. Click on the **Browse** button to add the ...\Data TOPO5000.style style file. To match the existing symbols with the values of the CND subtype field, press the **Match Symbols** button. Click on **Apply** and on **OK**.
3. Go to the **Bookmarks** menu and explore the spatial bookmarks that show you all the graphic errors of the linear symbols.
4. To resolve the drawing errors shown by the Forest and Forest 2 bookmarks, right-click on the RoadL layer and click on **Use Symbol Levels**. Inspect the results.
5. In the **Table Of Contents** section, right-click on the RoadL layer and navigate to **Properties** | **Symbology**. Uncheck the **<All other values>** value. Click on the **Advanced** button and select **Symbol Levels**.
6. In **Default View**, change the order of symbols, as shown in the following screenshot:

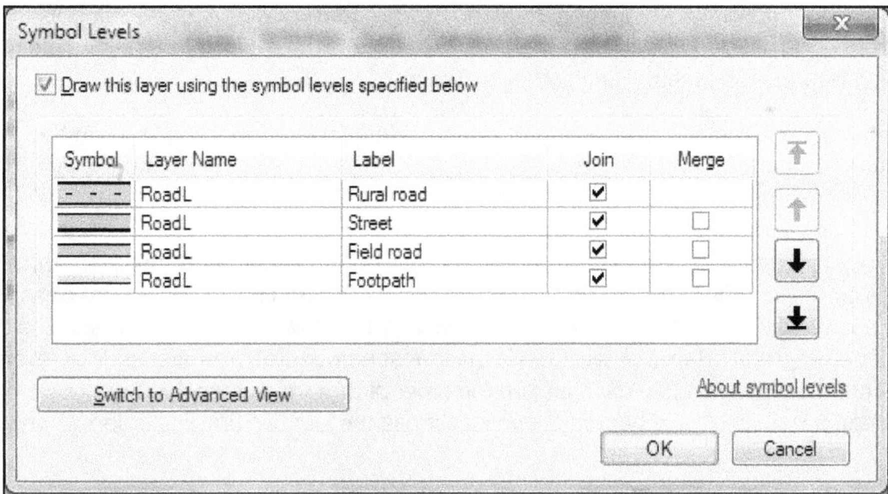

7. Click on **OK** twice to close the **Layer Properties** window. Inspect every bookmark to see the difference.
8. Return to the **Symbol Levels** dialog box and check the **Merge** column for Street and Field road.

9. Click on **OK** twice to close the **Layer Properties** window. Inspect every bookmark to see the results.
10. Return to the **Symbol Levels** dialog box and change the view mode by clicking on the **Switch to Advanced View** button.
11. In the **Symbol** column, click on the down arrow to the right of every symbol to see the number of levels.
12. For `column 1` (black layer), change the value: `0` for all symbols. For `column 2` (colored layer), change the values to `Footpath: 1`, `Field road: 2`, `Street: 3`, and `Rural road: 4`. For `column 3` (dashed line), change the value to `5`.
13. Click on **OK** twice to close the **Layer Properties** window. Inspect the results.
14. Navigate to **File** | **Save as** and save your map document as `MySymbolLevels.mxd` at `...\Data\MapDocuments`.

You can find the results at `...\Data\MapDocuments\Chapter5\SymbologLevelRoad.mxd`.

How it works...

By selecting **Use Symbol Levels**, you resolved road intersections of the same type.

In **Default View**, you saw two options: **Join** and **Merge**. **Join** joined the same symbols that were intersecting each other on the map from the same feature class. **Merge** joined different symbols that were intersecting each other on the map.

In **Advanced View**, you saw the number of symbol layers for every symbol. The number indicates the order of drawing layers in ArcMap. `Rural road` has three symbol layers. The black layer has the drawing order `1`, the colored layer has the drawing order `2`, and the dashed line layer has the drawing order `3`. This means that the black layer will be drawn first. All the other types of roads have only two symbol layers. Also, in **Advanced View**, there are three columns corresponding to the maximum number of layers identified in the feature class. The software is assigned by the default number to each symbol layer. These numbers define the order of drawing in ArcMap. The layers with the lowest number (`0`) draw first and the layers with highest number draw last (`5`). You changed the order of drawing at step 12. The `Rural road` will appear at the top of all other roads because it has the last two drawing orders (`4` and `5`).

There's more...

If you want to resolve the gaps between the segments of the same feature type, you should change the Line cap type. Follow these steps to fill in the gaps:

1. In your map document, `MySymbolLevels.mxd`, open the **Layer Properties** and select the **Symbology** tab.
2. Open the **Symbol Property Editor** dialog for the `Rural road` subtype (click on the **Edit Symbol** button).
3. In the **Layers** panel, select every symbol layer, and for the **Line Caps** option, check **Round**. Click on **OK** to close all open dialogs.
4. Inspect the results. Save your map documents.

Using representation

In the previous recipes, you worked with traditional symbology. In this recipe, you will use advanced techniques to symbolize geographical features on a map. The representation allows you to solve different cartographic challenges, such as the overlapping of different symbols. With multiple representations for the same feature class, you can obtain different thematic maps displayed at different scales while needing only one feature class entry in the **Table Of Contents** section. The representations are the property of the feature class. In the previous sections, you depended on the `.mxd` or `.lyr` file to keep the symbology for a feature class on a map. Both types of files save the path to the physical store of the data and the way you choose to symbolize the features. Representations have rules and rules have one or more symbol layers. A representation is stored in the feature class and is managed by two fields: `RuleID` and `Override`. `RuleID` stores the rules for a representation. `Override` stores the exceptions from the rules. Rules define how it will be symbolized as a feature or a group of features (subtypes). Geometry overrides separate symbols to change or move the symbols in order to decongest an overfull map area.

Getting ready

You will convert the existing symbology to representation for three feature classes: `Bridge`, `RoadL`, and `LandUse`. Also, you will create your own representation marker named **Traffic Signal** and save it in `TOPO5000.style`.

How to do it...

Follow these steps to create cartographic representations for `Bridge`, `RoadL`, and `LandUse`:

1. Start ArcMap and open the existing map document `Representation.mxd` from `<drive>:\PacktPublishing\Data\Representation`. `Representation.mxd` uses the `TOPO5k.style` style file from `<drive>:\PacktPublishing\Data\Representation`.
2. In general, a representation of a feature class refers to a given scale. Please remember that your product will be a topographic map at the scale `1:5,000`. That way, `Representation.mxd` has set `1:5,000` as the reference scale for the size of symbols or representations. The reference scale is a property of the data frame.
3. To define the reference scale for the data frame and symbols too, first type the scale `5000` in the **Map Scale** section. Second, right-click on the **Layers** data frame and navigate to **Reference Scale | Set Reference Scale**. Use **Zoom in** or **Zoom out** to see what happens to the size of the symbols. Notice that your symbol size relates to the display scale of the data frame. Symbols became smaller at small scales (for example, `1:15,000`) and bigger at large scales (for example: `1:500`).
4. To delete the reference scale for the data frame, in the **Table Of Contents** section, right-click on the **Layers** data frame and navigate to **Reference Scale | Clear Reference Scale**. Use **Zoom in** or **Zoom out** to see what happens to the size of the symbols without a reference scale for the data frame. Notice that your symbol size remains unchanged onscreen at different data frame scales.
5. Right-click on the **Bridge** layer and select **Create Symbology to Representation**. Accept the default parameters and uncheck **Add new layer to map symbolized with this representation**. Click on **Convert**.
6. In the **Table Of Contents** section, right-click on the **Bridge** layer and navigate to **Properties | Symbology | Representations**.
7. There are two sections: **Rules** and **Layers**. In the **Rules** section, there are three rules defined by the symbols of subtypes: **Examine all the options, You can add or delete the rules**, and **You can modify every rule by adding more layer symbol** (point, line, and polygon). You can change the graphic characteristics for all the existing layers. You can also control the **Symbol Levels** parameter if you click on the black arrow in the bottom-right corner of the **Layers** section.
8. To see the **Override field** section, select the **Fields** tab and check the **Override** parameter. Click on **OK**.
9. In the **Table Of Contents** section, right-click on the **Bridge** layer and select **Open Attribute Table**. Examine all the fields.

10. Repeat steps 6 to 8 for `RoadL` and `LandUse` layers.
11. You will create your own representation marker named `My Traffic Signal` using the **Marker Editor** tool. The zebra crossing will indicate the pedestrian crossing along the `Rural road` subtype feature.
12. In the **Table Of Contents** section, right-click on the `RoadL` layer and navigate to **Properties** | **Symbology** | **Representations**. Select the `Rural road` rule, and in the layers section, select the **Add new marker Layer** button, as shown in the following screenshot:

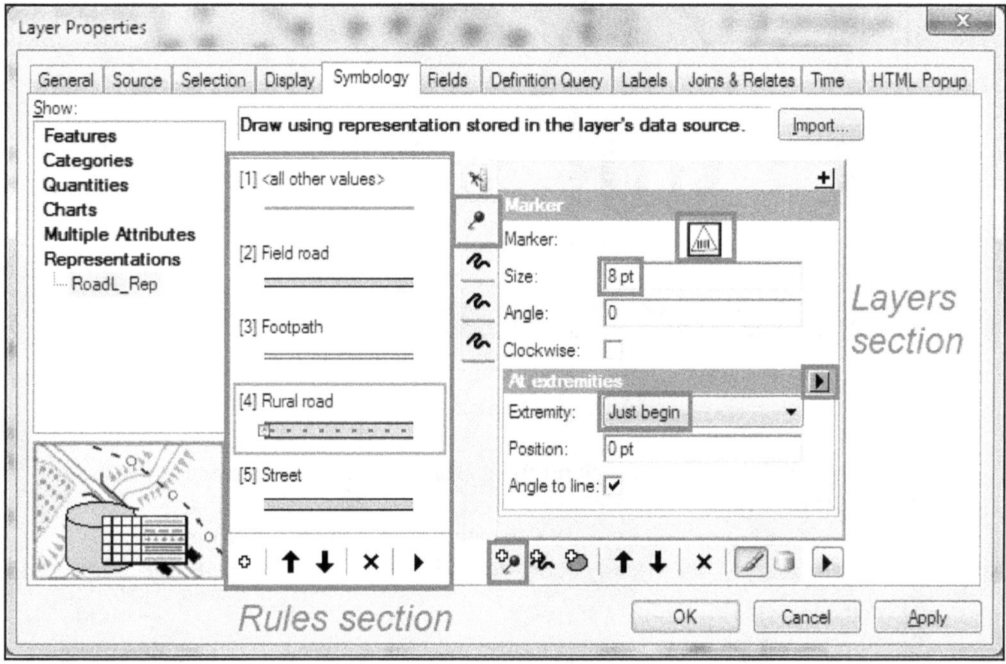

13. Select the default **Marker** (a black square), and in **Representation Marker Selector**, select **Square** from `ESRI.style`.

14. Select the **Properties** button and use the **Create Line** tool from **Marker Editor** to create a triangle with five vertical lines (zebra crossing), as shown in the following screenshot:

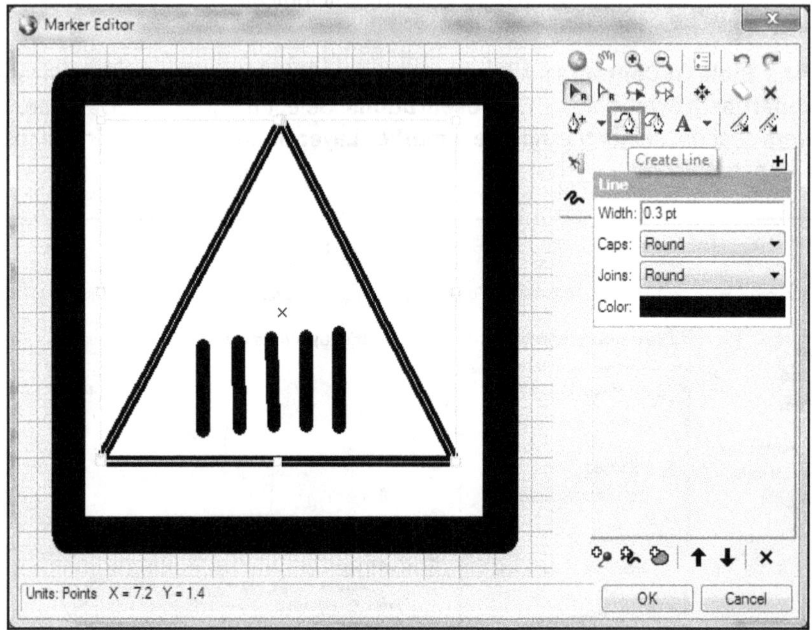

15. After you finish the symbol, click on **OK**. In **Representation Marker Selector**, save the custom symbol as `MyZebraCrossing` into `TOPO5k.style` in the `...\Data\Representation` folder. Select **OK** to return to the **Layer Properties** dialog. `Zebra crossing` has already been created in `TOPO5k.style\Representation Markers`.
16. Set Size to `8` **pt** for **Zebra** crossing marker into the Marker section. In the section Along line select black arrow from the right to open the **Market Placements** dialog. Navigate to the option **Line input** | **At extremities** and click on **OK**. For Extremity select Just begin. Click **OK** twice. Inspect the results in **Data View**.

    In the upcoming steps, you will create a geometry effect to the hydrography:

17. Double-click on the `LandUse` layer to open the **Layer Properties** dialog. Navigate to **Symbology** | **Representations** and select the **Hydrography** rule. In the `layers` section select its **Line layer** and click on the **Remove layer** button from the bottom of the section to remove the line symbol layer.
18. In the layers section, select the **Solid color pattern** layer and click on the black plus sign to open **Geometric Effects**. Navigate to **Polygon input** | **Smooth** and click on **OK** to return to **Layer Properties**.
19. Set **Flat tolerance** to `2` **pt** and click on **OK** twice. Inspect the `Hydrography` layer to see the results.

20. The **Smooth** effect doesn't influence the vertices or geometry. To check this, right-click on the `LandUse` layer, and navigate to **Selection** | **Make This The Only Selectable Layer**.
21. From the **Editor** toolbar, start **Edit session**, and with the **Edit** tool, select the vector, and select the **Edit vertices** bottom. Visualize the river to see the difference. If you want to remove the smooth effect applied to the river feature in Data View, stop the Edit session and navigate to Layer **Properties** | **Symbology** | **Representation**. Select the **Hydrography** rule and click on the black arrow from **Smooth** section to select **Remove Effect**.
22. In the next steps, you will create overrides for `River`. Go to **Bookmark**, select **Overrides** to see the river section. The symbols of the road and river from `LandUse` layer are overlapped.
23. Start the edit session. Add the **Representation** toolbar by navigating to **Customize** | **Toolbars** | **Representation**. You will use the tools from the **Representation** toolbar.
24. With **Direct Select Tool**, select the river. All its vertices are selected. With the *Shift* key pressed, click again on the river to unselect all its vertices. With the *Shift* key pressed, select one by one the eight vertices from the overlapping river, as shown in the following screenshot:

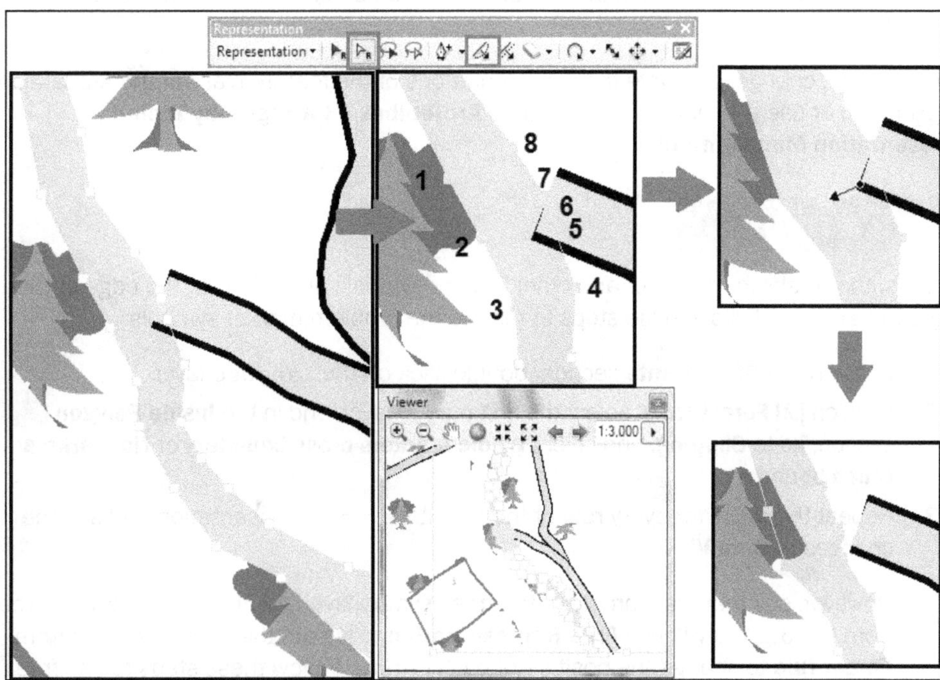

25. Select **Warp Tool**, click on the corner of the road symbol, and drag-and-drop to reshape the edge as shown in the preceding screenshot.

26. The override doesn't influence the geometry of the feature from `LandUse`. To check this, select the river and click on the **Edit Vertices** tool to see the vertices of the polygon feature.
27. Save and stop the edit session. Save your map document as `MyRepresentation.mxd` at `...\Data\MapDocuments`.

How it works...

Every time you create a representation for a feature class, a domain will be created in your file geodatabase. The domain controls representation rules from the **RuleID** field. The codes of the domain have the same values with the defined rules of representation. For example, the **Bridge** feature class uses the domain **Bridge_Rep_Rules** that has 4 values: **1** for **Bridge**, **2** for **Small wooden bridge**, **3** for **Unknown**, and **-1** for **Free Representation**.

In ArcMap, you can't convert, create, or modify cartographic representations in the edit mode (edit geometry of features). Instead, the overrides can be made during the edit session. When you have selected the river edge using **Select Tool** from the **Representation** toolbar, you have noticed that the selected representations look different than a selected feature using **Edit Tool** from the **Edit** toolbar. The reshaping of a representation doesn't influence the geometry of a feature. The shape or position of representation can be different from the original shape of the feature.

You can create, change, or delete the representations of the feature class in the ArcCatalog context menu or use the tools by navigating to **ArcToolbox** | **Cartography Tools** | **Representation Management**.

There's more...

The trees, thorny shrub, vineyard, and other marker symbols are cut off at the edge of the feature in `LandUse`. Follow these steps to resolve the problem marker symbols:

1. In the **Table Of Contents** section, double-click on the `LandUse` layer.
2. Click on **[2] Forest Rule**, select the first marker layer, and in the **Inside Polygon** section, go to **Clipping**, and select **Whole markers cross boundary** or **No markers touch boundary**.
3. Repeat the steps for every rule of the `LandUse_Rep` representation and see the changes in **Data View**.

   However, still there is a problem: the trees symbol overlaps the `Footpath` subtype from `RoadL`. You will use **Free Representation** to isolate the marker symbol from **Forest rule** to change the position of each symbol. Follow these steps to use free representation to move the trees away from the footpath:

4. Start the **edit** session. From the **Representation** toolbar, click on **Select Tool**, and select the forest.

5. Navigate to **Representation** | **Free Representation** | **Convert to Free Representation**. Keep the forest selected. Navigate to **Representation** | **Free Representation** | **Edit Free Representation**. Please remember that **Forest rule** has two **Marker** layers: `Coniferous` and `Deciduous`.
6. With the **Select Part** tool from the right-hand side, select the whole forest.
7. Select **Marker** (`Coniferous`), and with the **Select Part** tool activated, right-click on the selected forest and choose **Separate Symbols Layers**, as shown in the following screenshot:

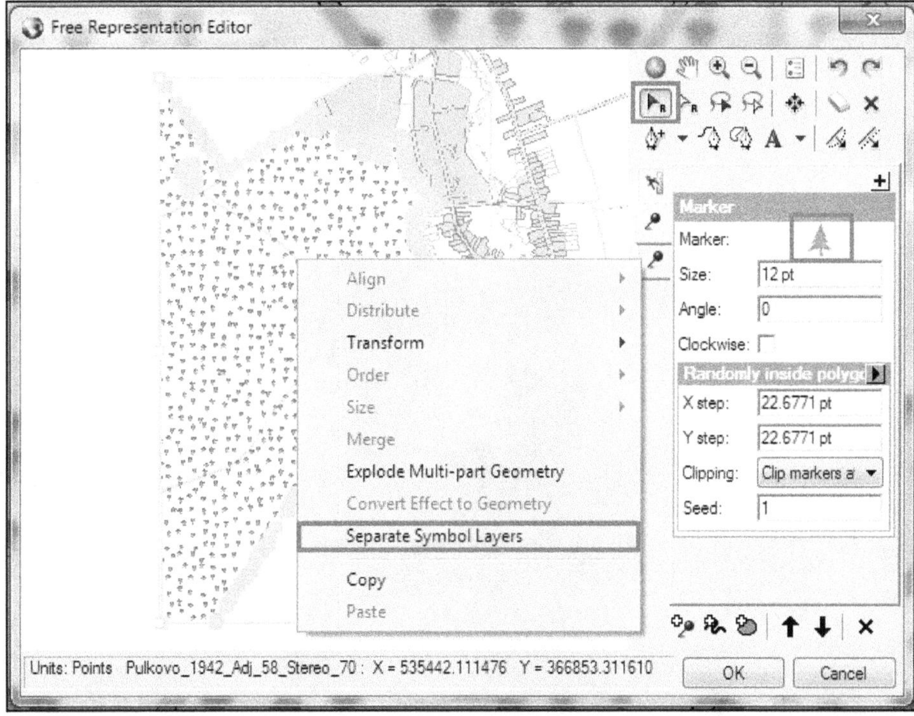

8. Right-click again and choose **Convert Effect to Geometry**. Right-click again and choose **Explode Multi-part Geometry**. Now, you can select every coniferous tree from the forest. You can move or change the size of every tree individually.
9. Repeat steps 6 to 8 for the `Deciduous` tree symbol.
10. Click on `OK` to close the **Free Representation Editor** window. **Save** and stop the **edit** session. Save your map document.

See also

- For information about creating and working with labels for point, polyline, and polygon features on a map, please refer to the next chapter, *Building Better Maps*

6

Building Better Maps

In this chapter, we will cover the following topics:

- Creating and editing labels
- Labeling with Maplex Label Engine
- Creating and editing annotation
- Using annotation and masking for contours
- Creating a bivariate map

Introduction

In map design, a cartographer is not only thinking about the symbol of geographic feature but also its attributes. The descriptive text for features can vary in graphic characteristics in order to emphasize the differences, hierarchy, and importance of the features. The overall appearance and message of the map can be significantly affected by text characteristics, such as fonts, style (for example, italic, bold, and so on), character spacing, leading (space between lines in a block text), callouts and halos, fields used for labeling size, color, shape, position relative to the features, or display scale of the text. You can add descriptive text to your map using dynamic labels, annotations, and graphic text.

The dynamic labels are based on attribute values and are stored as a property of a layer in a map document, as a part of a map layer (`.lyr`), layer package (`.lpk`), or map package (`.mpk`). The dynamic labels are created using the **Standard Label Engine** or the **Maplex Label Engine**.

The labels can be converted into annotations. The annotation makes your text editable (for example, font, size, position, text string, and so on). The annotation stores the text string, position, and display properties of the text. There are two types of annotations:

- Map document annotation (annotation group) stored in the map document (`.mxd`) as a property of the data frame
- Geodatabase annotation: standard annotation (not feature-linked) and feature-linked annotation

The graphic text is created for individual features using the **Text** tools from the **Draw** toolbar. The graphic text can be created based on the feature attribute but is not tied to the attribute value (for example, if the attribute is changed, the text is not updated). The graphic text is stored as a map annotation group in the map document.

In this chapter, you'll use advanced tools to create labels and annotations. In addition to this, you will display the height values with a visual effect for the contour lines. Finally, you will learn how to map quantitative data to create a bivariate map.

Creating and editing labels

When you are creating labels for an easy-to-read final map, you should do the following:

- Establish one or more attribute fields for labeling
- Define the position of labels relative to the features its describes
- Define the position relative to other features or labels in order to minimize overlapping (for example, label weight ranking)
- Define the order of drawing labels on a map (for example, label priority ranking)
- Define the symbols for labels
- Create different label classes for the same feature class
- Avoid too many duplicate labels
- Add custom text to the labels if necessary
- Set a scale range (to control the display of the labels on the map) and work with the reference scale (to fix the size of the labels at a specific scale)
- Work with the **Fast** label quality (**Labeling** toolbar) to explore the preliminary results, and use the **Best** label quality for final labeling

Getting ready

In this recipe, you will use the **Draw** toolbar to add a graphic text for the `TriangulationPoint` feature class, and the **Labeling** toolbar to create dynamic labels for the `Elevation` and `ContourLine` feature classes, as shown in the following screenshot:

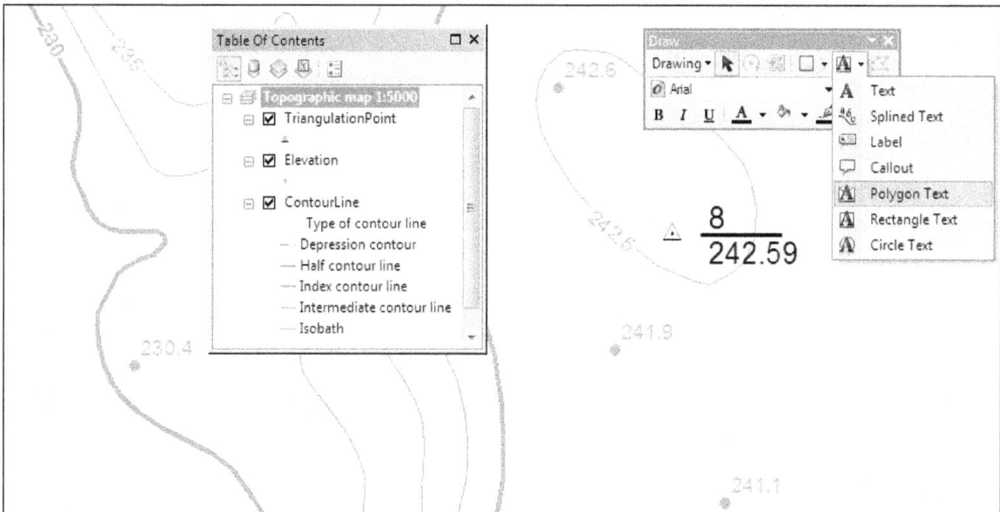

From the **Labeling** toolbar, you will work with **Label Manager** in the **Standard Label Engine** mode. The **Label Manager** dialog has two levels: feature class level and label class level. At the **Feature Class** level, you have two sections: **Add label class** and **Add label classes from symbology categories**. At the **Label Class** level, you have three main sections: **Text String**, **Text Symbol**, and **Placement Properties**. There are also three important buttons: **Scale Range**, **SQL Query**, and **Label Styles**.

How to do it...

Follow these steps to add a graphic text and dynamic labels:

1. Start ArcMap and open the existing map document, `Label.mxd`, from `<drive>:\PacktPublishing\Data\LabelingMaplex`.
2. Load the **Draw** and **Labeling** toolbars. You will add text for a triangulation point feature from the `TriangulationPoint` layer using the **Rectangle Text** tool from the **Draw** toolbar, as shown in the preceding screenshot.

3. Next to the point feature, draw a small rectangle. With the **Select Elements** tool from the **Tools** toolbar, right-click on the rectangle, and navigate to **Properties** | **Text**. In the **Text** section, type `8`, create a new line using the *Enter* key, and type `242.59` on the second text line. Navigate to **Change Symbol** | **Style References**, add . . . `\Data\TOPO5000.style` to the **Symbol Selector** list. Select the **Geodetic Network** text symbol. At this point, you can modify the symbol properties and save it in `TOPO5k.style` with a different name.
4. Click on **OK** and on **Apply** to update the changes. Select the **Frame** tab and set **Border**: **None**. Click on **Apply**. Select the **Size** and **Position** tabs, and set **Width**: `90` and **Height**: `65`. Click on **Apply** and on **OK** to return to **Data View**.
5. With the **Select Elements** tool, select and move the rectangle to the best placement from the cartographic point of view (for example, to the right-hand side of the point feature and slightly above the point).

   In the next steps, you will create dynamic labels for the `Elevation` and `ContourLine` layers. From the **Labeling** toolbar, you will work with **Label Manager** in the **Standard Label Engine** mode, as shown in the following screenshot:

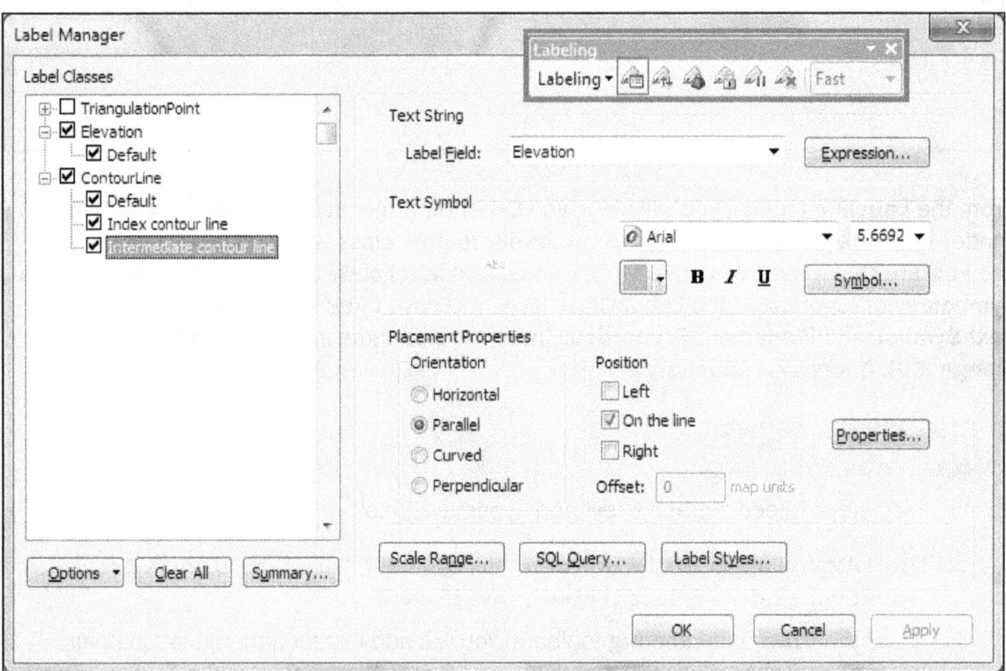

The **Label Manager** dialog has two levels: feature class level and label class level. At the **Feature Class** level, you have two sections: the **Add label class** and **Add label classes from symbology** categories. At the **Label Class** level, you have three main sections: **Text String**, **Text Symbol**, and **Placement Properties**. The steps are explained as follows:

6. Open the **Label Manager** dialog, check `Elevation` from the left-hand side of the dialog, and select the `Default` label class. Click on **Expression** and create the following expression: `FormatNumber([Elevation],1)`. Click on **Verify** to check the expression, and click on **OK**. Click on **Apply** to update the settings and on **OK** to return to **Data View**. Inspect the results.
7. Return to the **Label Manager** dialog, check, and select the `ContourLine` feature class. In the **Add label** classes from the symbology categories section, there are five default label classes that correspond to the `CNT` (`Type of contour line`) subtypes filed. Keep checked only `Index contour line` and `Intermediate contour line`. Click on the **Add** button. Read carefully the warning message, and select **No**.
8. Every new label class inherits the characteristics of the `Default` label class. As the new label classes are created based on the subtypes, the groups are already defined in the **SQL Query** dialog.
9. Select the `Default` label class, click on **SQL Query**, and type the following expression: `CNT <> 1 AND CNT <> 2`. Click on **Verify** to check the expression. Click on **OK**. The `Default` label class will create labels for all subtypes except `Index contour line (1)` and `Intermediate contour line (2)`.
10. Select the `Index contour line` label class and set the following parameters:
    - **Label Field**: Set this to `Elevation`
    - **Text Symbol**: Set this to `Arial`, `5` and `Bold`
    - **Symbol**: Set this to `Elevation`
    - **Edit Symbol** | **Mask** | **Halo**: Set the **Size** parameter 6 to `0.2` and **Fill Color** to `White`
    - **Placement Properties** | **Orientation**: Set this to `Parallel`
    - **Placement Properties** | **Position**: Set this to `On the line`
    - **Properties** | **Duplicate Labels**: Set this to **Place one label per feature part**
11. Accept the default settings for the remaining parameters. Click on **OK**.
12. Select the `Intermediate contour line` label class and repeat step 9, as shown in the preceding screenshot.

13. Click on **Scale Range** to define the display range of the scale for the intermediate contour labels.

    Because the precision of topographic datasets corresponding to the scale `1:5,000` and the intermediate contour are not labeled on a classical map, you will set a minimum and maximum scale to display the labels, as explained in the following steps:

14. Select **Don't show labels when zoomed**, and set **Out beyond**: `4990` and **In beyond**: `1000`. Click on **OK** twice to close the dialog windows.
15. Explore the results. To fix the dimensions of the labels for the scale `1:5,000`, select `1:5,000` on the **Standard** toolbar and right-click on the **Layers** data frame and navigate to **Reference Scale | Set Reference Scale**.
16. Use the **Lock Labels** and **View Unplaced Labels** options from the **Labeling** toolbar to fix the position of dynamic labels and see the unplaced labels.
17. Save your map as `MyLabel.mxd` at `...\Data\MapDocuments`.

You can find the results at `...\Data\MapDocuments\Chapter6\LabelFinal.mxd`.

How it works...

The graphic text created at step 3 is stored in the annotation default group as a property of the `Topographic map 1:5,000` data frame by navigating to **View | Data Frame Properties | Annotation Groups**. If you lose or damage the `.mxd` file, your individual graphic text will be lost too. The dynamic labels created are the property of the `Elevation` and `ContourLine` layers and can be stored in the `Label.mxd` map document or the `.lyr` file.

`FormatNumber()` is a VBScript function that returns a number. This function has one required parameter (expression) and four optional parameters. You used only the first optional parameter named `NumDigAfterDec` to define one decimal place for the numeric number format.

For the `ContourLine` layer, you have to use label classes to create three different label styles for three different groups of features.

A common mistake is to forget to define the group of features that will be labeled by the label classes.

If you want to see the mask of labels, in the **Table Of Contents** section, right-click on the **Layers** data frame, navigate to **Properties | Data Frame | Background**, and select a color.

See also

- To learn how to better control the dynamic labels using **Maplex Label Engine**, please refer to the next *Labeling with Maplex Label Engine* recipe

Labeling with Maplex Label Engine

In the previous recipe, you worked with the **Standard Label Engine** parameter. In this recipe, you will use the **Maplex Label Engine** parameter to create high-quality dynamic labels. The dynamic labels will be placed and oriented on the map based on criteria such as best fitting, density, or priority and weight rankings for features and labels that are overlapping.

The **Maplex Label Engine** parameter allows you to control the displayed text of the label using the **Abbreviation Dictionary** tool.

There are three types of abbreviations:

- **Translation**: The value of the label will be replaced with a custom one; the abbreviation is used whether there is space or not to display the label (for example, A can be used in place of Arable).
- **Keyword**: This will return an acronym formed by the first letter of each word from the label; the abbreviation is used only if there is not enough space to display the label (for example, DA in place of Dwelling Annex or GSG in place of Gibson Square Gardens).
- **Ending**: This abbreviates the last word from the label; the abbreviation is used only if there is not enough space to display the full label (for example, Oldhill St in place of Oldhill Street).

For instance, you can use the **Abbreviation Dictionary** tool to translate the label values into another language without changing the structure of the attribute table or field values.

Getting ready

In this recipe, you will create labels corresponding to a topographic map at the scale of `1:5,000` for the `LandUse` feature class, and labels corresponding to a cadastral map at the scale `1:500` for the `Buildings` feature class, as shown in the following screenshot:

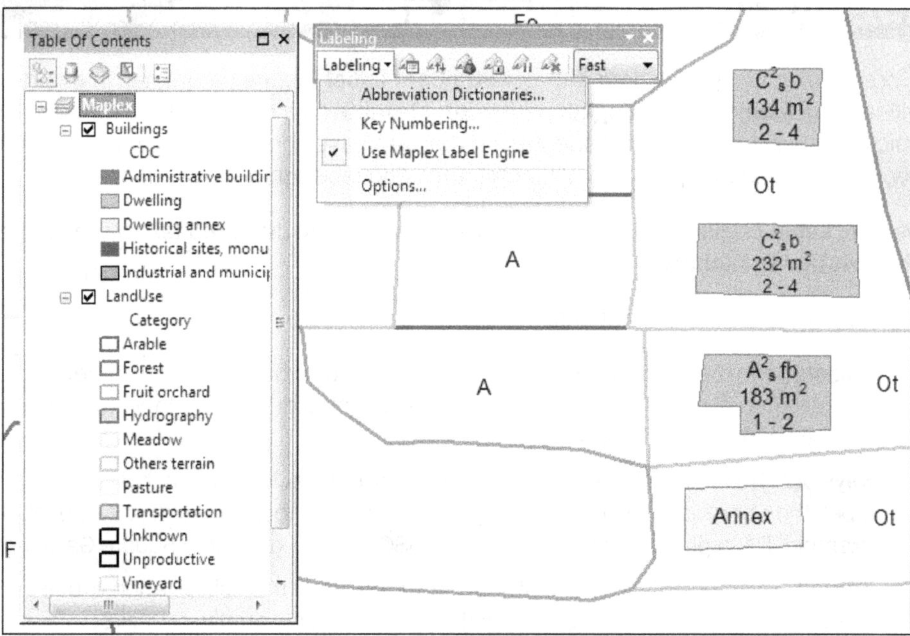

The following screenshot describes the structure of the `Dwelling` label class according with the cadastral map label style:

$$[Structure]^{[Stories]}_{[Basement]}[State]$$
$$\text{(gross building area) 'm}^2\text{'}$$
$$\text{[NumberFamilies] - [NumberPersons]}$$

How to do it...

Follow these steps to create dynamic labels using the **Maplex Label Engine** parameter:

1. Start ArcMap and open the existing map document `Maplex.mxd` from `<drive>:\PacktPublishing\Data\LabelingMaplex`.

2. Load the **Labeling** toolbar. From the **Labeling** toolbar, navigate to **Labeling Use | Maplex Label Engine**.
3. In the next steps, you will define and add abbreviations for the coded values of the LandUse: CAT (Category) and Buildings: CDC (Destination) fields.
4. Navigate to **Labeling | Abbreviation Dictionaries**. Select **New** and type MyLandUse. Select **Add Row** and for the new empty row, set the following: **Keyword**: 0, **Abbreviation(s)**: Unk, and **Type**: Translation.
5. Navigate to **Options | Append From File** and select the existing dictionary file LandUse.dic from ...\Data\LabelingMaplex. Click on **Open**. Now you have abbreviations for every subtype from the CAT (Category) subtype field.
6. Let's add the second dictionary for the Buildings layer. Select in **Options: Open From File** and load the existing dictionary file Building.dic from ...\Data\LabelingMaplex. Click on **Open**. Select **Rename** and type MyBuilding. Click on **OK** twice to close the dialog windows and return to **Data View**.
7. Open the **Label Manager** dialog, and check the Buildings and LandUse feature classes.
8. Select LandUse and for **Select symbology categories**, uncheck all symbology categories except Hydrography and Transportation. Click on the **Add** button. Read carefully the warning message, select **No** to keep the Default label class, and add the two additional classes.
9. Check and select the LandUse: Default label class, and set **Label Field**: CAT (Category). For **Text Symbol**, select **Symbol**, and choose the LandUse symbol from ...\Data\TOPO5000.style. Click on **OK**.
10. To exclude the Hydrography and Transportation features from being labeled with the Default label class, click on **SQL Query** and build the following expression: CAT <> 31 OR CAT <> 41. Uncheck **Display coded value description** to work with the coded value on your map. Click on **Apply** to update the settings in your map.
11. In the **Placement Properties** section, select **Position**: **Horizontal** and **Land Parcel Placement** to add horizontal labels that will not overlap the buildings inside the parcel polygons. Click on **OK**.
12. Select **Properties** to open the **Placement Properties** dialog. Click on the **Fitting Strategy** tab and check **Reduce font size**. Check **Abbreviate label**, select **Options**, and choose the MyLandUse dictionary name to replace the coded values of the CAT (Category) subtype field with the custom abbreviations. Click on **OK**.
13. Click on the **Strategy Order** button, use black arrows to add the **Abbreviate** label at the top and **Reduce font size** in second place. Click on **OK**.

14. Click on the **Label Density** tab, navigate to **Minimum feature size for labeling | Area**, and set `2000` **Map Units**. Explore the remaining tabs to see all the **Maplex** advanced options. Return to the **Label Manager** dialog box. Select **Scale Range** and check **Don't show labels when zoomed**. Type `10001` in the **Out beyond** section and `100` in the **In beyond** section. For this scale range, the parcel labels will be displayed on your map. Click on **Apply**.
15. Check and select the `LandUse: Hydrography` label class. Click on **Expression** and uncheck **Display coded value description**. Navigate to **Text Symbol | Symbol** and choose the **Stream** symbol from `TOPO5000.style`. Click on **OK**.
16. In the **Placement Properties** section, select **River Placement** with **Position**: `Curved`. For **Properties**, set the following parameters:
    - **Fitting Strategy | Abbreviate label**: `MyLandUse`
    - **Conflict Resolution**: `Never remove`
17. Click on **OK** and on **Apply**.
18. Uncheck the `LandUse: Transportation` label class. You will not add labels for roads, but you will use this label class later in this exercise to constrain the `Dwelling` label class to overlap with road features.
19. Select `Buildings` and for **Select symbology categories**, uncheck all symbology categories except the `Dwelling` category. Click on **Add** and on **No** for the warning message.
20. Check and select the `Building : Default` label class and set **Label Field**: `CDC (Destination)`. Uncheck **Display coded value description** to work with the coded value on your map. Click on **OK**.
21. For **Text Symbol**, select **Symbol**, and choose the `Building` symbol from `TOPO5000.style`. Click on **OK**.
22. To exclude the `Dwelling` features to be labeled with the `Default` label class, click on the **SQL Query** parameter, and build the following expression: `CAT <> 1`. Click on **Apply**.
23. In the **Placement Properties** section, select **Position**: **Horizontal** and **Regular Placement**. For **Properties**, set the following parameters:
    - **Fitting Strategy**: `Reduce font size`
    - **Abbreviate label**: `MyBuilding`.
    - **Strategy Order**: first place **Abbreviate label**, and second place **Reduce label in size**
24. Return to the **Label Manager** dialog. Navigate to **Scale Range | Don't show labels when zoomed**, and set **Out beyond**: `5000` and **In beyond**: `<None>`. The labels will be displayed starting from the scale `1:5,000` and larger (for example: `1:500`). Click on **OK**.

25. Check and select the `Building: Dwelling` label class, navigate to **Label Field** | **Expression** | **Load**, and choose the `BuildingLabel500.lxp` expression file from `...\Data\LabelingMaplex`. Uncheck **Display coded value description**.
26. Click on **Verify** to check the expression and click on **OK**.
27. Navigate to **Text Symbol** | **Symbol**, and set **Placement Properties** parameters as shown for the `Building` parameter in the `Default` label class. For the `Building` parameter in the `Dwelling` label class, do not check **Abbreviate label**.
28. Navigate to **Scale Range** | **Don't show labels when zoomed**, and set **Out beyond**: `1000` and **In beyond**: `100`. Click on **OK**.
29. In the **Label Manager** dialog box, click on **Apply** and on **OK**. It may take several seconds for the labels to be placed.
30. Set the reference scale for the labels at `1:500`. To see and analyze the position and dimensions of the labels, navigate to **Labeling** | **Lock Labels**, and use the **View Unplaced Labels** tools. The position of the labels should be locked and the unplaced labels displayed in red.
31. From the **Labeling** toolbar, set the **Label Priority Ranking** and **Label Weight Ranking** parameters, as shown in the following screenshot:

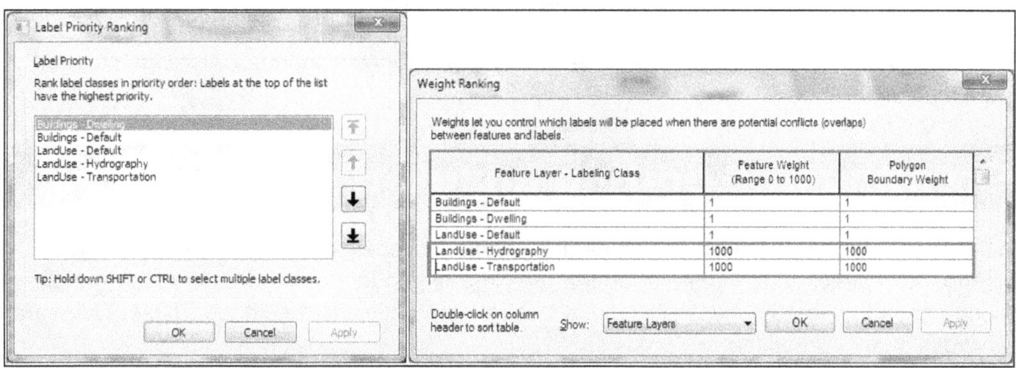

The value `1000` for `Hydrography` and `Transportation` means that no labels will be placed on the river and road polygons. The value `1` for `Buildings` tells the **Land Parcel Placement** style (from step 10) to avoid placing the parcel labels on any polygon features (buildings) inside the parcel polygons.

32. Clear the reference scale for your data frame, and explore the results on the map using the **Pan**, **Zoom In**, and **Zoom Out** tools.
33. Return to **Label Weight Ranking** and set the **Transportation** parameter to `0`. Notice how dwelling labels are drawn on the road now.
34. Save your map as `MyMaplex.mxd` at `...\Data\MapDocuments`.

You can find the results at `...\Data\MapDocuments\Chapter6\MaplexFinal.mxd`.

How it works...

You worked with two abbreviation dictionaries that renamed your subtype codes from the `Buildings` and `LandUse` feature classes without changing the values in attribute fields.

When you use **Label Priority Ranking**, you define which label classes have higher priority and which have lower priority in displaying on the map. When you use **Label Weight Ranking**, you define the label position relative to other features or labels in order to minimize overlapping in displaying on the map.

Priority and weight ranking influences the labeling speed because the ArcMap label engine evaluates the location of features before placing the labels, and based on label conflict detection and space availability, labels are rearranged every time you use **Full Extend**, **Pan**, **Zoom In**, and **Zoom Out**.

There's more...

As you might have already noticed, some labels are too small to read clearly, especially for dwellings. Go to the Windows menu and select **Magnifier** to see your labels at different scales. Explore the results on the map using your magnifier window. You can change the zoom, or you can set a specific scale by selecting the black arrow from the right-hand side of the **Magnifier** window. Select **Properties** and make your own changes. Also, you can add more magnifier windows.

Creating and editing annotation

In this recipe, you will convert your high-quality, dynamic labels created with **Maplex Label Engine** into a geodatabase annotation feature class. A geodatabase annotation feature class is another type of feature class stored in a feature dataset inside the geodatabase. There are two types of annotation feature classes:

- Not feature-linked annotation feature class (standard)
    - They are stored as independent pieces of text
    - Only if text has been derived from the attributes of a feature, annotations are not tied to the attribute value (if the attribute value is changed, the text is not updated)
    - If you move or erase a feature, the annotation will remain the same
- Feature-linked annotation feature class
    - They are generated based on feature attributes

- It is a composite relationship class with 1-M cardinality between a feature class as the origin table and an annotation feature class as the destination table, using the **OBJECTID** field as the primary key and the **FeatureID** field as the foreign key
- If you move a feature, the annotation will move with the feature
- If you erase a feature, the annotation will be erased too
- If you change the attribute value for a feature, the feature-linked annotation text will be changed

Getting ready

Firstly, you will improve the position of labels for `Elevation` and `ContourLine` using **Maplex Label Engine**. After that, you will create a standard annotation feature class and two feature-linked annotation feature classes for the reference scale `1:5,000`.

For the `Buildings` feature class, you will create a feature-linked annotation feature class for the reference scale `1:500`, as shown in the following screenshot:

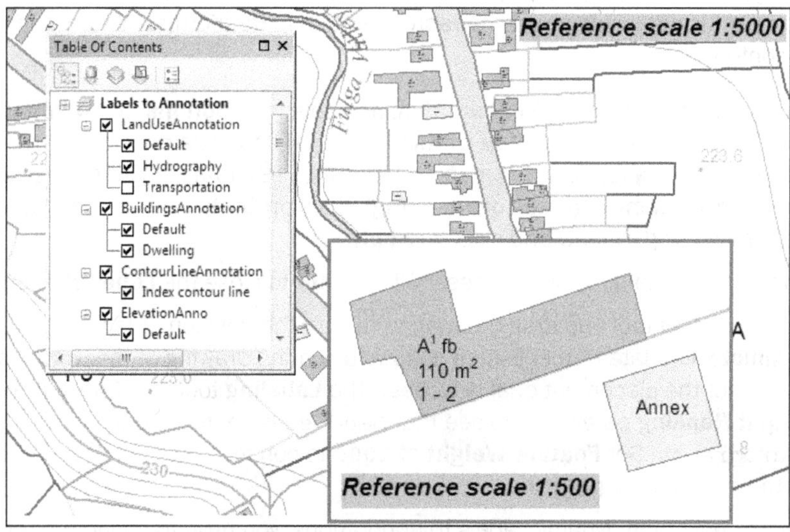

In the last part, you will edit the annotation features to correct the position, dimension, and other graphic characteristics of the annotation text.

How to do it...

Follow these steps to convert dynamic labels to geodatabase annotations:

1. Start ArcMap and open the existing map document `Annotation.mxd` from `<drive>:\PacktPublishing\Data\Annotation`.
2. Load the **Labeling** toolbar and check whether the **Use Maplex Label Engine** option is selected. Open **Label Manager** and set the `Elevation` parameter to the `Default` label class for the feature class. In the **Placement Properties** section, set **Position** as `Best Position`. Navigate to **Properties** | **Conflict Resolution**, and check **Background Label (placed first)** and **Never remove (place overlapping)**. Click on **OK** and on **Apply**.
3. For the `ContourLine` layer, right-click on the `Default` label class, and select **Delete Class**. Delete also the `Intermediate` label class. Select the `Index contour line` label class and navigate to **Placement Properties** | **Contour Placement**. Navigate to **Properties** | **Label Position** | **Options**, and check **Uphill alignment** and **No Laddering**. Click on **OK**.

   Using the **Uphill alignment** option, you will label contour lines in accordance with topographic principles about the representation of landforms and terrain features by contours.

4. For **Label Density**, check **Repeat label**, and select its **Options**. Set **Minimum Repetition Interval**: `1000` **Map Units** to have a reasonable density of labels per contour line for a topographic map at the scale `1:5,000`. For **Conflict Resolution**, check **Never Remove (place overlapping)**. Click on **OK** to return to the **Label Manager** dialog window and click on **Apply**.
5. Inspect the remaining label classes. Click on **OK** to close the **Label Manager** option.
6. Select the **Pan** tool and drag the map to redraw the position of labels in the **Label to Annotation** data frame. Even if it slows down the drawing of the dynamic labels, set the placement quality to **Best** (the **Labeling** toolbar). Check the **Label Weight Ranking** parameter to see that `LandUse` is set to `Transportation` and `Hydrography`. Set **Feature Weight** of `1000` to constrain the labels not to be placed on the road or river polygons.

   Before converting dynamic labels to annotations, you have to define the reference scale for the annotation. An annotation uses the reference scale for the current data frame. If the reference scale for the data frame is not defined, the annotation will use the display scale of the current data frame by navigating to **Standard** | **Map Scale** and set it to `5000` before converting labels to annotations. Let's create a standard annotation feature class by performing the following steps:

7. Set the reference scale for the data frame to the scale `1:5,000`. In the **Table Of Contents** section, right-click on the `Elevation` layer, and select **Convert Labels to Annotation**. Set the parameters, as shown in the following screenshot:

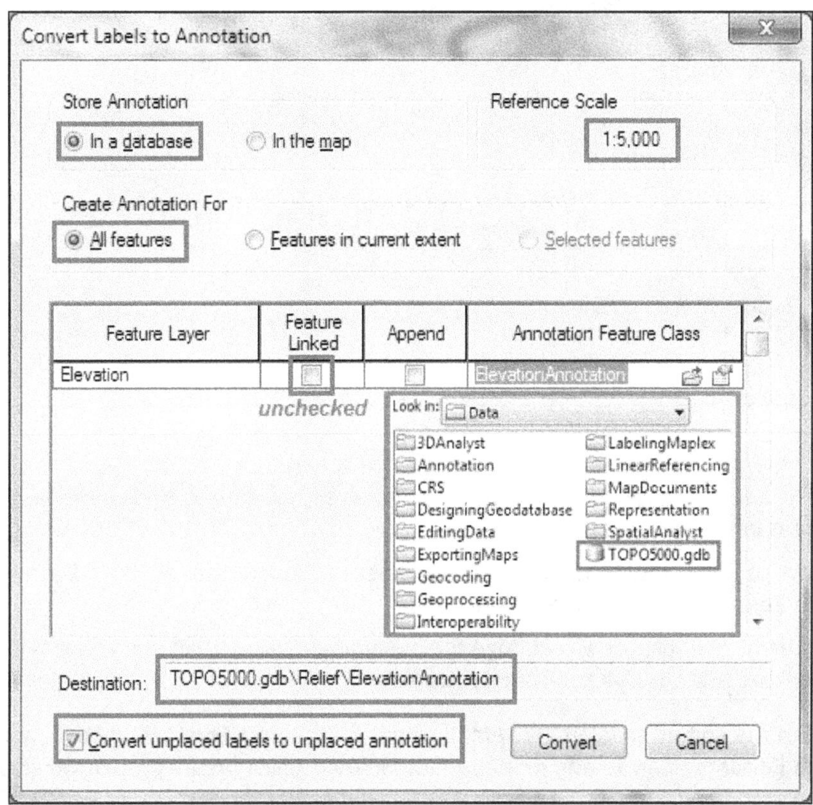

Because of the overlapping conflicts, some labels will be hidden. The **Convert unplaced labels to unplaced annotation** option will transform the unplaced labels too. The **In the map** option means that your annotation will be stored as `Annotation Group` associated to the `Elevation` layer. The annotation will be stored in the map document (`.mxd`) as a property of your current data frame.

8. Click on **Convert**. The `ElevationAnnotation` layer has been automatically added in the **Table Of Contents** section. Notice that dynamic labels of the `Elevation` layer have been deactivated in **Label Manager**.

Let's create three feature-linked annotations for the `ContourLine`, `LandUse`, and `Buildings` layers, as shown in the following screenshot:

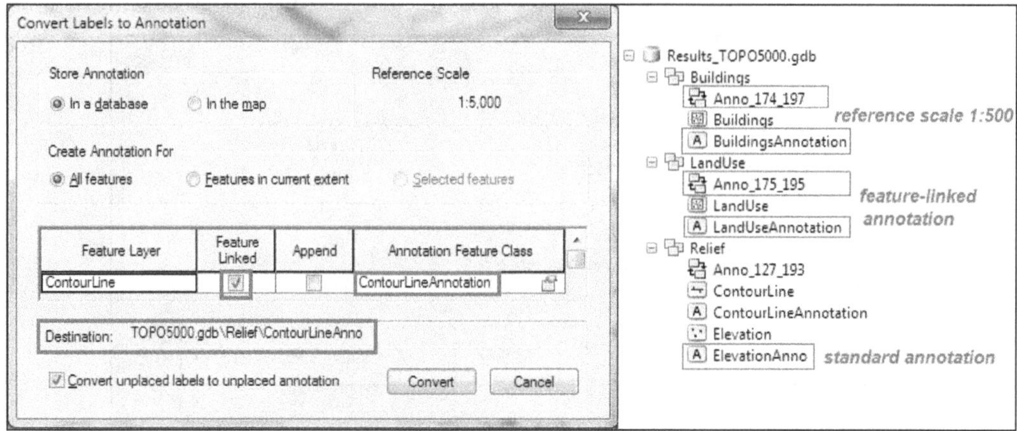

9. For the `Buildings` layer, set the reference scale to `1:500`. After you create the `1:500` annotation for the `Buildings` layer, navigate to **Reference Scale** | **Clear Reference Scale**.
10. Now, you have four annotation feature classes. Explore the results in **Data View** using the **Zoom In**, **Zoom Out**, and **Pan** tools.

    In the next steps, you will change the position and the size of one dwelling annotation from the `BuildingsAnnotation` layer:

11. Start the editing session. By default, you have the **Edit Annotation Tool** parameter on the **Editor** toolbar. To add more annotation tools, click on the drop-down arrow from the right-hand side of the **Editor** toolbar, and select **Customize**. Select **Commands** and for **Show commands containing**, type `anno`. Click and drag the **Unplaced Annotation** and **Construct Annotation with A Leader Line** tools to the **Editor** toolbar. Click on **Close**.
12. Click on the **Unplaced Annotation Window** tool and in the **Show** section, select `BuildingsAnnotation` as the `Dwelling` annotation class.

    The **Unplaced Annotation** window shows three columns: **Text**, **Class** (annotation class), and **FID Linked** (the `OBJECTID` value of the linked dwelling feature that is related to the annotation feature).

13. Uncheck **Visible Extent** and check **Draw** to see the unplaced annotations in a red outline rectangular. Click on **Search Now** and right-click on the first unplaced annotation to choose **Zoom to Feature**. Right-click again on the selected annotation and choose **Place Annotation** to change the annotation status as **Placed**, as shown in the following screenshot:

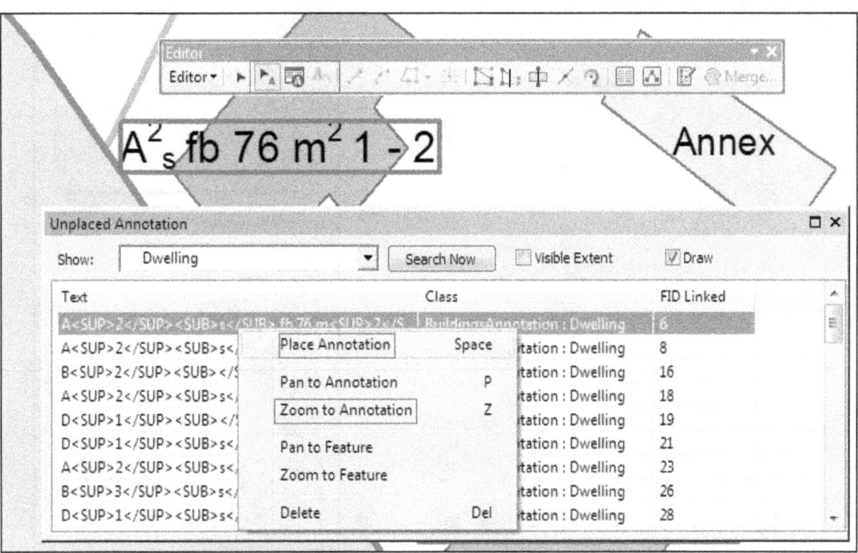

14. Your annotation list can be slightly different. You can set the **Placed** status for unplaced annotations in the `BuildingsAnnotation` attribute table. The status is stored in the **Status** field that has assigned the `AnnotationStatus` domain.

    In the next steps, you will edit the annotation, as shown in the upcoming screenshot:

15. Select the previous annotation with the **Edit Annotation Tool** parameter and click on the **Attributes** button. In the **Attributes** dialog, click on the **Expand All Relationships In Branch** button. Explore the attribute fields of the annotation feature and the related-dwelling feature from the `Buildings` feature class.
16. Select the related building feature and change the `Stories` value to `10`. The text from the annotation feature class is immediately updated because the annotation is linked to the dwelling feature.

17. Select the annotation feature and click on the **Annotation** tab. Split the single text line into three different lines. Select the **Leader** button and choose **Simple Line Callout**. Click on **OK** and on **Apply**. Select the annotation and move outside the polygon boundary to see the leader line.

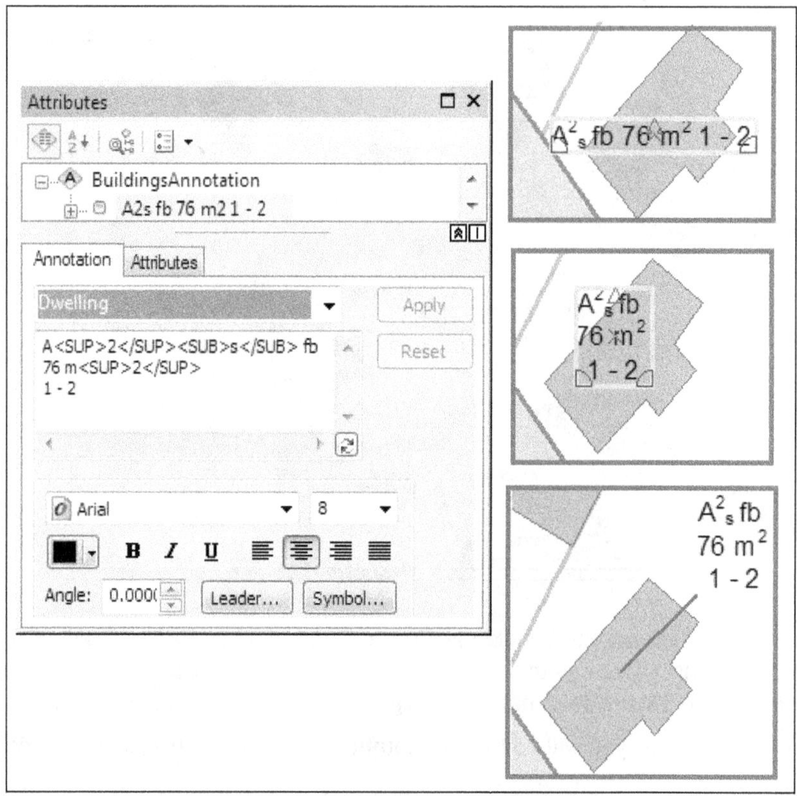

Let's add a new annotation to your building. To quickly view the `OBJECTID` value from the `Buildings` layer, navigate to its **Layer Properties** | **Display** and set **Field:** `OBJECTID` and **Show MapTips using the display expression**.

18. From the **Editor** toolbar, select the **Create Features** tool. Set `BuildingsAnnotation` as the `AB Dwelling` feature, choose **Construction Tools**, and set it as the **Leader** tool. In the **Annotation Construction** window, type `demolished in 2012`. Your mouse pointer has a small cross with the custom text. Click your pointer on the building to create the annotation as shown in the following screenshot:

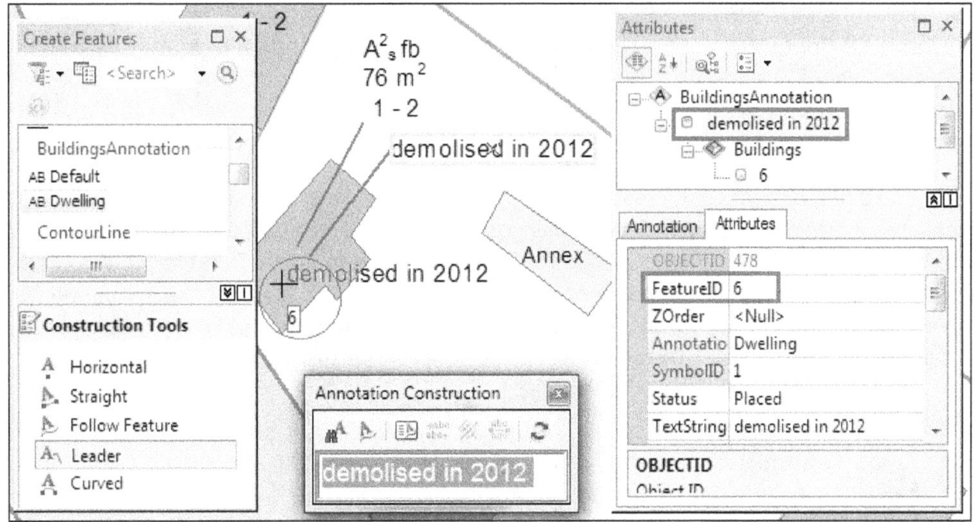

19. To create a feature-linked annotation, you have to edit the `FeatureID` field with the building `OBJECTID` value. Save the edit by navigating to **Editor** | **Save Edits**.

    Let's test the relationship between the `Buildings` and `BuildingsAnnotation` layers, as follows:

20. With the **Edit** tool, select and right-click on the related building and choose **Delete**. The buildings and the two related annotations from `BuildingsAnnotation` will be erased too.
21. Undo the last action by navigating to the **Standard** toolbar | **Undo Delete Feature** tool. Select the building and move to a new location. The two related annotations will be moved too.
22. Save your map as `MyAnnotations.mxd` at `...\Data\MapDocuments`.

You can find the results at `...\Data\MapDocuments\Chapter6\AnnotationFinal.mxd`.

How it works...

If you have two data frames in a map document and erase one of them, you will lose all map annotation groups that are associated with that data frame. If you save the labels in the map document as an annotation group, the unplaced dynamic labels will be reported as **Overflow Annotation**. To keep an overflow annotation in a map document, right-click on the annotation, and choose **Add Annotation**.

While converting labels to the annotation feature class, the annotation is automatically added in the **Table Of Contents** section. Notice that all annotations (standard or feature-linked) have annotation classes created from the label classes. These annotation classes are stored in the geodatabase as subtypes in the `AnnotationClassID` subtype field. Examine in ArcCatalog the properties of the standard and feature-linked annotation feature classes you just created.

If you made a mistake, you can erase the existing annotation class and create another one. You can also create many annotation feature classes for the same feature class. This can be useful if you are creating different thematic maps at different scales for the same dataset. If you decide to combine all annotation feature classes in one, you will have two options: check **Append** from the **Convert Labels to Annotation** dialog, or use the **Append Annotation Feature Classes** tool by navigating to **ArcToolbox** | **Data Management Tools** | **Feature Class**.

An annotation created from standard dynamic labels cannot combine with an annotation created with Maplex Label Engine. Both annotations must use labels that are created with the same label engine. Two feature-linked annotation feature classes created for the same feature class can be combined in a single one. A new, empty annotation feature class can be created with ArcCatalog, at the geodatabase level or the feature dataset level and edited in ArcMap.

The annotation feature class is stored as a polygon feature class. The polygon representing the outline of the annotation can be resized, rotated, flipped, curved (edit the **Base Line Sketch** parameter), and converted into multiple parts. The curve annotation has vertices that can be deleted, added, or moved. To make all these changes, select annotation, and right-click to choose the edit options.

There's more...

You might have noticed in the **Table Of Contents** section a new symbol on the left-hand side of the annotation classes when you use **Zoom In** or **Zoom Out**. This square with a double base indicates that the annotation class will not be drawn because it is out of the scale range already defined for the dynamic label class and inherited by the annotation class. You can change this restriction by right-clicking on the annotation class and selecting the **Visible Scale Range** parameter.

See also

- Explore the `ContourLineAnnotation` layer with the **Magnifier** tool at the reference scale of `1:5,000`. There are some values with small segments between the digits. To learn how to use annotation in this cartographic issue, please refer to the next *Using annotation and masking for contours* recipe.

Using annotation and masking for contours

According to the commonly used rules in a topographic map, all contour lines should be interrupted beneath the elevation information. In this recipe, you will learn how to use the geodatabase annotation to display the elevation value with a visual effect for contours.

Getting ready

In this section, you will display on a topographic map at the scale of `1:5,000` the elevation value for the contour lines and elevation points as shown in the following screenshot:

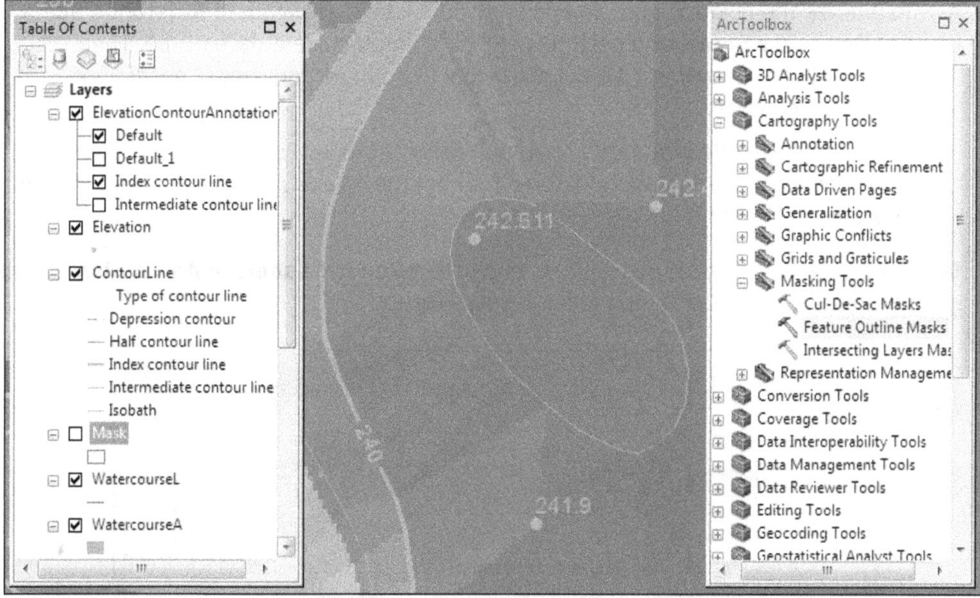

Firstly, you will create an annotation feature class that stores the point and index contour line elevation. Secondly, you will use the **Feature Outline Masks** tool by navigating to **ArcToolbox** | **Cartography Tools** | **Masking Tools** to obtain an intermediary polygon feature class that contains the mask (a polygon created around the text) for the annotations. Finally, based on the mask layer, you will create a visual discontinuity for index contour lines in accordance with the position of the annotation for contour lines and elevation points.

How to do it...

Follow these steps to create labels for contour lines and elevation points:

1. Start ArcMap and open the existing map document `AnnoMaskAdvancedDraw.mxd` from `<drive>:\PacktPublishing\Data\Annotation`.
2. Navigate to the **Labeling** | **Label Manager** dialog box. Inspect the label settings for the `Elevation` and `ContourLine` layers.
3. Uncheck the white halo for the `ContourLine: Index contour line` label class by navigating to **Text Symbol** | **Symbol** | **Edit Symbol** | **Mask** and set it to `None`. Keep clicking on **OK** until all the dialog windows are closed.
4. To analyze the position and dimensions of the labels, navigate to **Reference Scale** | **Set Reference Scale**, and set it to `1:5,000` for the current data frame. Then navigate to the **Labeling** | **Lock Labels** tool to hold the convenient position of the displayed labels.

   Let's create a single standard annotation called `ElevationContourAnnotation` that will store the annotation features for both the `Elevation` and `ContourLine` layers by performing the following steps:

5. Right-click on the `Elevation` layer and select **Convert Labels to Annotation**. Set the parameters, as shown in the following screenshot:

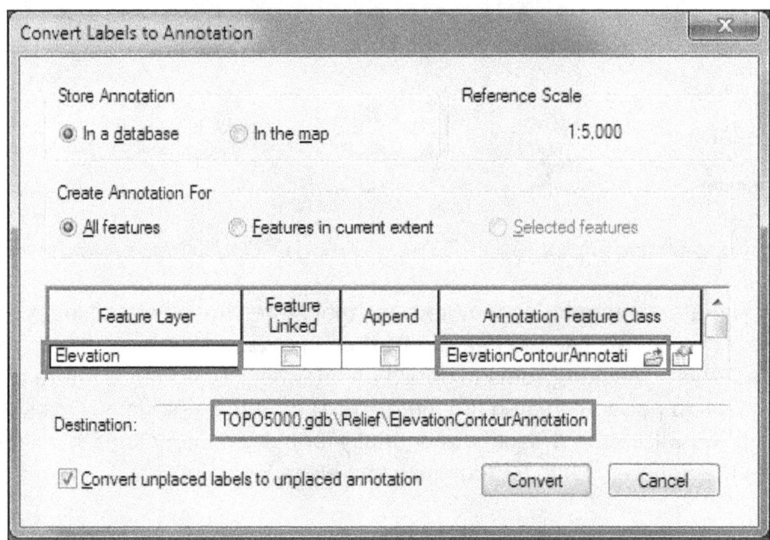

6. Click on **Convert**. Right-click on the `ContourLine` layer and select **Convert Labels to Annotation**. Set the parameters, as shown in the following screenshot:

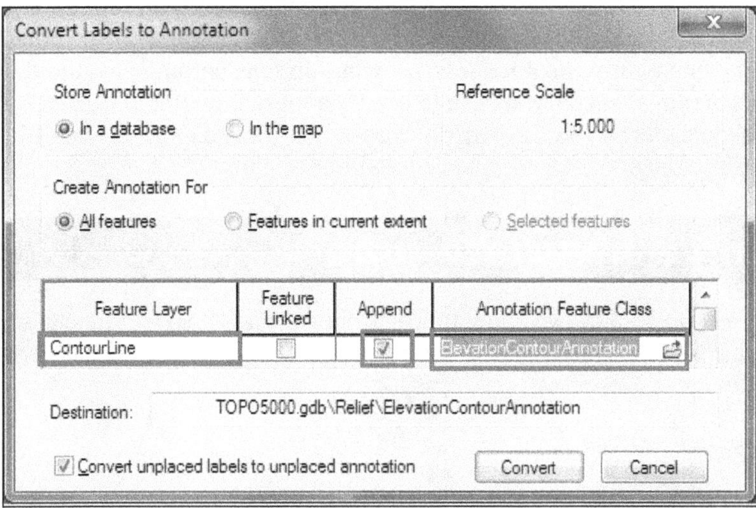

7. Check the **Append** option to store the annotation features in the existing `ElevationContourAnnotation` geodatabase annotation. Click on **Convert**.
8. Check and correct the position of the annotations in **Data View** if it's necessary.
9. Open **ArcToolbox** and navigate to **Cartography Tools** | **Masking Tools**. Double-click on the **Feature Outline Masks** parameter, and set the following parameters:
    - **Input layer**: `ElevationContourAnnotation`
    - **Output Feature Class**: `TOPO5000.gdb\Relief\Mask`
    - **Margin (optional)**: `2.5 Meters`
10. Leave unchanged the remaining default options. Click on **OK**.
11. In the **Table Of Contents** section, turn off the `Mask` layer. Right-click on the current **Layers** data frame and select **Advanced Drawing Options**. Check **Draw** using specified masking options. For **Masking Layers**: `Mask`, check **Masked Layers**: `ContourLine`. For **Masking Layers**: `WatercourseA`, check **Masked Layers**: `WatercourseL`. Click on **Apply** and on **OK**.
12. Explore the results. Save your map as `MyMaskAnnotations.mxd` at `...\Data\MapDocuments`.

You can find the results at `...\Data\MapDocuments\Chapter6\AnnoMaskAdvancedDrawFinal.mxd`.

How it works...

When an `ElevationContourAnnotation` feature annotation is moved, the corresponding feature from `Mask` will not automatically move. You have to move the polygon mask to align with the annotation feature. The **Advanced Drawing Options** parameter is the property of data frame, and for this reason, you have to keep the mask layer in the current data frame from your map document (`.mxd file`). Saving the `ContourLine` as a `layer` file doesn't help to keep the visual effect for contour line features.

Creating a bivariate map

In this recipe, you will display two quantitative attributes of a single feature class. The result of displaying two attributes at the same time for a polygon feature class is called a **bivariate map**.

Getting ready

In this section, you will use the `Census` feature class to create a quantitative bivariate map that analyzes two variables from the census data: `NoEducation` (number of persons with no education) and `Unemployment` (unemployment rate).

How to do it...

Follow these steps to create a bivariate map using the **Quantity by Category** option:

1. Start ArcMap and open a new map document. In the **Table Of Contents** section, right-click on the data frame called **Layers**, select **New Group Layer**, rename it by selecting the new group layer, and type `Multiple Attributes/Quantity by category`.
2. Right-click on the group layer and select **Add Data**. Load the `Census2010` feature class from `...\Data\TOPO5000.gdb\Boundaries`.
3. To display only the subset of features that have the census information, right-click on the `Census` layer, and navigate to **Properties** | **Definition Query** | **Query Builder**. Build the following expression: `PopulationOver10 <> 0`. Click on **Apply**.
4. Click on the **Symbology** tab and navigate to **Multiple Attributes** | **Quantity by category**. In the **Value Fields** section, select the first field `Unemployment` and the second field `NoEducation`. Click on the **Add All Values** tab and uncheck the **<All other values>** option.

Let's group the **Values** option in four classes based on the `Unemployment` field values, as follows:

5. Select all values **4** (**4, 18166**; **4, 13173**; **4, 13879**; **4, 8462**), right-click on the selected values, and choose **Group Values**, as shown in the following screenshot:

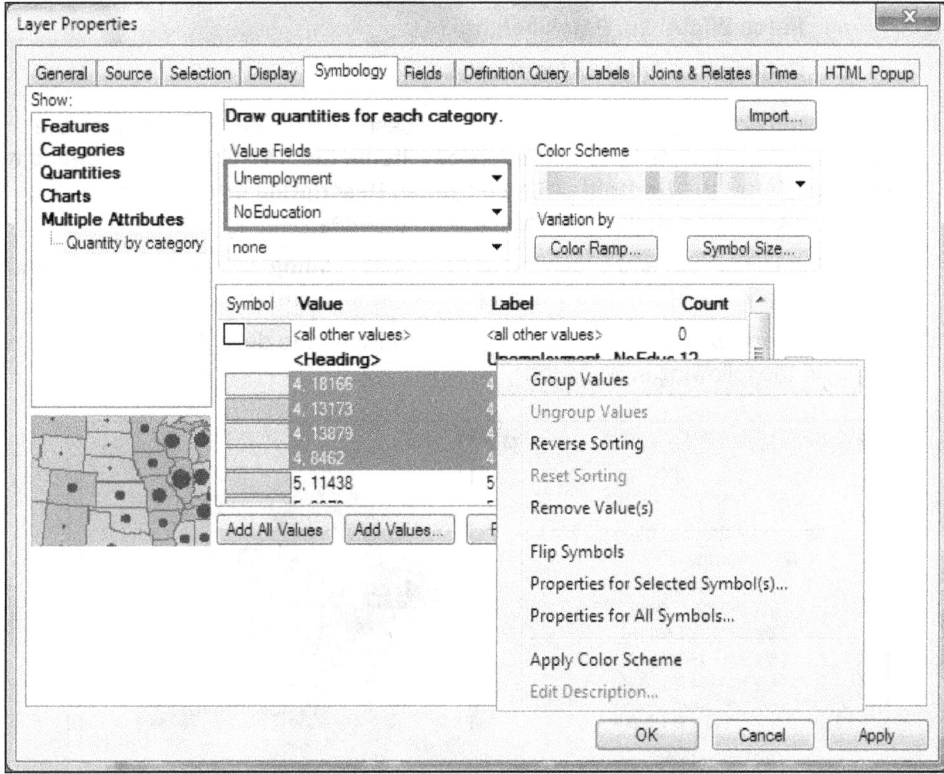

6. Repeat step 5 to group values **5**, **6** and (**7-8**). You obtained four new classes. Click on **Apply**. Set **Color Scheme:** `Errors` color ramp and click on the **Value** field to choose **Reverse Sorting**.
7. Navigate to **Variation by** | **Color Ramp**. Set **Value:** `NoEducation`, navigate to **Classification** | **Natural Breaks (Jenks)** | **Classes**, and set it to `4`. Click on **OK**. For **Color Ramp**, choose `Cyan-Light to Blue-Dark`. On the **Value** field, click on and choose **Reverse Sorting**. On the **Symbol** field, click on and choose **Flip Symbols**.
8. Click on **OK** to close the dialog box and on **Apply** to see the updates in **Data View**. Edit **<Heading> Label:** `Unemployment/NoEducation`. Click on **Apply** and on **OK**.

9. Navigate to **View menu** | **Layout View** to add a legend for your bivariate map. Navigate to **Insert menu** | **Legend**. In the **Legend Wizard** dialog window, set the following parameters:
    - **Set the number of columns in your legend**: `4`
    - **Legend Title**: erase the text and leave empty
    - **Patch Width**: `30`; **Patch height**: `30`
    - **Columns**: `0`; **Patches (vertically)**: `0`
10. The **Legend** option generated on your map doesn't look good. Right-click on the **Legend** option, and navigate to **Properties** | **Items tab** | **Style button** | **Legend Item Selector**. Select the **Horizontal Single Symbol Description Only** parameter, and click on **OK**. Click on **Apply** to see the legend updates.
11. Select the **Layout** tab and check the **Right to left reading** option to switch the matrix. Click on **Apply** to see the matrix update. Click on **OK** to return to **Layout View**.
12. You can insert text by navigating to **Insert menu** | **Text** to describe the two axes, as shown in the following screenshot:

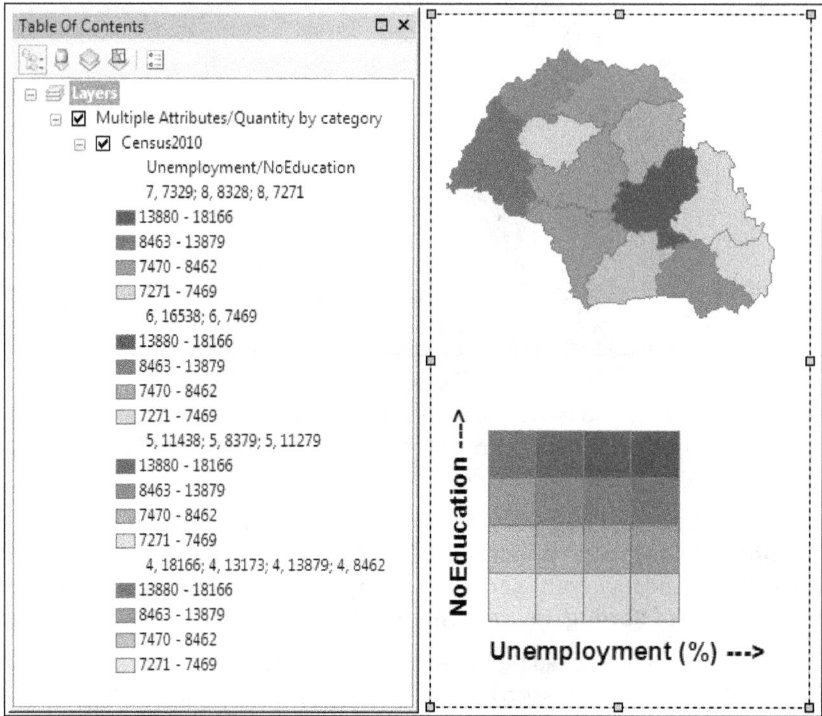

13. Navigate to **Standard** toolbar | **Select Element** to select the text or legend elements. Explore the results. Save your map as `MyBivariateMap.mxd` at `...\Data\MapDocuments`.

You can find the results at `...\Data\MapDocuments\Chapter6\BivariateMap.mxd`.

How it works...

You grouped the displayed features in four categories of unemployment rates (7-8 percent, 6 percent, 5 percent, and 4 percent). Every category of unemployment rate groups the number of persons with no education into four quantitative classes based on the natural grouping of `NoEducation` field values. Based on the sixteen groups, you created a legend color matrix with four rows and four columns.

The four categories of unemployment rate are symbolized by the Error ramp scheme.

The four quantitative classes representing the number of persons with no education are symbolized with the `Cyan-Light to Blue-Dark` color ramp (step 7).

The **Quantity by category** classification creates the final symbol, that of colors, by keeping the Hue from the first symbology (Error ramp scheme) and adds Saturation and Value from the secondary symbology (`Cyan-Light to Blue-Dark`). You can check this using the **Eye Dropper** tool and the **Color Selector** tool (RGB and HSV color space).

See also

- For more information about working in **Layout View**, please refer to *Chapter 7, Exporting Your Maps*

7
Exporting Your Maps

In this chapter, we will cover the following topics:

- Creating a map layout
- Exporting your map to the PDF and GeoTIFF formats
- Creating an atlas of maps
- Publishing maps on the Internet

Introduction

Before we start creating a map or a poster containing a map body, think about the following aspects that can influence your final product:

- Audience (for example, the technical team)
- Objectives (for example, the topographic map)
- Map scale (for example, `1:2,000`)
- Visual balance (for example, the main map body as a central theme of the map)
- Technical limits (for example, printer map size)

In the next four recipes, you will design, prepare, and export a map. You will share the contents of the map with your colleagues using a map package format through the **ArcGIS Online** platform. You can find the results of this chapter at `<drive>:\PacktPublishing\Data\ExportingMaps\MyResults`.

Creating a map layout

A **map layout** is a final map product created in **Layout View** that can be shared with other users. A map layout is designed for digital view or print. **Layout View** is the map design environment in ArcGIS.

The main map elements are map body, graticule, scale bar, scale text, north arrow, legend, title, and map border. Some details regarding the map projection, contour interval (for a topographic map), source of data, name of the organization, and some explanatory notes will support the main message of a map.

Getting ready

In this recipe, you will design a thematic poster that presents a topographic map at the scale `1:5,000`. Your poster will contain three data frames and a title, scale, legend, and dynamic text. The following table describes the shape and size of an old paper map:

Scale	**Trapezoid**		**Grid**
	Latitude	**Longitude**	
`1:5,000`	1' 15"	1' 52, 5"	500 m
`1:2,000`	37.5"	56.25"	200 m

The following screenshot represents the dimensions and geographic coordinates (latitude and longitude) for a sheet of a topographic map at the scale `1:5,000`:

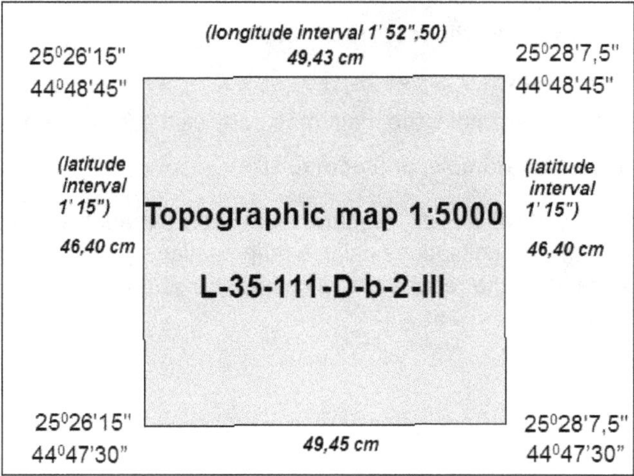

You have the final poster in PDF format at <drive>:\PacktPublishing\Data\ExportingMaps\FinalMapLayout.pdf. Please use this example while you work with the exercise from the next section.

How to do it...

Follow these steps to create a map layout using ArcMap:

1. Start ArcMap and open the existing map document MapLayout.mxd from <drive>:\PacktPublishing\Data\ExportingMaps.
2. To prepare the map data for print, you have to create a map layout in **Layout View**. Switch from **Data View** to **Layout View** by selecting **View** | **Layout View**.

   A new toolbar has automatically been added: **Layout** toolbar. The tools on this toolbar can be used only in **Layout View**. To select, move, and resize the map elements on your map layout, you will use the **Select Elements** tool from the **Tools** toolbar. A selected map element will be outlined in blue and will have eight selection handles. If you place the mouse pointer over one blue selection handle, you will be able to resize the map element. Right-click on any selected map elements to change its properties.

3. Before starting work on your map layout, define the virtual page size. From **File** | **Page and Print Setup** | **Map Page Size**, uncheck **Use Printer Paper Settings**. For **Page**, select **Custom** as **Standard Sizes**. For **Width**, type 80, and for **Height**, type 115. For **Orientation**, select **Portrait**. Click on **OK**.

   Layout View has two rulers: at the top and on the side of the virtual page. Let's adjust them:

4. Navigate to **Customize** | **ArcMap Options** | **Layout View**. You can open **ArcMap Options** directly by right-clicking on one of the rulers and selecting **Options**. Set the parameters as shown in the following screenshot:

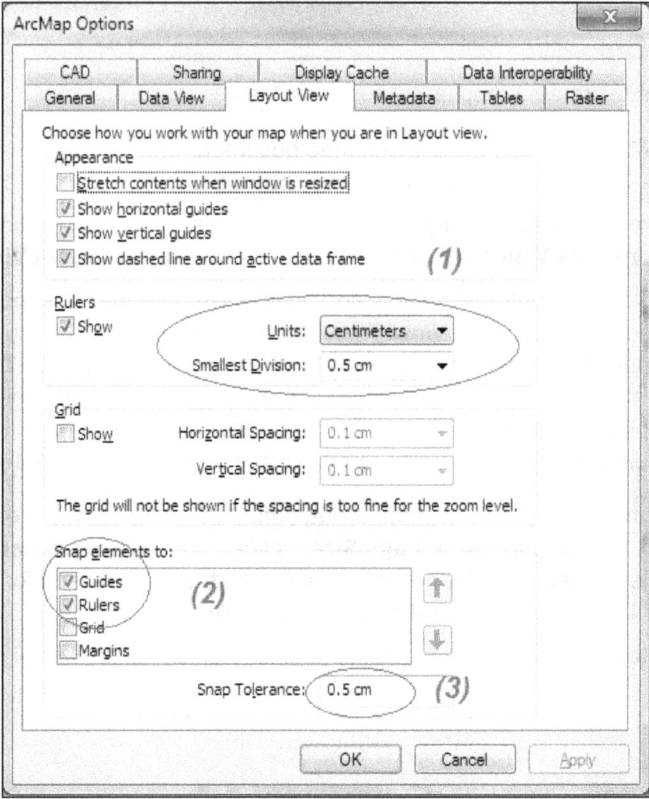

5. Click on **Apply** and on **OK**. To add guides, right-click on the vertical ruler from the left-hand side, and select **Set Guide**. Drag the light-blue ruler guide at `110` centimeters. Use **Zoom In**, **Zoom Out**, and **Pan** on the **Layout** toolbar to see the smaller division of the ruler. On the vertical ruler, add three more ruler guides at: `111`, `40`, and `5`. On the top ruler, add three guides at `5`, `6`, and `75`, as shown in the following screenshot:

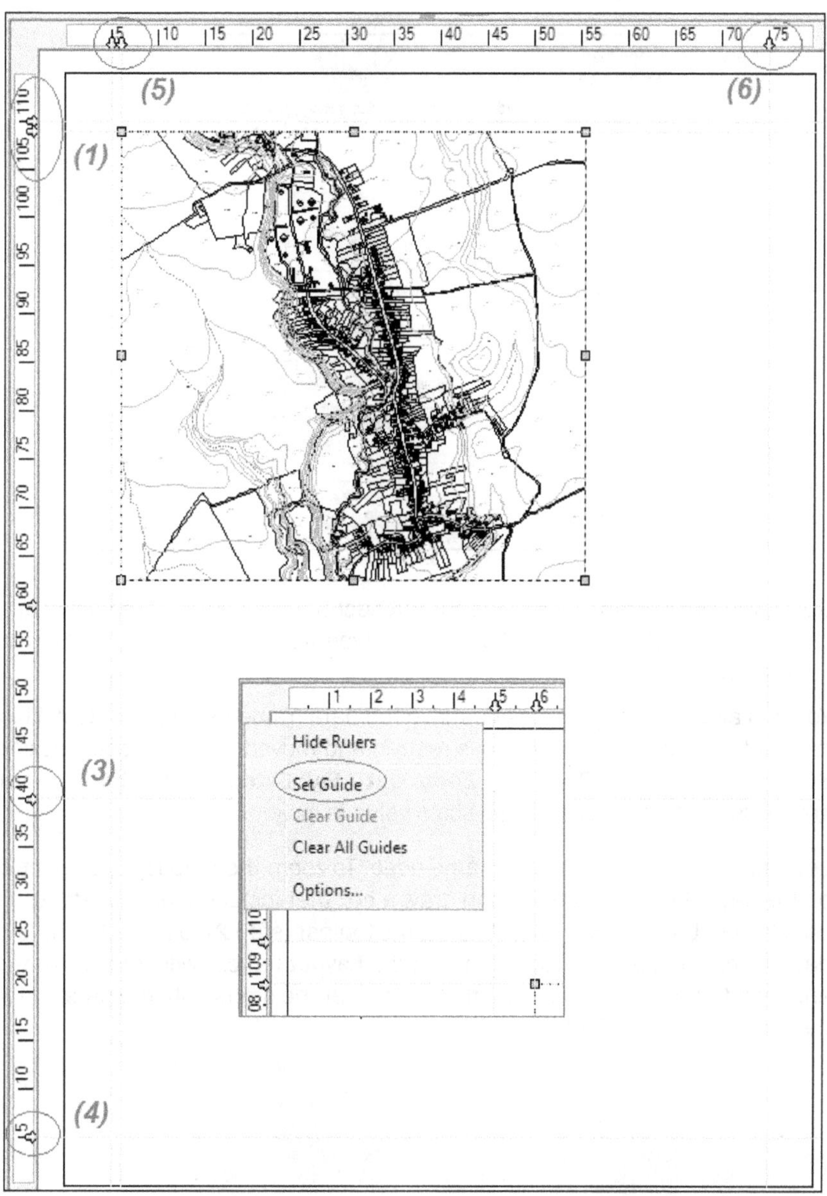

Resize the `Topographic Map 1:5,000` data frame and define a fixed scale for your data frame.

6. Select the data frame with the **Select Elements** tool. Right-click on the selected data frame and choose **Properties** | **Size and Position**. Set **Size**, **Position**, **Anchor Point**, and **Element Name**, as shown in the following screenshot:

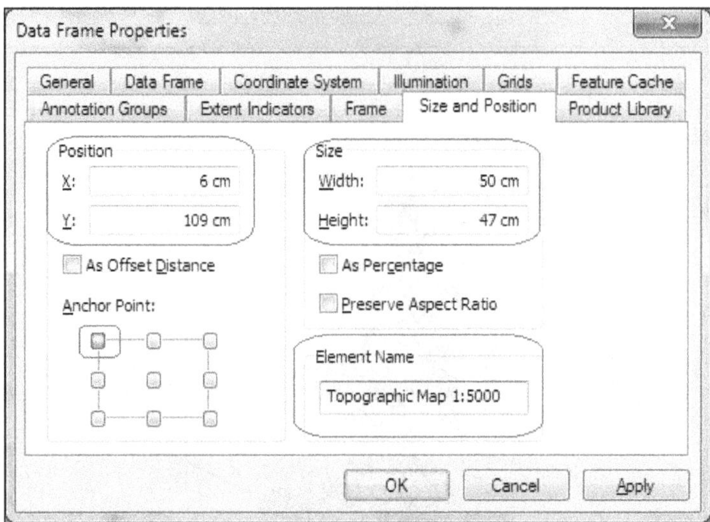

7. Select the **Data Frame** tab. In the **Extent** section, click on the drop-down arrow and choose the **Fixed Scale** option. For **Scale**, type `5000` and click on **Apply**. Click on **OK** to close the window.

   You can also manually resize and move the data frame using the **Select Elements** tool. Notice that your data frame is resized and moved. Also, the scale box on the **Standard** toolbar and **Zoom In** / **Zoom Out** / **Full Extent** on the **Tools** toolbar are gray. This indicates that the map scale value is locked.

8. Let's align the data to the data frame edge. To zoom the map layout, select **Zoom In** on the **Layout** toolbar (step **1**) and draw a box that includes the top left corner of the data frame. Use the **Pan** tool on the **Tools** toolbar (step **2**) to move the data in the data frame (step **3**). The zoom tools on the **Layout** toolbar will not change the scale and map the extent of your data frame. Repeat the steps until the data is aligned to the edge of the data frame.

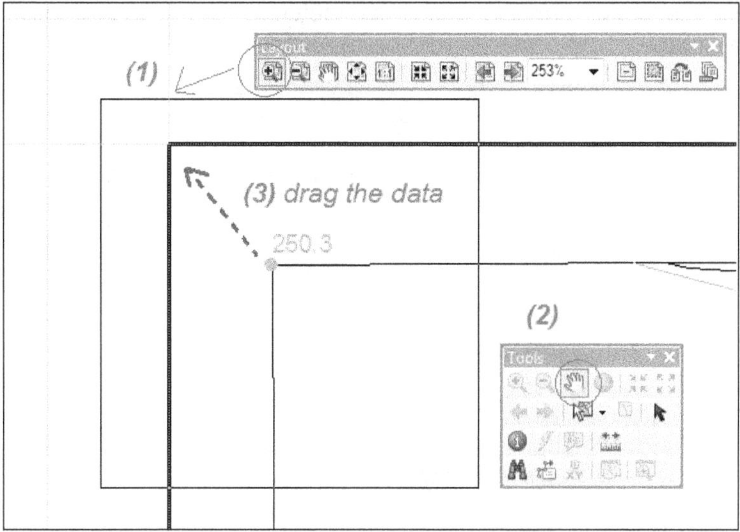

In the next steps, insert the title of the map:

9. Navigate to **Insert** | **Title**. Type `Topographic Map 1:5,000` and click on **OK**. Select the title and drag it above the data frame. Right-click on the title and navigate to **Properties** | **Text**.
10. Select **Change Symbol** and set **Size**: `88`. Select **Save As** and set the following parameters: **Name**: `Map Title 5000`, **Category**: `LABELS`, and **Style**: `<drive>:\PacktPublishing\Data\TOPO5k.style` custom style.
11. Click on **Finish** and on **OK**. To unselect the title, click outside the map layout.

    From time to time, be sure to save the map document as `MyMapLayout.mxd` at `...\Data\ExportingMaps`.

    In the next steps, you add a scale text and a bar to your map:

12. Navigate to **Insert** | **Scale Text** and select the `Absolute Scale` symbol, and click on **OK**.
13. Select and drag the scale element on the right-hand side of the map. Notice that the value is the fixed scale value of the data frame. Right-click on the selected map element and set **Properties** | **Format** | **Size**: `30`. Save the text style. Click on **OK** to close the dialog.
14. Navigate to **Insert** | **Scale Bar** and select the `Alternating Scale Bar 1 Metric` symbol, and click on **OK** to accept the default properties and close the dialog box. Select and drag the map element to the right-hand side of the map below the numerical scale. Double-click on the scale to open **Alternating Scale Bar Properties**.

15. For the **Scale and Units** tab, set the parameters, as shown in the following screenshot:

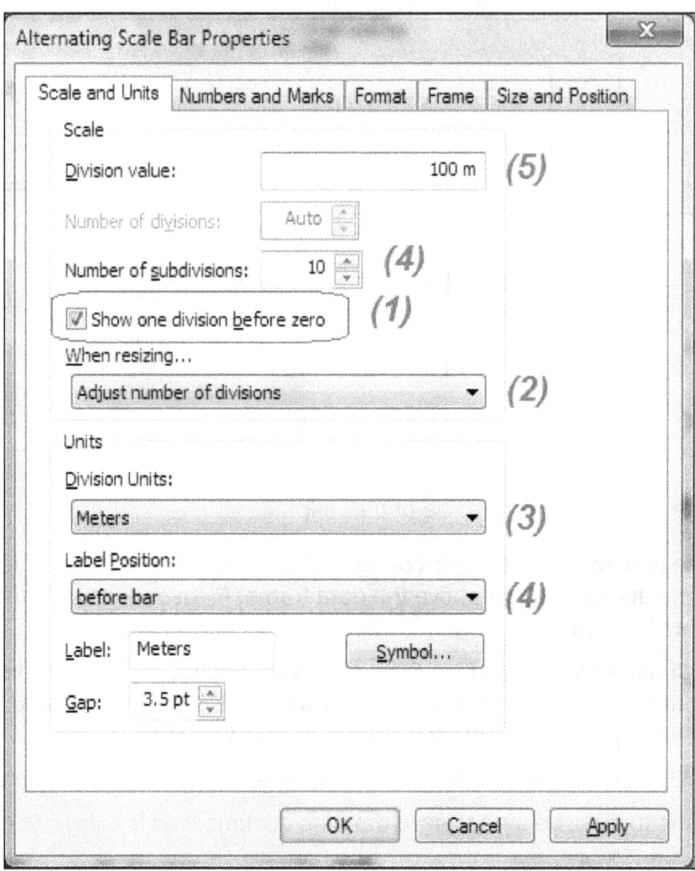

16. Click on **Apply**. For the **Number and Markers** tab, set the parameters as shown in the following screenshot:

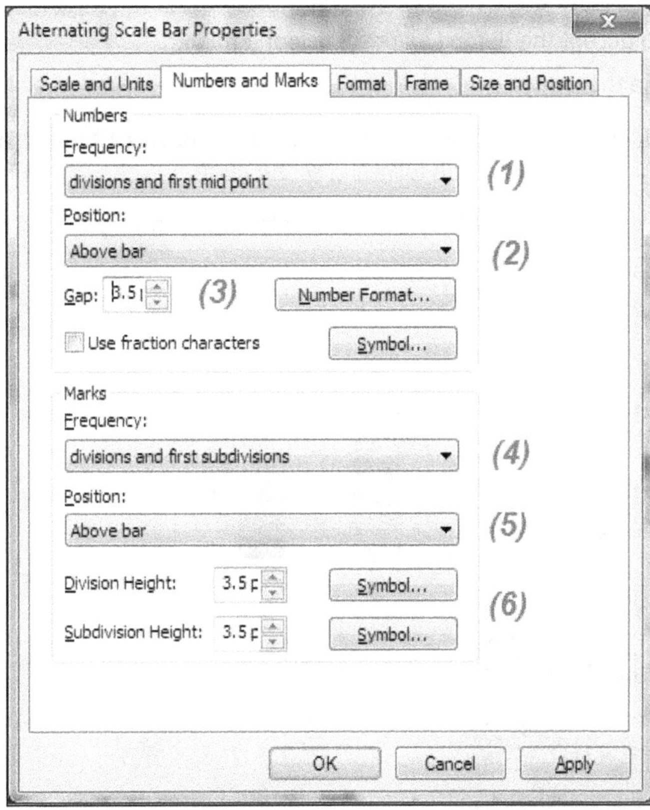

17. Select the **Format** tab, and set the parameters **Text Size**: `14` and **Bar Size**: `6`.

18. Click on **Apply**. Save your scale style by navigating to **Style** | **Style selector** | **Save**: `My 1:5,000 Scale Bar`. Move the scale style to the custom style `TOPO5k.style`.

19. For the **Size and Position** tab, set the following parameters: **Position X**: `58` and **Position Y**: `105`, **Size Width**: `16`. Click on **Apply** and on **OK**.

    In the next steps, you will insert the legend:

20. Navigate to **Insert** | **Legend**. By default, all layers are included in your legend (on the right-hand side). Change the order of layers in the Legend **Items** section: `Elevation`, `ContourLine`, `WatercourseL`, `WatercourseA`, `Buildings`, and `LandUse`.

21. For **Set number of columns in your legend**, type `1`. Click on **Next** and accept the default values for the next two panels. In the symbol patch panel, select `WatercourseA`, and for **Patch Area**, select the `Water body` symbol. Select the `Buildings` layer and set **Patch Area**: `Urbanized Area`. Click on **Next** on the final panel and accept the default values. Click on **Finish**.

    Let's refine the legend:

22. Right-click on the selected legend, navigate to **Properties** | **Layout**, and change the **Gaps** values, as shown in the following screenshot:

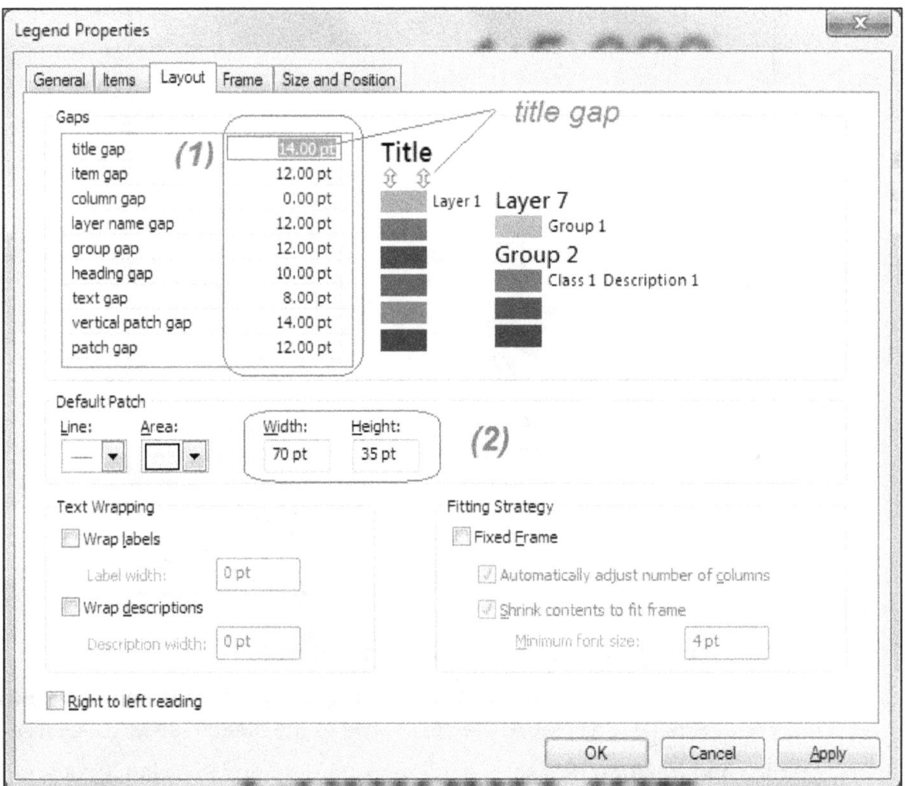

23. Select the **Size and Position** tab, and set the following: **Position X**: `60` cm and **Position Y**: `72` cm. Click on **Apply** to save the changes. Those are the main settings for a legend.

    You can continue improving the legend by selecting the **Items** tab. You can change the label properties for every layer from the list. Let's test some options:

24. Select the `ContourLine` layer. For the **Font** section, choose **Apply to the heading**. Set **Symbol** | **Size**: `22`. Click on **Apply**.

25. You can save the style of the legend for `ContourLine`. Make sure that the `ContourLine` layer is still selected before saving your style. Click on the **Style** button. Save your legend as `ContourLine Legend`. Keep clicking on **OK** to close all the dialog windows.
26. To move your new legend style (the `Legend Items` folder) from the default personal style file to `TOPO5k.style`, use **Customize** | **Style Manager**.

    Let's insert a border around the map using a custom border style from `TOPO5k.style`:

27. Select the data frame using the **Select Elements** tool. Go to **Insert** | **Neatline** and set the parameters as shown in the following screenshot:

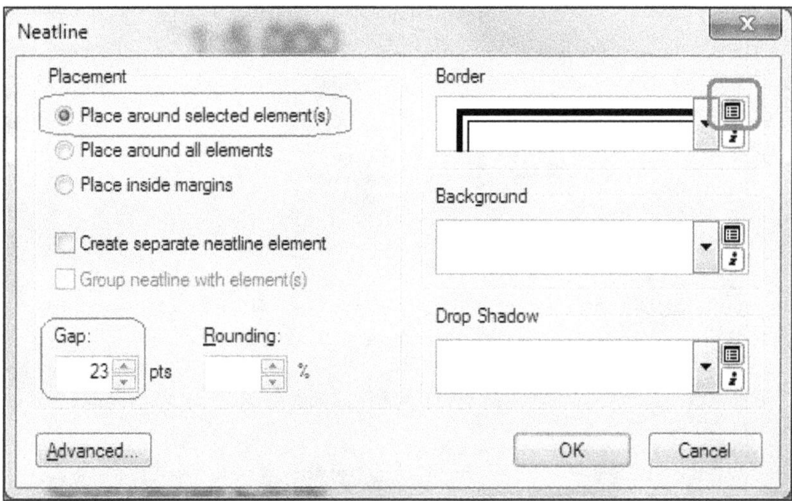

28. In the **Border** section, click on the **Style Selector** button. In the **Border Selector** window, click on **More Styles** to check whether `TOPO5k.style` is checked. Select the `Map Frame 5000` symbol. Click on **OK**. To change the properties of the neatline, go to **Data Frame Properties** | **Frame** | **Border**.

    Add a graticule (parallels and meridians) and a grid (rectangular coordinates). The grid will be superimposed on the graticule.

29. In the **Table Of Contents** section, right-click on the `Topographic Map 1:5,000` data frame, and navigate to **Properties** | **Grids** | **New Grid**. Check **Graticule**, and for **Grid Name**, type `Geographic coordinates`. Click on **Next**. Check the **Graticule and labels** option. In the **Intervals** section, set **Place parallels every** | **Min**: `1` and **Sec**: `15`. Set **Place meridians every** | **Min**: `1` and **Sec**: `52.5`. Click on **Next**. Check only the **Major division ticks** option. Go to the next panel. Check **Place a simple border at edge of graticule** and **Store as a fixed grid that updates with changes to the data frame**. Click on **Finish**.

We want to reproduce the old map values of the geographic coordinates. For this reason, we defined such small intervals for meridians and parallels. You will refine the interval values and define your own origin in the next steps:

30. In **View** | **Data Frame Properties** | **Grids**, select **Geographic coordinates**, and click on **Properties** | **Intervals.** Set the parameters as shown in the following screenshot:

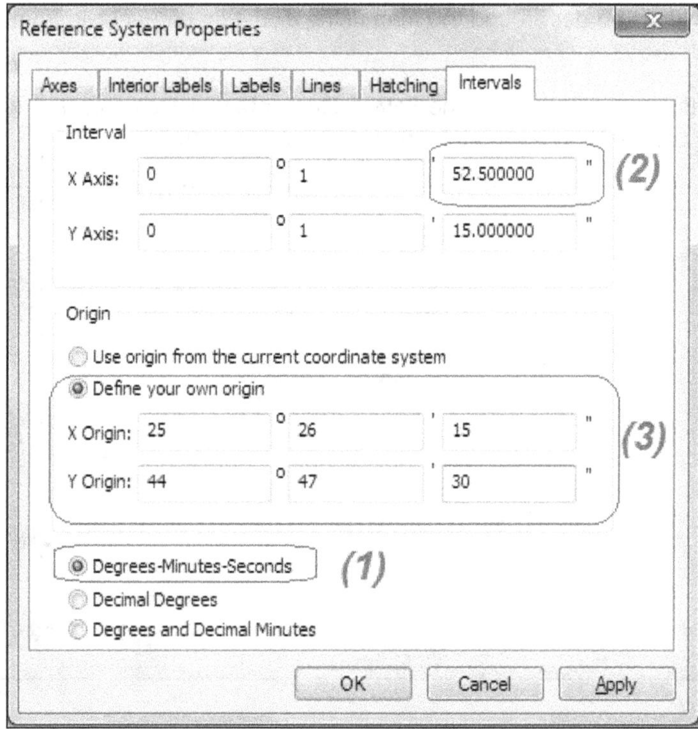

31. Click on **Apply** to save the changes. Select the **Axes** tab. For **Major Division Ticks**, add the line symbol named `Grid` from `TOPO5k.style`. Check **Display ticks outside** with **Tick size**: `23` pts. Select the **Labels** tab. For **Label Offset**, type `1` pts. Select **Additional Properties** and choose as **Label type** the **Stacked** option. Click on **OK** to close all windows. Inspect the results. Align the data to the data frame edges to have geographic coordinates for all corners of the data frame.
32. In **Data frame Properties** | **Grids**, add **New Grid**. Check **Measured Grid**, and for **Grid Name**, type `Projected coordinates`. Click on **Next**. Check the **Grid and labels** option. In the **Intervals X Axis and Y Axis** section, type `500` with unit as **Meters**. Click on **Next** twice to accept the default settings and click on **Finish**.

33. In **Data frame Properties** | **Grids**, select **Projected coordinates**, and click on **Properties** | **Intervals**. For **Origin**, select **Define your own origin**, and type `500,000` for both **X Origin** and **Y Origin**. Select the **Labels** tab. For **Label Offset**, type `1` pts. For **Label Orientations**, check only **Left** and **Top**. Select **Additional Properties** and choose **Specify the number of digits in a group**: `3`. Select **Number Format**, check **Number of significant digits**: `6`, and select **Show thousands separators**. Click on **OK** until all windows are closed.

    Save your grids as `My Geographic coordinates` and `My Projected coordinates`.

34. In **Data frame Properties** | **Grids**, select the reference grid and click on **Style** | **Save**.
35. Use **Style Manager** to move your symbols from the *default personal style* file (the `Reference Systems` folder) to `TOPO5k.style`.

    Let's insert some information regarding the coordinate reference system on your map:

36. Navigate to **Insert** | **Dynamic Text** | **Coordinate System**. Move the text box below the legend. Change the font size to `18`.

    A map layout can have one or more data frames with different coordinate reference systems, scales, and extent. Add two data frames in your map layout.

37. Navigate to **Insert** | **Data Frame**. In **Data Frame Properties**, rename the new data frame to `Romania Pulkovo 1942`. For **Size and Position**, set the following elements:
    - **Position X**: `5` cm; **Position Y**: `60` cm
    - **Size Width**: `25` cm; **Size Height**: `17` cm
    - **Anchor Point**: **top left**
    - **Define a fixed scale**: `3,000,000`
38. Add the `Counties.lyr` layer in your data frame from `...\Data\ExportingMaps\LayerFiles`. In the map layout, add your data at the center using Pan tool from the Tools toolbar.
39. Create a new data frame named: `Europe ETRS89`. For **Size and Position**, set the following elements:
    - **Position X** : `5` cm; **Position Y** : `40` cm
    - **Size Width**: `70` cm; **Size Height**: `35` cm
    - **Define a fixed scale**: `10,000,000`

40. Add the `Europe.lyr` layer from `...\Data\ExportingMaps\LayerFiles`. Add your data in the center of the data frame.

    Let's define an extent indicator for this data frame. The data frame `Europe ETRS89` will be the reference map, and the `Romania Pulkovo 1942` data frame will be the inset map.

41. In **Data Frame Properties**, select the **Extent Indicators** tab, and move the `Romania Pulkovo 1942` data frame to the right-hand side section **Show extent indicator for these data frames**. Click on the **Frame** button if you want to change the default symbol. Check the **Use simple extent** option. Click on **Apply** and on **OK** to close the window.

    You obtained a red rectangle around a country from your `Europe` data frame. If you change the dimension of the `Romania Pulkovo 1942` data frame, the red rectangle will be updated to the new extent.

    Continue to add more map elements on your map layout to fill the center gap or to improve the final poster.

    In the next steps, you will save your map and set up the printer:

42. Save the map document as `MyMapLayout.mxd` at `Data\ExportingMaps`.
43. Go to **Page and Print Setup**, and for **Printer Setup** | **Name**, choose a large format printer (plotter).
44. For **Paper Size**, select a proper printer paper size in accordance with the size of your map layout. Make sure **Orientation** | **Portrait** is checked. For **Map Page** | **Size**, check **Use Printer Paper Settings** and **Show Printer Margins on Layout**. Uncheck Scale Map Elements. Click on OK.
45. Navigate to **File** | **Print**, and for **Printer Engine**, choose **Windows Printer**. Set **Output Image Quality** to **Normal**. Click on **OK**. Before starting to print, go to **File** | **Print Preview** to check the final details.

How it works...

It is good practice to use a fixed scale for a data frame when you are working on a map layout.

Dragging the data in a map layout will not change the coordinates of your data.

By default, all inserted map elements will appear on the center of the graphic area.

You have noticed that the edges of your extent are curved. This is because the `Romania Pulkovo 1942` data frame has defined a *projected coordinate system* and the `Europe ETRS89` data frame has defined a *geographic coordinate system*.

If you have the same coordinate system (geographic or projected) for both data frames, then the extent will be straight.

The total length of your scale bar should not exceed a third of the width of your data frame. The **When resizing** | **Adjust number of divisions** option will update the number of divisions and will keep the distance represented by each division when you decide to change the length of your scale from `1:5,000`.

You have already worked with an active data frame in *Chapter 3, Working with CRS*. If you select the data frame in the map layout, the name of the activated data frame is indicated with bold letters in the **Table Of Contents** section. When you work with multiple data frames in a map layout, always look at the state of data frames in the **Table Of Contents** section.

There's more...

A template contains a designed map layout with one or more data frames. A template can contain data as basemap layers. Every new map document is based on a normal template that contains a single empty data frame. The normal template is the **Blank Map** saved as `Normal.mxt` at `%AppData%\ESRI\Desktop10.2\ArcMap\Templates`. A map document `mxd` can be used as a map template.

You will create *your own template* to save time when you create another map at the scale `1:5,000` with the same components. Your custom template will contain `Europe.lyr` as a basemap layer and all map elements from the previous section:

1. Start ArcMap and open an existing map document `Template.mxd` from `<drive>:\PacktPublishing\Data\ExportingMaps`. The map document contains three data frames. Only the `Europe ETRS89` data frame contains data. Notice the source of data. Close ArcMap.
2. In Windows Explorer, right-click on the `Template.mxd` **ArcGIS ArcMap** document, and choose **Copy**. Go to `%AppData%\Roaming\ESRI\Desktop10.2\ArcMap\Templates` and create a new folder, `Custom templates`. In this empty folder, right-click and choose **Paste**.

Because your template map contains a basemap layer, save `Template.mxd` from `%AppData%\Roaming\ESRI\Desktop10.2\ArcMap\Templates` with relative pathnames. Open your template map with ArcMap, set the data source, and select **File** | **Map Document** | **Properties**. Check the **Store relative pathnames to data source** option.

3. Open ArcMap. In the **ArcMap-Getting Started** dialog, go to **New Maps** | **My Templates** | **Custom Templates**, and choose **Template**. Click on **OK**.
4. In the **Table Of Contents** section, right-click on the `Topographic Map 1:5,000` data frame, and select **Activate**. Add the `LandUse` feature class from `...\Data\TOPO5000.gdb\LandUse`.
5. Notice the update from the map layout. Save your map document as `MyNewMap.mxd`.

If you want *all users of system* to use a standardized template, copy `TemplateAllUsers.mxd` to a new folder named `MyOrganization` at `<ArcGIS_HOME>\Desktop10.2\MapTemplates`.

Next time, when you open ArcMap, you will get the following structure:

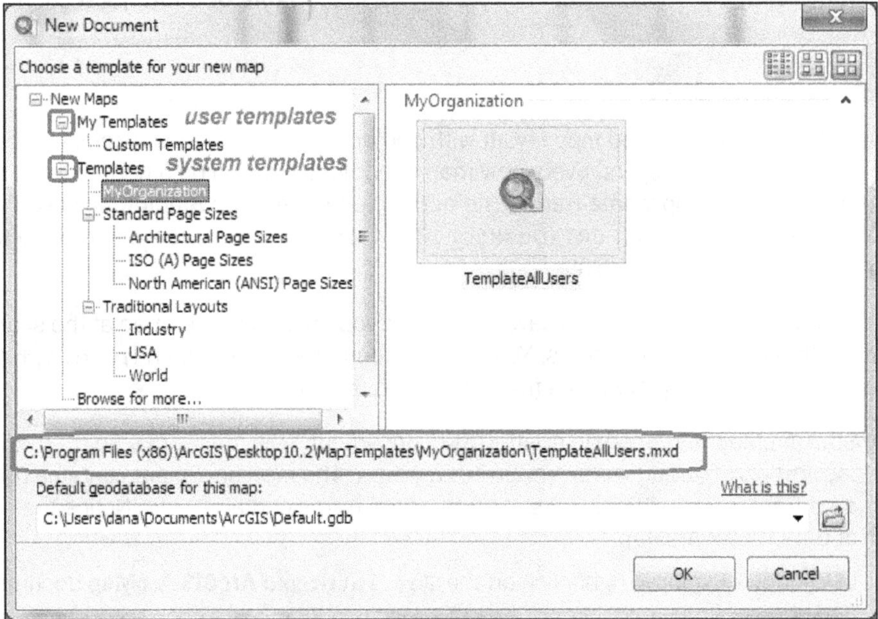

Notice that the `TemplateAllUsers.mxd` template doesn't have data. The template contains only three empty data frames and specific map elements.

Exporting your map to the PDF and GeoTIFF formats

Even if your colleagues don't have ArcGIS, you still can share your map with them. You can export your map in:

- **Raster format**: This includes bitmap, **Tagged Image File Format** (**TIFF**), and **Joint Photographic Experts Group** (**JPEG**)
- **Vector format**: This includes **Enhanced Metafile** (**EMF**), **Encapsulated PostScript** (**EPS**), **Adobe Illustrator** (**AI**), and **Portable Document Format** (**PDF**)

Getting ready

PDF is a vector-based file format that supports the information coming for vector and raster sources. Use **Adobe Reader** to view the results of this recipe. You can download the software for free from `www.adobe.com`.

TIFF is a raster-based file format used mainly to store raster images. **GeoTIFF** is TIFF with the coordinate reference system embedded. It is used to store data such as aerial photography, satellite images, terrain digital elevation models, or scanned old maps.

Before starting, please refer to:

- *Online Acrobat Help / Geospatial PDFs* at `http://helpx.adobe.com/acrobat/using/geospatial-pdfs.html`
- The GeoTIFF website at `http://geotiff.osgeo.org`

How to do it...

Follow these steps to export your map to a PDF format using ArcMap:

1. Start ArcMap and open an existing map document `FinalMapLayout.mxd` from `<drive>:\PacktPublishing\Data\ ExportingMaps\MyResults`.
2. Select **File** | **Export Map**. For the **Save As** type, select the **PDF (*.pdf)** format. For **File name**, type `MyTopographicMap300dpi`.
3. Select the **General** tab. For **Resolution**, keep the default resolution: `300` **dots per inch** (**dpi**). Set **Output Image Quality** to **Normal**.

4. Select the **Format** tab. Select **RGB** for **Destination Colorspace** because you intend to use a laser printer for a quick preview of the results. If you intend to use commercial offset printing, you should use **CMYK**. See the *There's more...* part of the *Modifying symbols* recipe of *Chapter 5, Working with Symbology*. Make sure **Embed All Document Fonts** is checked. Accept all other default options.
5. Select the **Advanced** tab. For **Layers and Attributes**, select **Export PDF Layers and Feature Attributes**. Check **Export Map Georeference Information** to access the georeference information from the PDF file. Click on **Save**.
6. Repeat the steps for a PDF with `72` dpi and save it as `MyTopographicMap72dpi`.
7. Open the files with Adobe Reader. To see the layers and attributes of the features from your map, go to **View** | **Show/Hide** | **Navigation Panel** | **Layers** and **Tree View**. To see the geographic coordinates, go to **Tools** | **Analyze**, and select the **Geospatial Location** tool.

To save your map layout to a **Layout GeoTIFF** (`*.tiff`) file, you will need the **Esri Production Mapping** extension. This extension will allow you to check the following options: **Write World File** and **Write GeoTIFF Tags**. You also have a **Production** tab, where you can choose which data frame will define the coordinate reference system of the GeoTIFF file. GeoTIFF supports only one CRS at one point of time. Remember that your map layout contains three data frames with two different coordinate reference systems. If you don't have the extension, you can save as a TIFF, as follows:

1. Navigate to **File** | **Export Map**. For **Save As**, select the **TIFF (*.pdf)** format. For **File name**, type `MyTopographicMap72dpi`.
2. Select the **General** tab. For **Resolution**, type `72` dpi. Accept the default number of pixels. Accept all other default parameters. Click on **Save**.
3. Repeat the steps for a TIFF file with `300` dpi and save it as `MyTopographicMap300dpi`.

Compare the file size in Windows Explorer of the four files and notice the quality of maps.

How it works...

When you export your map to any format allowed, the **Export Map** | **Clip Output to Graphics Extent** option eliminates the white empty margin around your map elements.

In relation to the **Embed All Document Fonts** option from step 4, please be aware about the fonts used. Some fonts do not support embedding in a PDF file. Embedding means that PDF uses the same fonts as your map layout even if the file is opened on a platform that does not have the map's fonts (for example, Esri fonts). You can verify that your font is embeddable at **Microsoft typography** at `http://www.microsoft.com/typography`.

You can control what layers and their attributes should be transferred to the PDF format. In the **Table Of Contents** section, you can unselect the layers. To reduce the number of attributes for the visible layers, uncheck the attributes in **Layer properties** | **Fields**. An example of a PDF with restricted layers and attributes is `RestrictedTopographicMap.pdf` from the `...\ExportingMaps\MyResults` folder.

The PDF vector format is used to display web maps. When you use **Pan** and zoom in your software, the displaying is quite fast. This format is not quite the best solution for print publishing because PDF could make small changes in font characteristics.

The TIFF format will build the map pixel by pixel. This format is used for print publishing. A high resolution will create a high-quality map with a price: a large file size (300 dpi/over 300 MB). When you use **Pan** and zoom in your graphics software, the displaying will take a while for the `MyTopographicMap300dpi` map. The labels and line symbols from the `MyTopographicMap72dpi` map are slightly jagged. However, this will not affect the quality of a **PowerPoint** presentation, for example.

Creating an atlas of maps

An atlas of maps is a collection of maps. Those collections of maps use spatial dataset series and other additional information. In accordance with *INSPIRE Metadata Implementing Rules Technical Guidelines*, a series *is a collection of related datasets that share the same product specification*.

The **Data Driven Pages** function allows you to generate a series of maps starting from a map layout and an index layer. An index layer is used to specify the dimension and the number of pages from the map book. Each feature from the index generates one output map. You can use a *regular grid*, a *strip map that follows a linear feature*, or any *feature layer* from the data frame as an index layer.

The general steps in generating a series of map pages are:

1. Creating a map layout.
2. Creating **Grid Index Features**.
3. Setting up the **Data Driven Pages** function.
4. Exporting the pages in PDF format.

When the **Data Driven Pages** function is enabled in your map layout, there will be a new tab in the **Export Map** window. The **Pages** tab will allow you to control the **Data Driven Pages** export.

Getting ready

You will create a series of maps at the scale `1:2,000`, starting from the map extent of the scale `1:5,000`. The map layout from the following section will contain two data frames with the same coordinate reference system. Every data frame contains the same layer structure. The `Topographic Map 1:5,000` data frame will be unchanged in every output page. The map pages will be generated based on the `Topographic Map 1:2,000` data frame content.

Before starting the exercises, please open the `Map_L-35-111-D-b-2-III-1.pdf` file from the `ExportingMaps\MyResults` folder. This PDF file is one of the final results in this recipe.

How to do it...

Follow these steps to create a map series at scale `1:2,000` using **Data Driven Pages**:

1. Start ArcMap and open an existing map document `DrivenPages.mxd` from `<drive>:\PacktPublishing\Data\ExportingMaps`. Make sure that the data frame `Topographic Map 1:2,000` is your current data frame or active (shown in bold in the **Table Of Contents** section).

   Notice the red extent indicator from the `Topographic Map 1:5,000` data frame. It's indicating the extent of the `Topographic Map 1:2,000` data frame.

2. Firstly, you will create a grid index in a new feature class. Open **ArcToolbox**, go to **Cartographic Tools** | **Data Driven Pages**, and double-click on the **Grid Index Features** tool, as shown in the following screenshot:

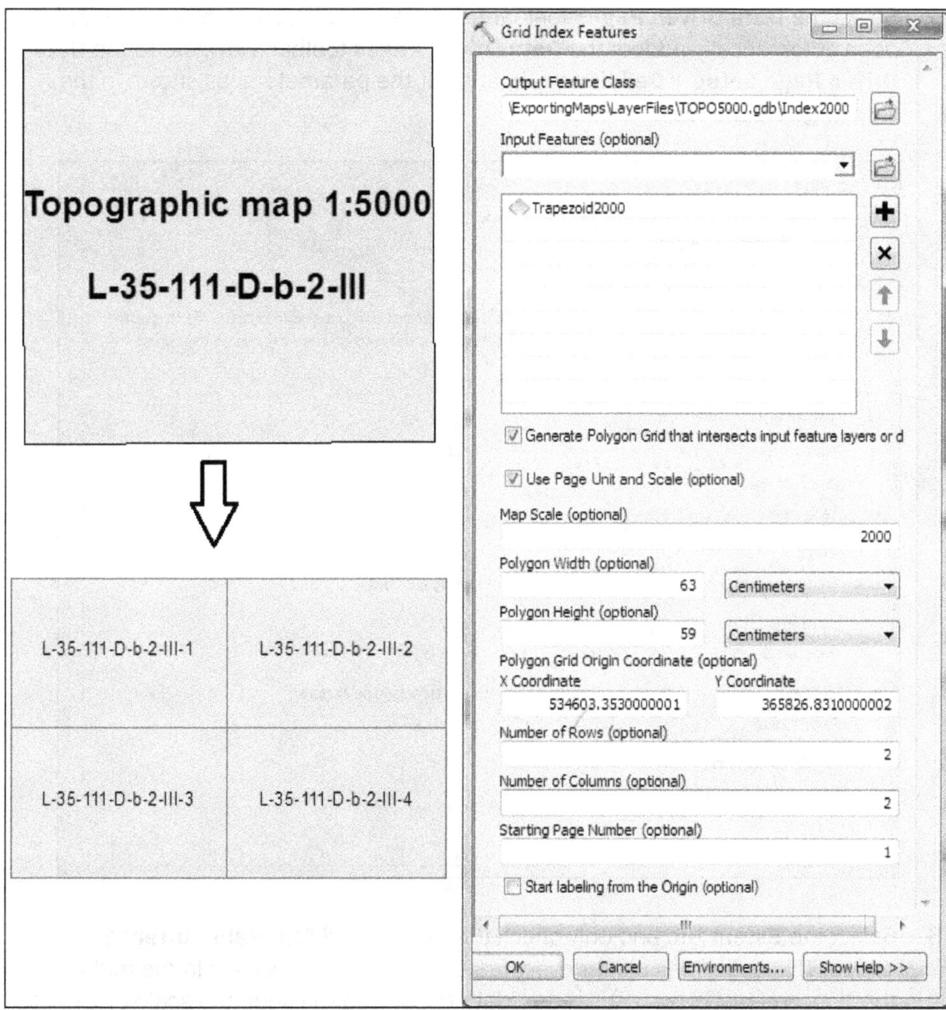

3. For the **Output Feature Class**, go to `ExportingMaps\LayerFiles\TOPO5000`, and type `Index2K`. For **Input Features**, select `Trapezoid5000`. Check **Generate Polygon Grid** and **Use Page Unit and Scale**. Be sure that **Map Scale** is `2000`. For **Number of Rows** and **Number of Columns**, type `2`. For **Starting Page Number**, type `1`. Accept the default values for the rest of the parameters. Click on **OK**.
4. Add `Index2k` in your data frame. For the `Index2k` layer, change the symbol to **Hollow** with **Outline Color**: `Mars Red`. Open the `Index2k` attribute table. Select the **PageName** field and right-click on it to select **Field Calculator**. Select the **Load** button, go to `ExportingMaps\LayerFiles`, and select the `Index2000.cal` file. Click on **Open** and on **OK**.

5. Select the **Data Driven Pages** toolbar from the **Layout** toolbar. A new toolbar has been automatically added: the **Data Driven Pages** toolbar. Navigate to the **Data Driven Page Setup** | **Definition** tab, and set the parameters, as shown in the following screenshot:

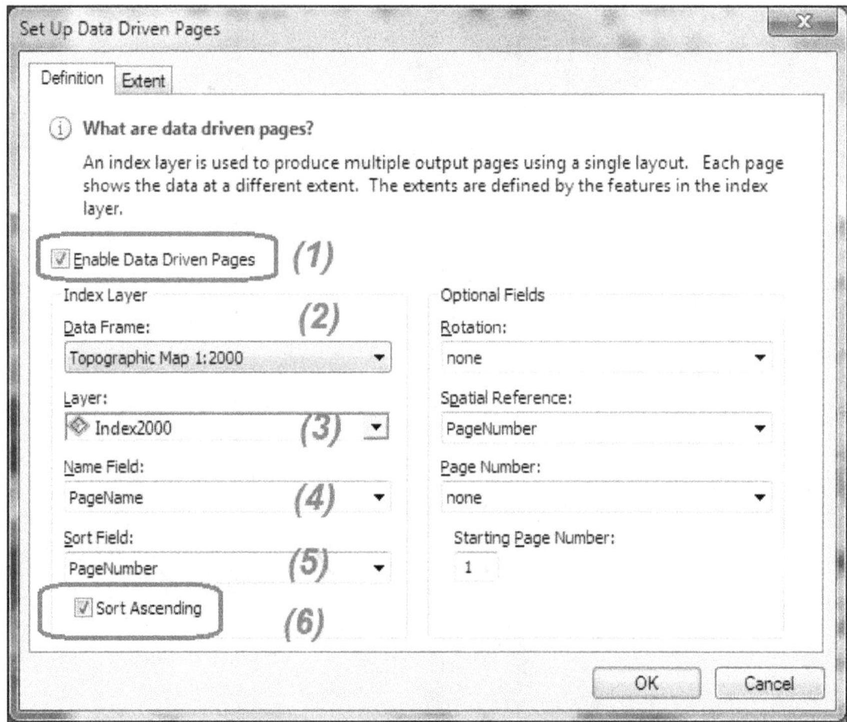

6. Select the **Extent** tab and only check the **Center And Maintain Current Scale** option. This option will center every feature from the index layer in the middle of the `Topographic Map 1:2,000` data frame and will keep the scale of `1:2,000`. Click on **OK**.

7. In **ArcToolbox**, go to **Cartographic Tools** | **Data Driven Pages**, and double-click on the **Calculate Adjacent Fields** tool. For **Input Features**, select the `Index2k` layer. For **Field Name**, select `PageName`. Click on **OK**. To inspect the result, use the **Open Attribute Table** tool from the **Data Driven Pages** toolbar.

   Let's add the index name for your maps:

8. Navigate to **Page Text** | **Data Driven Page Name**. The inserted map element will appear at the center of the graphic area. Right-click on the small rectangle selected and select **Properties** | **Change Symbol**. For **Size**, type `75`, and click on **OK** twice to close the dialog boxes. Drag the index name above the data frame's title. Use the blue arrows from the **Data Driven Pages** toolbar to see how the data frame and the index name are being updated.
9. Save the map document as `MyFinalDrivenPages.mxd` at `<drive>:\PacktPublishing\Data\ExportingMaps`.

   You will save your maps in four different PDF files:
10. Navigate to **File** | **Export Map**. For the **Save As** type, select the **PDF (*.pdf)** format. For **File name**, type `TopographicMap`. Navigate to **Options** | **General**. For **Resolution**, type `100` dpi. Select the **Pages** tab. Check **All (4pages)**. For **Export Pages As**, choose **Multiple PDF Files (page names)**. Select the **Advanced** tab. For **Layers and Attributes**, select **Export PDF Layers Only**. Check **Export Map Georeference Information**. Click on **Save**.
11. With Windows Explorer, explore the results from the four PDF files.

How it works...

If you have multiple data frames in your map layout, you should activate the data frame for which you want to create a series of layout pages. The parameters **Polygon Width**: `63` cm and **Height**: `59` cm from step 3 represent the size of the data frame. Having a fixed scale extent for the data frame ensures you do not miss the **Map Scale (optional)** value.

To generate a correct index for scale `1:2,000`:

- You used a reference layer named `Trapezoid5000`, which represents the extent of a map at scale `1:5,000`
- You have specified two rows and two columns in the index layer that will cover the extent of the reference layer

These are the two final remarks about the previous exercise:

- Take into account that various standards or norms specify that symbols characteristics (for example, size) do not remain constant as the scale changes. For the `Topographic Map 1:2,000` data frame, we used the symbols and annotation feature classes from the previous data frame (scale `1:5,000`) for ease of work. Normally, you should create the symbols and annotation for the scale `1:2,000`.
- We assumed that datasets have a positional accuracy corresponding to the scale `1:2,000`. Therefore, we used the same dataset for both data frames.

You can display your data at any scale you want, but don't overlook the positional accuracy of the original dataset. It is a right decision to mention this aspect on your final map.

Publishing maps on the Internet

The most common methods to share your maps are:

- **Layer package** (`lpk`)
- **Map package** (`mpk`)
- **Tile package** (`tpk`)
- Web map (a basemap layer and different data layers)
- Exported maps (for example, PDF file)

A layer package contains one or more layers (`.lyr` file) and the copy of the original data. Please remember that a layer file includes only the symbol and label definition, and the reference to the source data.

A map package contains a map document (`.mxd` file), the copy of the original data, and other documents (for example, a PDF file). A map package is used mainly for the vector format in order to edit, query, identify, or analyze data. You will create a map package in the next recipe.

A tile package contains the vector and raster formats. A tile package is used as a basemap (background reference information) and is structured in a set of tiles or images (for example, 256 x 256 pixels).

For more information about packaging and web maps, please refer to *ArcGIS Help (10.2)* online by navigating to **Desktop** | **Mapping** | **Sharing data through packaging** and also navigating to **Desktop** | **Mapping** | **Using web maps and GIS services**.

Getting ready

To publish your map on ArcGIS Online, you need:

- An Internet connection
- A web browser
- An Esri Global account

In your web browser, type `http://www.arcgis.com` to create an Esri Global account. Navigate to **Sign-up now** | **Create a Public Account**. Type all necessary information and click on **Review and Accept the Terms of Use and Create my account**.

How to do it...

Follow these steps to create a map package and share it on ArcGIS Online:

1. Start ArcMap and open the `SharingResults.mxd` map document from `<drive>:\PacktPublishing\Data\ExportingMaps`.

   Before starting to package our map, let's check the descriptive information about the `SharingResults.mxd` map document.

2. From the **File** menu, select `Map Document Properties`, and change the name of **Author**. Add more **Tags**, such as `topo maps`, `basemap`, `relief`, and `topography`. Make sure that **Store relative pathnames to data sources** is checked. Click on **OK**.

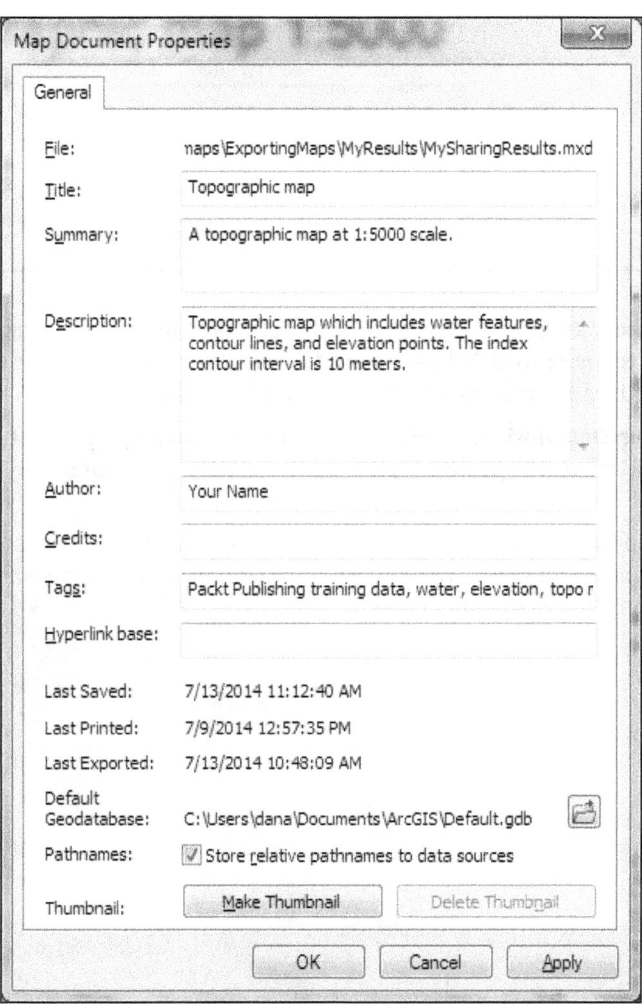

3. In the **File** menu, for **Sign In**, enter your user name and password to sign in to ArcGIS Online. Navigate to **Share As** | **Map Package**, as shown in the following screenshot:

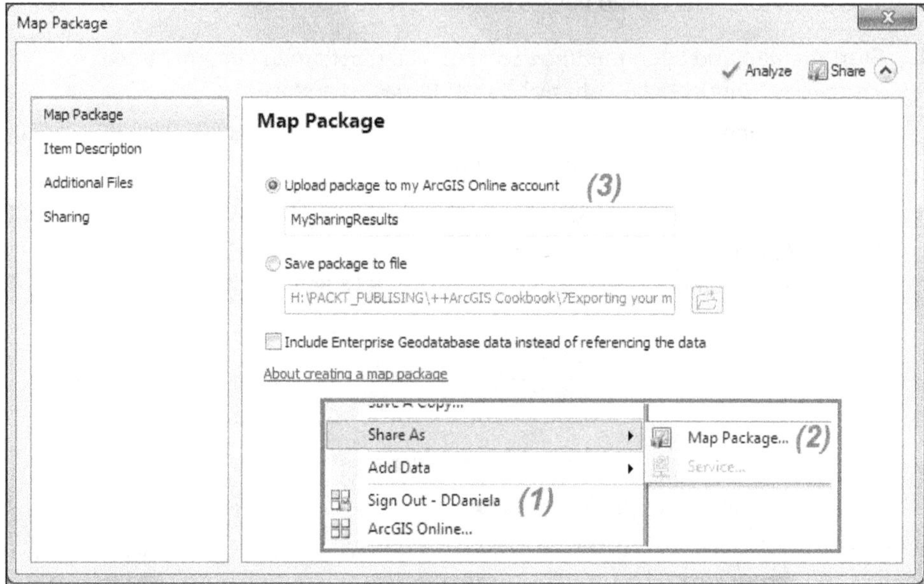

4. In **Map Package**, specify where to save your map package by checking the **Upload package to my ArcGIS Online account** option. You can change the name of your package. Uncheck **Include Enterprise Geodatabase data**.
5. Select **Item Description**. For **Access and Use Constraints**, please type `Packt Publishing will permit you to use this database by contacting contact@packtpub.com`.
6. Select **Sharing**. Check the **Everyone (public)** option. You can also share the data only with members of a group. Validate your map for any errors by selecting the **Analyze** button from the upper-right corner:

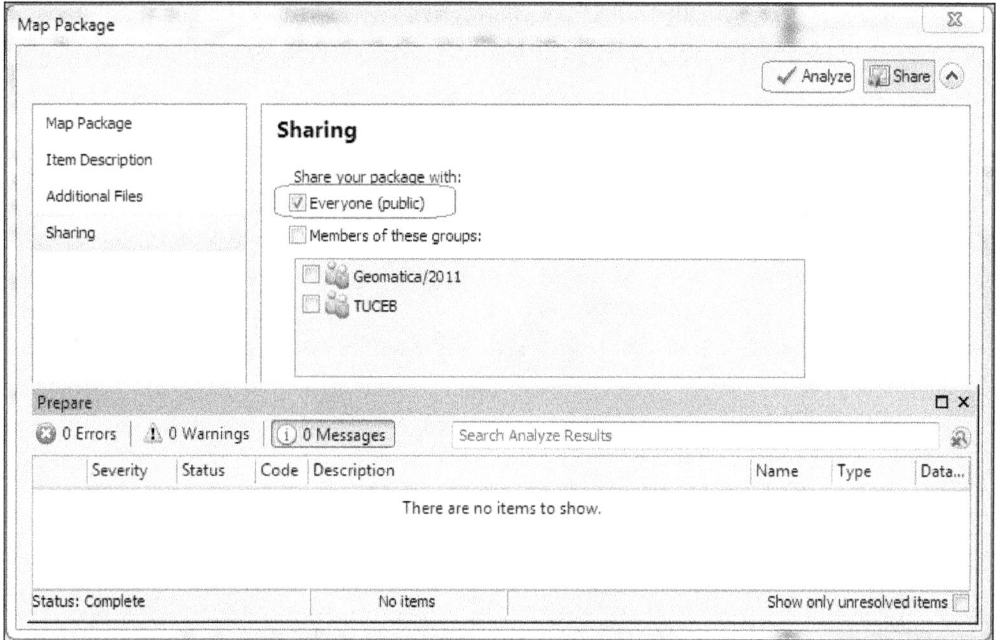

7. Click on **Share** to create and share the `MySharingResults` package. Click on **Yes** to save the map document. You will see the following messages:

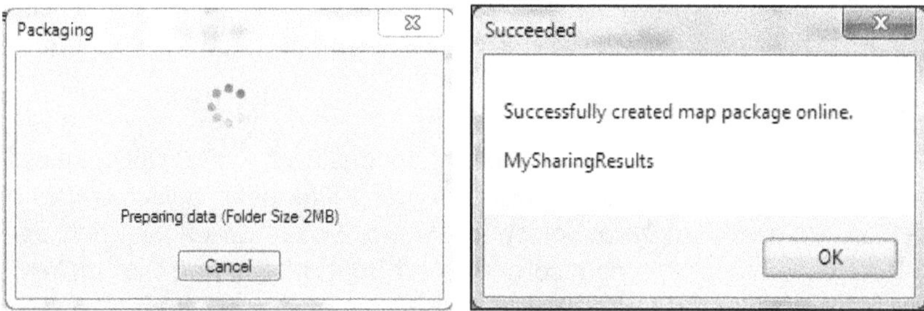

8. Open your web browser and go to `www.arcgis.com`. Sign in with your Esri Global account, and select **My Content**, as shown in the following screenshot:

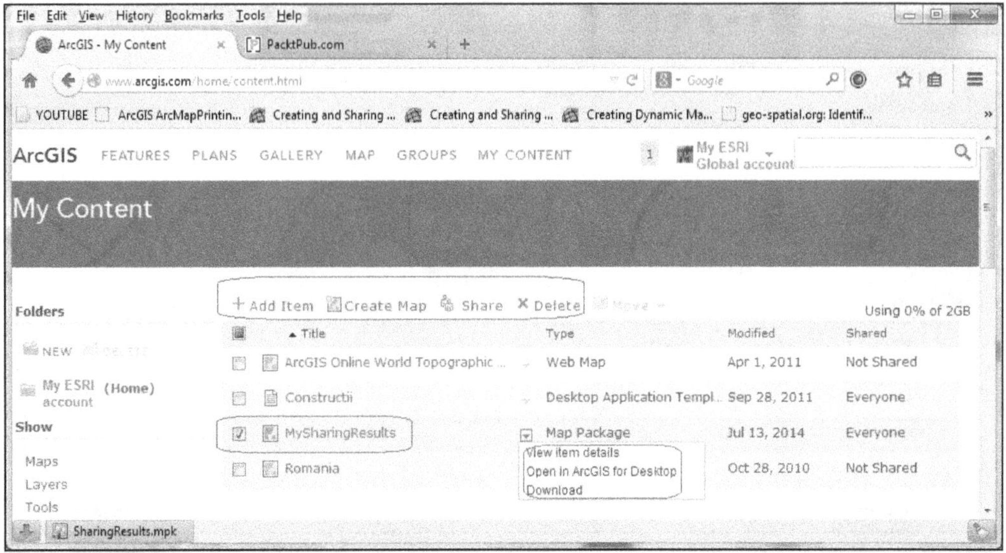

9. **My Content** lists your **Map Package**. You can see item details of your package, or you can download or open it in ArcGIS for Desktop.
10. Explore by yourself all options from ArcGIS Online.

How it works...

- After you have finished exploring different options from ArcGIS Online, please delete the package from your online account. From the **File** menu, select **ArcGIS Online**. Click on **My Maps and Data**. Select the `MySharingResults` map package and click on **Details**. In the bottom-right corner, select **Delete Item**, and click on **Yes**. Return to your browser and check whether the map package was deleted from ArcGIS Online.

See also

- If you want to learn more about how to manage map documents and layers' and how to automate map production and printing, please refer to *Chapter 3, Managing Map Documents and Layers* and *Chapter 5, Automating Map Production and Printing* from the book *Programming ArcGIS 10.1 with Python Cookbook, Eric Pimpler, Packt Publishing* (2013)

8

Working with Geocoding and Linear Referencing

In this chapter, we will cover the following topics:

- Preparing data for geocoding
- Geocoding addresses
- Geocoding with alternative names
- Creating and editing routes
- Creating and editing events
- Working with complex routes
- Analyzing events

Introduction

Geocoding is the process of converting the address information into spatial data that can be displayed on a map using an *address locator*. In accordance with the online ESRI GIS Dictionary, an address locator is *a dataset that stores the address attributes, associated indexes, and rules* that define the process of geocoding.

As an ArcGIS for Desktop user, you can store the address locator in a `.loc` file format stored in a folder or in an existing geodatabase as `Locator`. You can also share your address locator with everyone in your organization:

- As a locator package (`.gpk`) format with ArcGIS Online
- As a geocode service with ArcGIS for Server

The ESRI GIS Dictionary defines linear referencing *as a method for storing geographic data using a relative position along an already existing line feature.*

For more details about the geocoding and linear referencing concepts, please navigate to **Guide Books** | **Geocoding** | **Linear referencing** sections from *ArcGIS Help (10.2)* online at `http://resources.arcgis.com/en/help/main/10.2`.

Preparing data for geocoding

The geocoding process requires *reference data* that has address information stored in fields that correspond to the required elements of an address locator style. Reference data can contain point, polyline, or polygon geometry. Geocoding also requires *a nonspatial table with addresses* to geocode.

Getting ready

In this section, you will prepare reference data and a nonspatial table with addresses for the geocoding process.

Firstly, you will standardize the reference data according to the **US Address-Dual Ranges** address locator style, as shown in the following table:

HouseNum*	PreDir	PreType	StreetName*	SufType	SufDir
68	-	-	Dacia	Ave	-

The standardization process will split an address into standard fields required by a specific address locator style and will translate the address information into standard abbreviations. The `US Address-Dual Ranges` style considers that the odd numbers are on one side of a street and the even numbers are on the other side.

The dual ranges style works with polyline features as a data reference. Therefore, you will use the `Street` polyline feature class from the `VeloGIS.gdb` geodatabase.

Secondly, you will prepare a nonspatial table imported into your geodatabase. The table contains customer addresses taken from a larger database of an imaginary delivery service named `Farm Food Delivery`. In this table, you should have the address information in one field, as shown in the following screenshot:

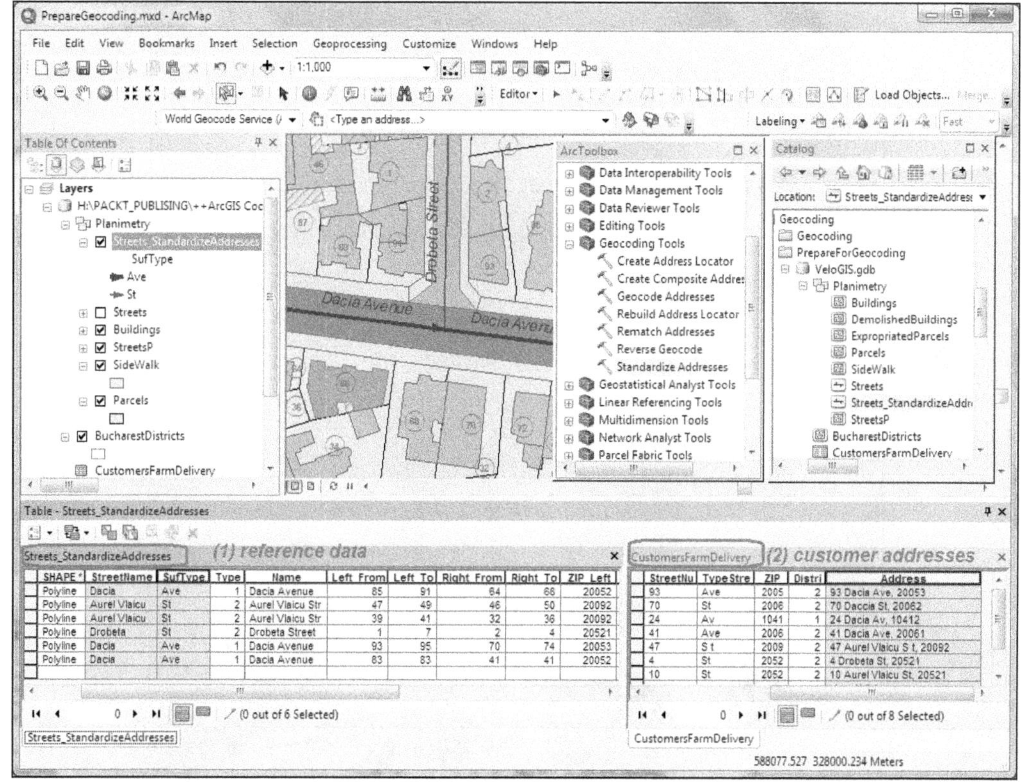

How to do it...

Follow these steps to examine and standardize the reference data:

1. Start ArcMap and open an existing map document `PrepareGeocoding.mxd` from `<drive>:\PacktPublishing\Data\Geocoding`.
2. The reference layer for geocoding is the `Streets` feature class from `...\Data\Geocoding\ VeloGIS.gdb\Planimetry`.
3. Let's examine its attribute table. Right-click on the `Street` layer and select **Open Attribute Table**. The `Left_From`, `Left_To`, `Right_From`, and `Right_To` fields contain the street number ranges on the left-hand side and the right-hand side of the streets. The `ZIP_Left` and `ZIP_Right` fields store the **ZIP code**. The `Name` field contains the name and type of the street.

4. You have to break down the name of the street into two elements required by the `US Address-Dual Ranges` address locator: street name and street type. Open **ArcToolbox** and go to **Geocoding Tools**. Double-click on the **Standardize Addresses** tool to open the dialog box and set the parameters, as shown in the following screenshot:

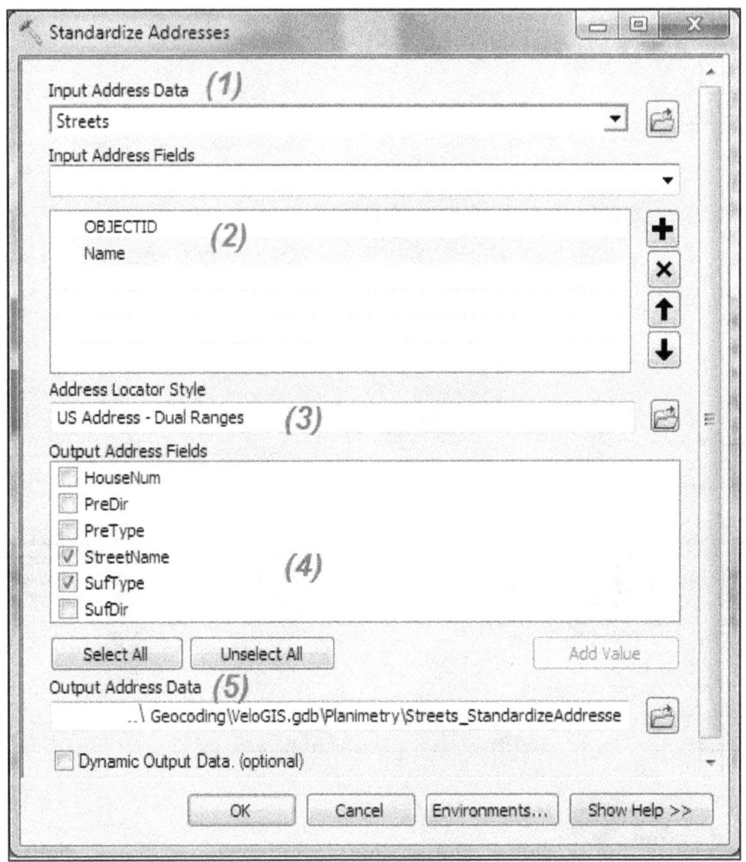

5. For **Input Address Data**, select the `Streets` layer. For **Input Address Fields**, select `OBJECTID` and the `Name` field. Select `US Address-Dual Ranges` as **Address Locator Style**. Check only `StreetName` and `SufType`. For **Output Address Data**, go to `...\Geocoding\VeloGIS.gdb\Planimetry` and type `Streets_StandardizeAddresses` for the name of the dataset. Click on **OK**.

6. You have obtained a new layer named `Streets_StandardizeAddresses`. Open its attribute table. The layer has two more fields: `ADDR_SN` (alias `StreetName`) and `ADDR_ST` (alias `SufType`). To toggle between the name and alias of the fields, select **Table Options**, and check or uncheck the **Show Field Aliases** option. Notice that the `ADDR_ST` field stores the corresponding abbreviations for street type. Your reference data is ready for geocoding your customer delivery addresses.

   Continue the following steps to examine and prepare the `CustomersFarmDelivery` nonspatial table:

7. In the **Table Of Contents** section, make sure that the **List by Source** tab is selected. Right-click on the `CustomersFarmDelivery` table and click on **Open**. The table contains the name of customers and their addresses in separate fields. The `US Address-Dual Ranges` address locator requires that all street-related information be stored in a single field.
8. You will create a new field. Start the edit session to undo the edits in the event of a mistake. In the `CustomersFarmDelivery` table, select **Table Options** | **Add Field**. For the **Name**, type `Address`. For **Type**, select `Text`, and accept the default length of characters. Click on **OK**.
9. You will populate the values of the **Address** field by concatenating the following fields: `StreetNumber`, `StreetName`, `TypeStreet`, and `ZIP`. We will consider the street number as the house number.
10. Right-click on the **Address** field and select **Field Calculator**. You will use an existing expression. Select **Load** and select `the US Address-Dual Ranges.cal` file from `...\Data\Geocoding`. Click on **Open** and then on **OK** to calculate the records. Inspect the result. Close the edit session. Your nonspatial table is ready for geocoding.
11. Save the map document as `MyPrepareGeocoding.mxd` at `...\Data\Geocoding`.

You can find the results at `...\Data\Geocoding\ PrepareForGeocoding`.

How it works...

As reference data, you can use feature classes or shape files. At step 5, you used two fields for address: `OBJECTID` and `Name`. The `OBJECTID` field will represent the numeric value for house numbers.

In the address table, you should have the entire address stored in one field. In the reference data, the address should be mostly in separate fields. The street polylines should connect to one another only at their endpoints. Before starting to use the reference data, check the connectivity of polylines using the following topological rules: **Must not have dangles** and **Must not intersect or touch interior**.

Geocoding addresses

The geocoding process requires four steps:

1. **Preparing reference data**: The following are the details of this step:
    - The point, line, or polygon geometry
    - It contains address information compatible with an address locator style
2. **Choosing an address locator style**: The following are the details of this step:
    - It has a different number of required and optional fields
    - Depending on the address information from your resource data, choose the most suitable template to create an address locator
3. **Creating an address locator**: The following are the details of this step:
    - Rules for interpreting addresses (for example, odd street numbers are on the left-hand side of the street)
    - List of standard street components (for example, abbreviations such as `Boulevard` = `Blvd`)
    - Snapshot of reference data (for example, centerline streets)
4. **Geocoding or matching addresses**: The following are the details of this step:
    - Finding individual addresses
    - Creating point features based on a nonspatial address table

Getting ready

In the next exercise, you will create an address locator named `Streets_Locator`. Based on the address locator, you will geocode the `CustomersFarmDelivery` address table. Finally, you will try to improve the geocoding results by manually fixing the missing address matches.

How to do it...

Follow these steps to create an address locator and geocode the address table:

1. Start ArcMap and open `MyPrepareGeocoding.mxd` to continue to work in the same geodatabase from `<drive>:\PacktPublishing\Data\Geocoding\VeloGIS.gdb`.

    If you did not succeed in finishing the previous step, please open an existing map document `Geocoding.mxd` from `...\Data\Geocoding\PrepareForGeocoding`. This map document uses a duplicate geodatabase named `VeloGIS.gdb` stored in the `PrepareForGeocoding` folder.

2. Open the **Catalog** window. Expand the `VeloGIS.gdb` geodatabase. Right-click on the geodatabase and navigate to **New** | **Address Locator**. Click on **Show Help** to read the descriptions for every section of the **Create Address Locator** window. Set the parameters, as shown in the following screenshot:

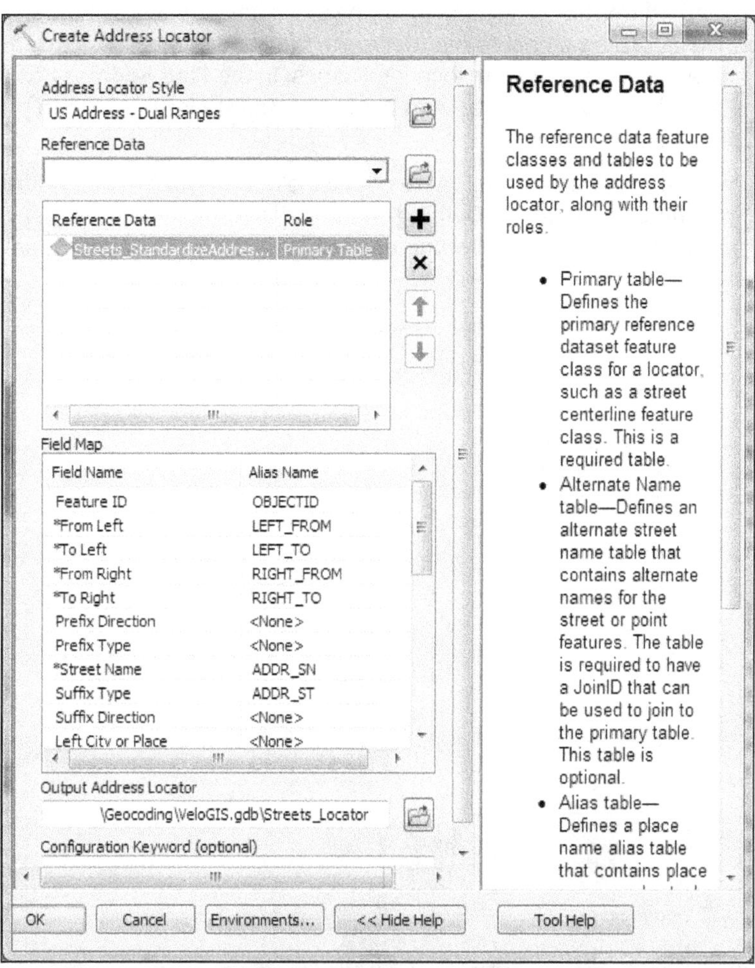

3. For **Address Locator Style**, choose `US Address-Dual Ranges`, and click on **OK**. For **Reference data**, select the `Street_StandardizeAddresses` layer. To correct the error, select `Primary Table` for **Role**. In the **Field Map** section, check whether the required fields (fields with an asterisk) are mapped with the corresponding fields from the reference data. For **Output Address Locator**, go to `VeloGIS.gdb`, and save your locator as `Streets_Locator`. Click on **OK** and then on **Save**.
4. `Streets_Locator` has been added to `VeloGIS.gdb`.

5. Let's test the address locator. Firstly, add the **Geocoding** toolbars by navigating to **Customize** | **Toolbars**. Click on the drop-down list and select **Manage Address Locators**. Click on the **Add** button and select `Streets_Locator`. Click on **Add** and on **Close**.
6. In the **Find Address** box, type the following individual address: `68 Dacia Ave`. Press *Enter*. The address was not found. Retype the full address: `68 Dacia Ave, 20061`, and press *Enter*. A green point location flashes. In the **Find Address** box, right-click on the address, and choose the **Add Point** and **Add Callout** options to place a point symbol and a callout label.
7. Let's make the address locator more permissible when you type an address in the **Find Address** box. In the **Catalog** window, right-click on `Streets_Locator`. Navigate to **Properties** | **Geocoding options** and change the options, as shown in the following screenshot:

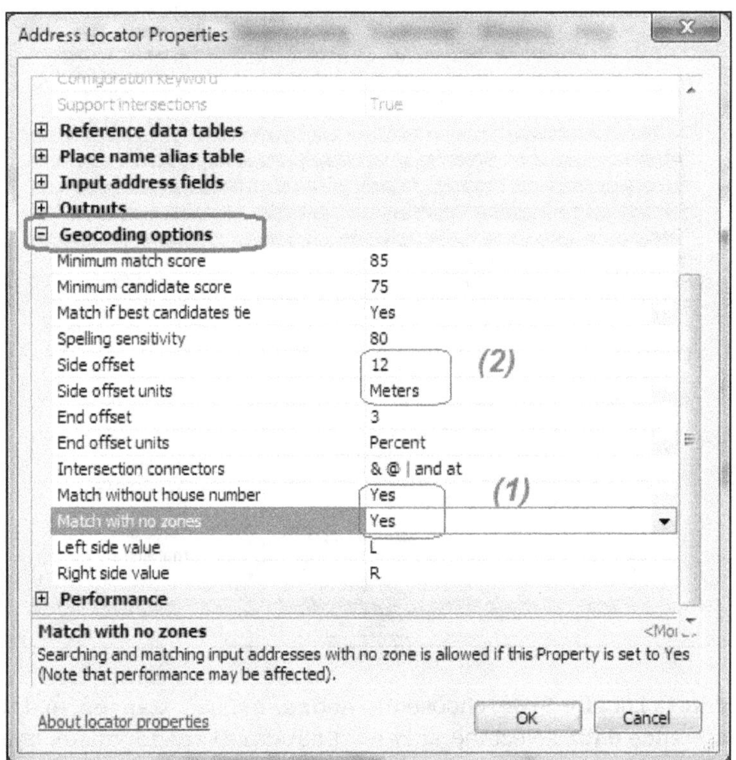

8. On the **Geocode** toolbar, click on the drop-down list of **Select Address Locator** and select **Manage Address Locators**. Select `Street_Locator` and click on **Remove**. Click on the **Add** button and select again `Streets_Locator`. Click on **Add** and on **Close**. This will refresh the settings for your address locator.

9. In the **Find Address** box, retype `68 Dacia Ave`, and press *Enter*. Now the address was found without the ZIP code `20061`.
10. Find a *location at intersections* by typing: `Dacia Ave & Drobeta St.`
11. Find two *relative locations* by typing:
    - `20 meter east from 68 Dacia Ave`
    - `20 meter south from Dacia Ave & Drobeta St`
12. Continue to follow the steps to geocode addresses from the `CustomersFarmDelivery` nonspatial table, as shown in the following screenshot:

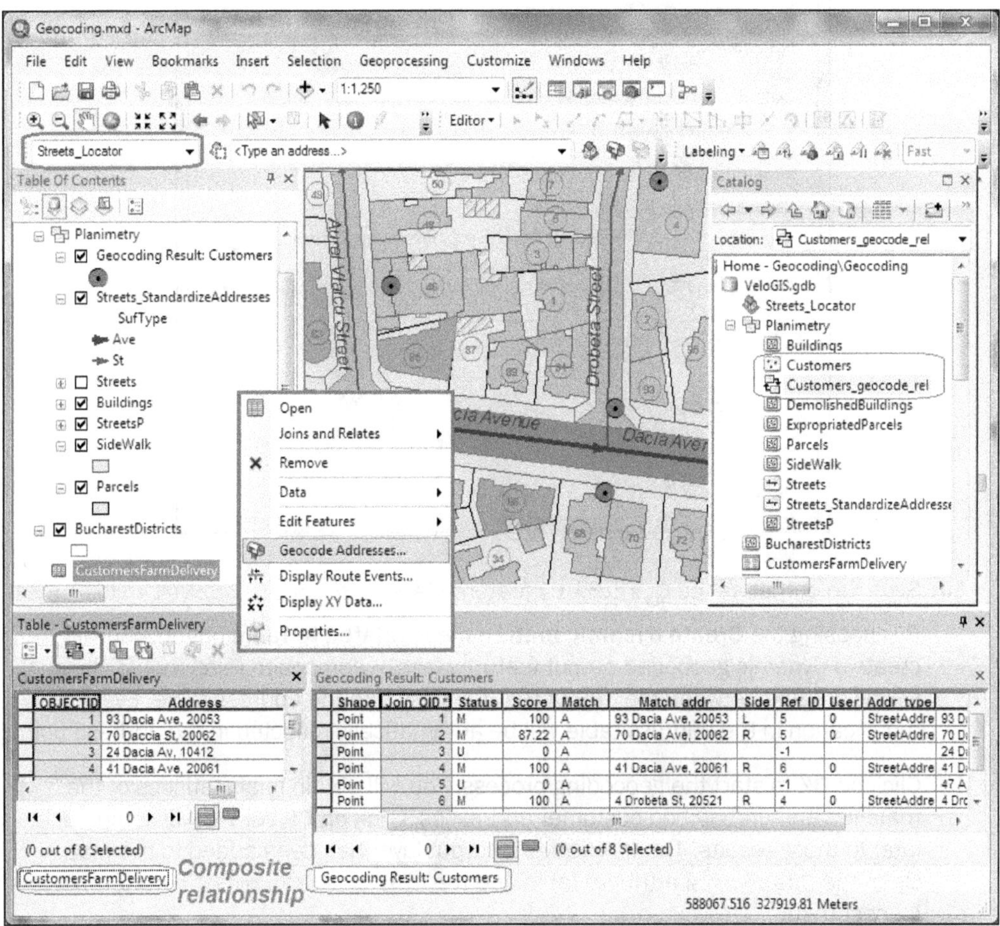

13. Make sure that `Street_Locator` is selected in the **Select Address Locator** section from the **Geocode** toolbar. In the **Table Of Contents** section, right-click on the `CustomersFarmDelivery` table, and click on **Geocode Addresses**. Select `Streets_Locator` and click on **OK**. In the **Geocode Addresses: Streets_locator** window, set the parameters as shown in the following screenshot:

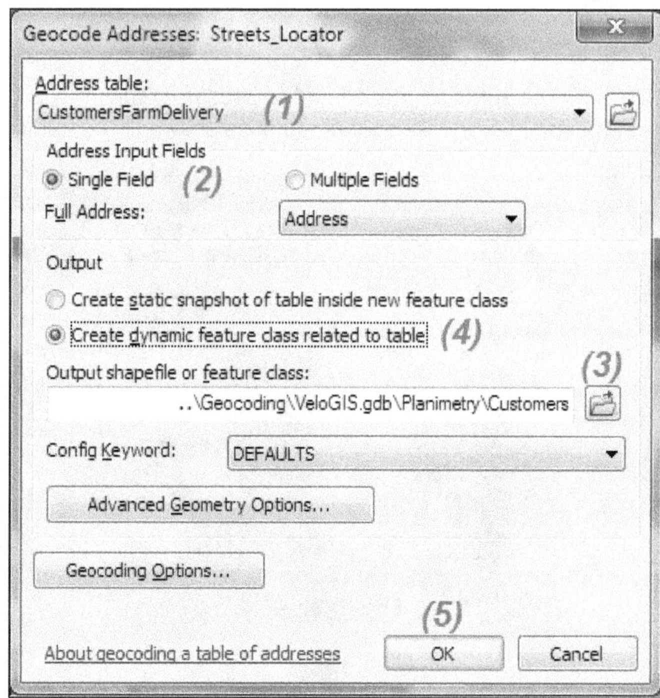

14. Save the output as the `Customers` feature class in `VeloGIS.gdb\Planimetry`.

    By choosing the **Create dynamic feature class related to table** option, you will create a dynamic geocoded output feature class. A composite relationship class will be created between the output feature class and the address table. Every modification in the address table will be automatically reflected in the address points.

15. Click on **OK** to start the geocoding process. You will obtain final statistics of the matched and unmatched customer addresses. Click on **Close**. We will come back later to those results. The `Customers` output layer has been added to the map. Open the `Customers` attribute table and the `CustomersFarmDelivery` table to inspect them.

You have noticed some additional fields in the `Customers` attribute table. Please navigate and refer to the **Guide Books | Geocoding | Locating addresses | About geocoding a table of addresses** section from *ArcGIS Help (10.2)* online to learn more about those new fields. Because your statistic results are not quite perfect, you will manually rematch the addresses, as shown in the following screenshot:

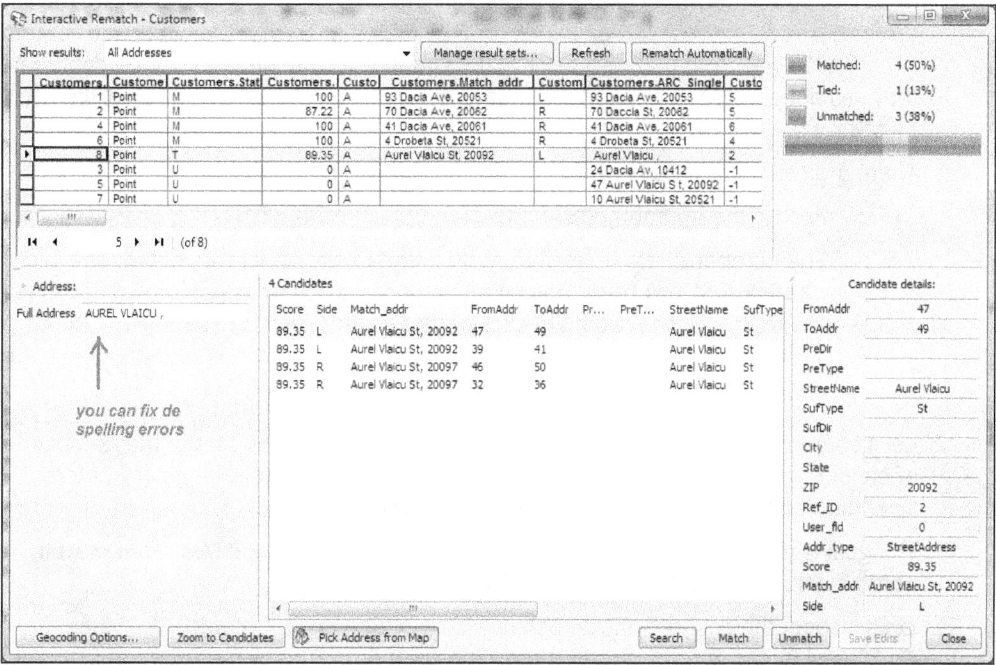

16. In the **Table Of Contents** section, select **Geocoding Result**: `Customers`. From the **Geocoding** toolbar, select the **Review/Rematch Addresses** tool. Right-click on the **Customers.Status** field and select **Sort Ascending** to sort the values.
17. In the **Geocoding results** table, select the row with **Customers.ObjectID**: `8` and with **Customers.Status**: `T`. In the **Address** panel, the address is incomplete because it contains only the name of the street. You have four candidates for this address in the **Candidates** section. You should choose one of them to change the **Status** value.
18. Select the first candidate and read the candidate details. Click on **Match** and on the **Refresh** button. Notice that **Customers.Status** has changed from `T` (tied) to `M` (matched) and that the **Customers.Match** type has changed from `A` to `M` (manually match).

19. If you want to have a score of `100`, firstly, select again the first candidate, and click on **Unmatch**. Secondly, change **Full Address** from `AUREL VLAICU` to `AUREL VLAICU ST`. The score for all four candidates has been changed to `100`. Select the first candidate, and click on **Match** and **Refresh**. Right-click on the **Customers.Status** field again and select **Sort Ascending** to sort the values.
20. In the table, select the row with **Customers.ObjectID**: `5` and with **Customers.Status**: `U` (unmatched). In the **Address** panel, for **Full Address**, correct the type of street from `S T` to `ST`. Click on the empty space of the **Candidates** section. Select the candidate that has appeared and click on the **Match** button. Notice that **Customers.ObjectID**: `5` has **Status**: `M` and **Score**: `100`.
21. There are two unmatched addresses left. You have two options:
    - Accept the unmatched status.
    - The number of the house is wrong. In the table, select the record, and click on **Pick Address from Map**. Drag the **Interactive Rematch** dialog box to see the map. On the map, right-click on the chosen location, and select **Pick Address**. A new point has been added and **Customers.Match** has been changed to `PP` (Pick by Point). Click on **Refresh**.
22. Let's change the `87.22` score for **Customers.ObjectID**: `2` and control the number of candidates by selecting **Geocoding Options**. Change the following parameters: **Spelling sensitivity** : `95` and **Minimum Candidate score** : `60`.
23. The goal is to see more candidates for an address but still keep **Minimum match score** higher. Click on **OK**. Click on **Refresh**.
24. Let's inspect the consequences. Select the row **Customers.ObjectID**: `2`. Notice that **Full Address** has two errors and there is no candidate. This is because you increased **Spelling sensitivity** to `95` and the spelling errors are not accepted by the new **Spelling sensitivity** value.
25. If you correct the errors by changing `DACCIA` to `DACIA`, and `ST` to `AVE`, the previous candidate will show up. Click on the **Unmatch** button to unmatch the existing address and then click on **Refresh**. Select the candidate and again click on **Match**. Notice that the **Score** value has been changed to `100`.
26. Select the row with **Customers.ObjectID** of `4` and `5`. Notice that address has one more candidate with a score below `70`. Select again **Geocoding Options** and change **Minimum Candidate score** to `70`. Click on **OK**. By increasing **Minimum Candidate score**, the second candidate for the **Customers.ObjectID** of `4` and `5` have disappeared.
27. Close the **Interactive Rematch** dialog box. The `Customers` layer has been updated according to your last settings.

    Let's test the composite relationship between those two tables.

28. Start the edit session to change an address from the `CustomersFarmDelivery` table and see the changes in the `Customers` layer, as shown in the following screenshot:

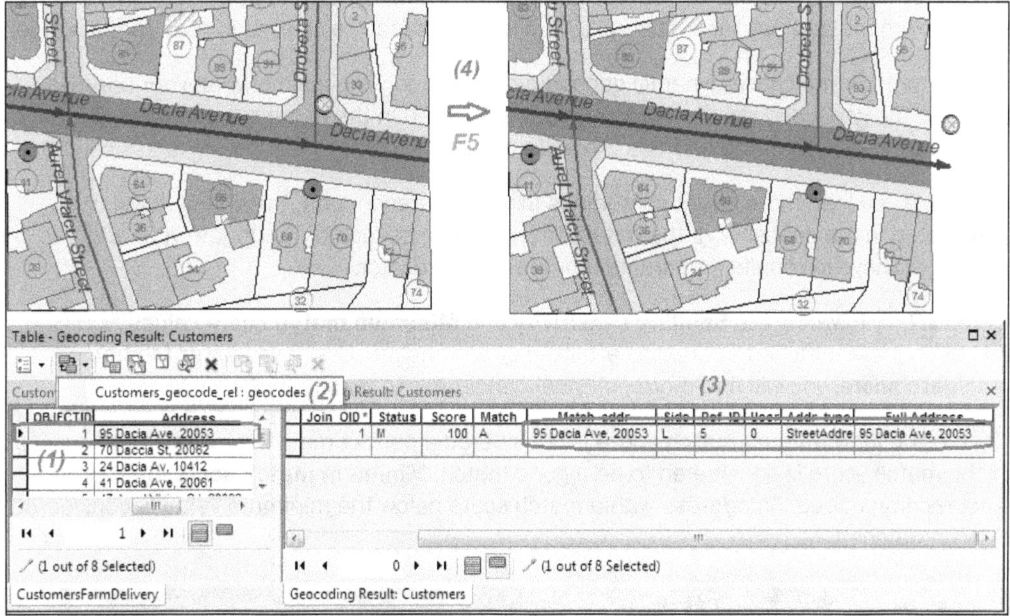

29. In the `CustomersFarmDelivery` table, select the record with **OBJECTID** as `1`. In the **Address** field, change the number of the street to `95`. Click on the **Related Tables** button and select the `Customers_geocode_rel: geocodes` relationship. The point feature from the `Customers` layer is now selected, and the `Match_addr` and `Full Address` values are automatically updated. Press the *F5* key to redraw the map. The point was moved according to the new address.
30. Add a new customer in the `CustomersFarmDelivery` table. Let's suppose that `Customer 9` will have the following address: `50 Aurel Vlaicu St, 20097`. Fill the empty cells with the corresponding address information.
31. Look at your map while you are selecting **View** | **Refresh**. A new point has been added. Notice that the `Customers` attribute table has been updated with a new record. Save the edits and stop the editing session.
32. Save your map document as `MyGeocoding.mxd`.

You can find the results of this recipe in the **Geocoding** folder at `...\Data\Geocoding`. Please do not modify the geodatabase from this folder. It will be used in the next recipe.

How it works...

The point and callout created at step 6 are graphic elements stored in your map documents. To move or delete them, use the **Select Elements** tool and the *Delete* key. The point features created from the address table at step 13 are stored as feature classes in your geodatabase.

Also, at step 13, you chose dynamic geocoding. This option is possible when your address table and the geocoded resulted feature class are stored in the same geodatabase. Dynamic geocoding refers only to the address table changes.

Later on, if you load more street centerlines in your reference data, the changes will not be reflected in the snapshot of the data from your address locator. To update your address locator, you have to *rebuild* it or create *a new* address locator.

At step 22, by lowering the **Spelling sensitivity** and **Minimum match score** values, you will obtain more matched addresses after the geocoding process. If you lower **Minimum candidate score**, you will find more potential candidates to match.

A `100` value for **Minimum match score** is considered a perfect match. A range from `80` to `99` for the match score is considered to be a good match. **Minimum match score** lower than `60` is not recommended. An address with a match score below the minimum value is considered to be unmatched.

Geocoding with alternative names

To locate an address, you have two more alternatives:

- Name aliases (for example, the `91 Dacia Ave` address is known as `Elementary School`)
- Alternative names (for example, `Dacia Avenue` is a small section of the road called `E60`; `E60` is part of the European Road Network)

The address locator with a name alias needs a supplementary table that must contain two required fields: the alias place name and the corresponding address. Your address locator will use this input table to search the address of the alias place name.

The address locator with alternative names needs an alternate name table that will contain specific fields in accordance with the address locator style (for example, prefix and suffix type, street name, and prefix and suffix direction). The alternate name table and the primary reference feature attribute table must contain the common `Join ID` field. Those two tables will join in a `Many to Many` relationship based on the common field.

Getting ready

You will geocode the PointOfInterest address table using a composite address locator, as shown in the following screenshot:

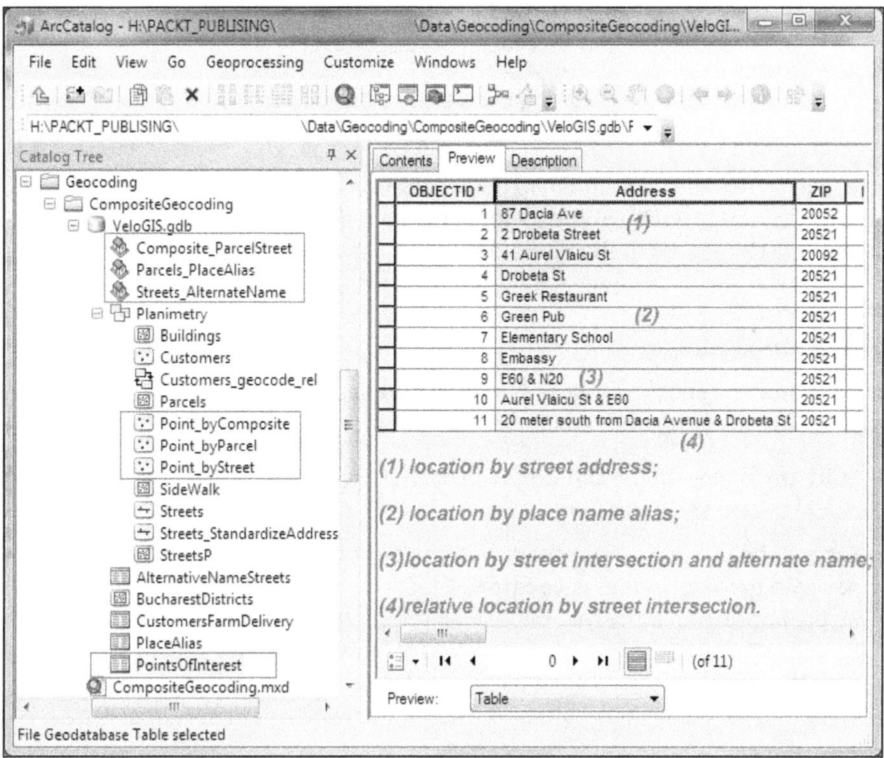

The composite locator will combine the functionality of two individual address locators:

Address locator	Parcels_PlaceAlias	Streets_AlternateName
Address locator style	US Address-Single House	US Address-Dual Ranges
Reference data	Parcels (polygon features)	Street__StandardizeAddresses (polyline features)
Alternative searches	Place aliases based on the PlaceAlias table	Alternate names for streets using the AlternativeNameStreets table
The PlaceAlias and AlternativeNameStreets tables are already created in the VeloGIS.gdb geodatabase.		

If you worked on all previous exercises in VeloGIS.gdb from <drive>:\PacktPublishing\Data\Geocoding, skip step 1 from the next section, and prepare your Street__StandardizeAddresses feature class for geocoding by:

- Adding a Join ID [Short integer] field
- Filling the Join ID field with the following values: 100 for street Aurel Vlaicu; 101 for street Dacia; and 102 for street Drobeta

Add in your map document the following tables: PointsOfInterest, PlaceAlias, and AlternativeNameStreets.

Otherwise, please use the data and map document from <drive>:\PacktPublishing\Data\Geocoding\Geocoding. The folder contains a duplicate VeloGIS.gdb geodatabase that will help you to succeed in the next steps.

How to do it...

Follow these steps to geocode addresses with place aliases and alternate names using a composite address locator in the ArcCatalog context menu:

1. Start ArcCatalog. In the **Catalog Tree** section, go to <drive>:\PacktPublishing\Data\Geocoding\Geocoding.
2. Expand the VeloGIS.gdb geodatabase. Right-click on the geodatabase and navigate to **New** | **Address Locator**. Select US Address-Dual Ranges as the address locator style. Choose as reference data the Street_StandardizeAddresses feature class for **Primary Table**, and choose as reference data the AlternativeNameStreets table for **Alternate Name Table**. Perform the steps shown in the following screenshot:

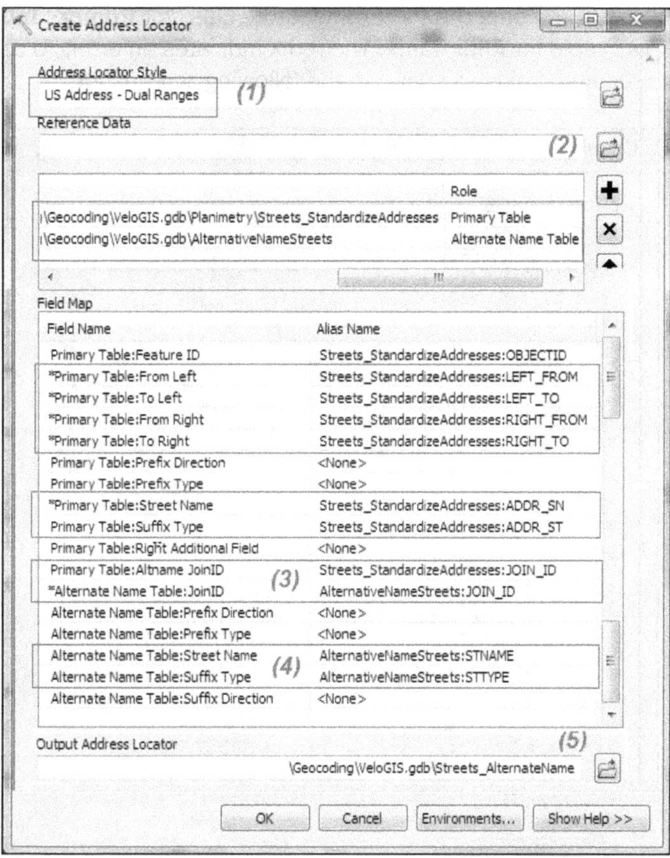

3. Most of the fields will match. Be careful about the following fields: for **Primary Table/ Alternate Name Table: JoinID**, choose the common field `JOIN_ID`. Save the address locator in your geodatabase. Right-click on `Streets_AlternateName` and navigate to the **Properties** | **Geocoding** options. Select **Yes** for the **Match without house number and Match with no zones** options. Click on **OK**.
4. Create a new **Address Locator.** Select `US Address-Single House` as the address locator style. This address locator style allows you to find an address point using polygon features as reference data.

5. Choose as reference data the `Parcels` feature class for **Primary Table** and the `PlaceAlias` table for **Alias Table**. The red X indicates an error. To correct the error, match the required fields as shown in the following screenshot:

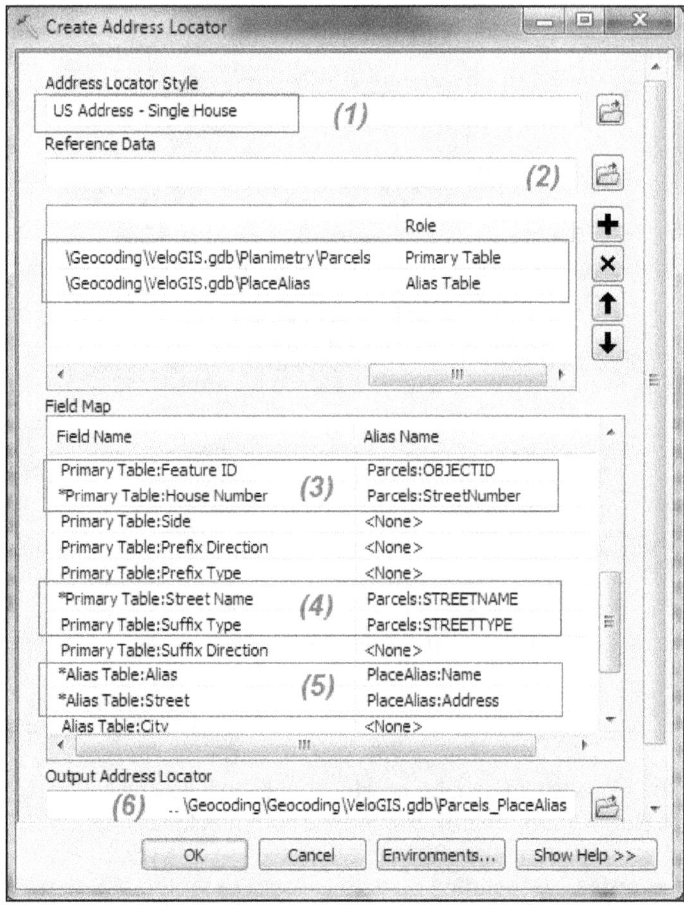

6. Change the properties of the new `Parcels_PlaceAlias` address locator by selecting **Yes** for the **Match without house number** and **Match with no zones** options. Click on **OK**.

7. Let's create a composite address locator. Right-click on the geodatabase and navigate to **New** | **Composite Address Locator**. For **Address Locator**, add `Parcels_PlaceAlias` and `Streets_AlternateName` address locators. Name the **Output Composite Address Locator** as `Composite_ParcelStreet`. Click on **OK**. Right-click on `Composite_ParcelStreet` and click on **Properties**. Inspect the default settings.

8. In the **Participating Address Locators** panel, select `Parcels_PlaceAlias` to activate its individual properties. Navigate to **Properties | Geocoding options**, and select **Yes** for the previous house and zones matching options. Click on **OK**. Similarly, change the properties of the second address locator.
9. Let's geocode the addresses. In ArcCatalog, right-click on the `PointOfInterest` table, and select **Geocode Addresses**. In the **Choose an Address Locator to use** pop-up window, choose `Composite_ParcelStreet`, and click on **OK**. Continue to set the following parameters:
   - **Address table**: `PointOfInterest`
   - **Address Input Fields**: `Single Field`
   - **Single Line Input**: `Address`
   - **Output feature class**: `...VeloGIS.gdb\Planimetry\ Point_byComposite`
10. Click on **Geocoding Options**, and for **Intersections | Connectors**, type `&`, and click on **OK**. Your statistic results are not quite perfect and you should manually rematch the addresses.
11. Try to repeat the last two steps to geocode the `PointOfInterest` table based on the individual address locators:

Output feature class	`Point_byParcel`	`Point_byStreet`
Address locator	`Parcels_PlaceAlias`	`Streets_AlternateName`

12. Use ArcMap to see the three-point feature classes. Symbolize the new layers and examine the results. Improve the geocoded address results from the `Point_byComposite` feature class.

You can find the results of this recipe in the `CompositeGeocoding` folder at `...\Data\Geocoding`.

How it works...

A composite address locator geocodes an address table using the functionality of multiple address locators in one geocoding process. The next table shows the advantages and disadvantages of using a simple address locator or a composite one:

Address locator	Advantages	Disadvantages
`Streets_AlternateName`	This supports intersections. There is a high number of matching addresses.	There is low position accuracy.
`Parcels_PlaceAlias`	There is high position accuracy.	This does not support intersections: **Properties** | **About the locator** | **Support intersections**: `False`
`Composite_ParcelStreet`	There is high position accuracy. This supports intersections. There is a high number of matching addresses.	Trying to have a high number of matching addresses, your composite locator will use the most accurate locator first (parcel) and the least accurate second (street). In the end, you will obtain a low accuracy result.

By choosing a composite address locator, you can improve the geocoding results, but you cannot compensate for the quality of address information from your reference data: *garbage in, garbage out*.

Creating and editing routes

A route is a linear feature that has a measurement system (for example, distance or time) and a unique identifier. Routes are stored in a route feature class (standalone feature class or in a feature dataset). The vertices of a route contain the *x* and *y* coordinates and a measure value or *m* value. Measure units are independent of map units. Measure values do not have to begin at zero and should be increased in the same direction.

Getting ready

In this section, you will create a route feature class that will store three simple routes that represent three different bus lines from your town: *bus 133*, *bus 79*, and *bus 135*. Your routes will be created from scratch using the **Trace** tool.

How to do it...

Follow these steps to create and edit simple routes in ArcMap:

1. Start ArcMap and open the existing map document `Routes.mxd` from `<drive>:\PacktPublishing\Data\LinearReferencing`. Open the **Catalog** window. Expand the `VeloGIS.gdb` geodatabase.
2. You will create a new route feature class. Right-click on the geodatabase and navigate to **New** | **Feature class**. For **Name**, type `BusRoute_m`. For **Type**, choose `Line Feature`. Under **Geometry Properties**, check **Coordinates include M values. Used to store route data**. Go to the next panel. Navigate to **Add Coordinate System** | **Import**, go to `VeloGIS.gdb`, and select the `Street` feature class. Accept the default settings from the next two panels. In the attribute table panel, add a new field that will store the unique identifiers of the routes. For **Field Name**, type `RouteID`. For **Data Type**, select `Short Integer`. Click on **Finish**. The empty `BusRoute_m` layer has been added in the **Table Of Contents** section.

   That name of the `BusRoute_m` route feature class suggests the measure units (letter `m` refers to meter). It's good practice to document the measurement system and measure units in feature-class metadata.

3. In the **Catalog** window, right-click on `BusRoute_m`, and select **Item Description**. Select the **Edit** button, and in the **Overview** | **Item Description** | **Description (Abstract)** section, type: `Bus routes. The measurement system is distance. The measure units are meters.`. Click on **Save** and then on **Exit**.

Let's add three routes in the `BusRoute_m` layer, as shown in the following screenshot:

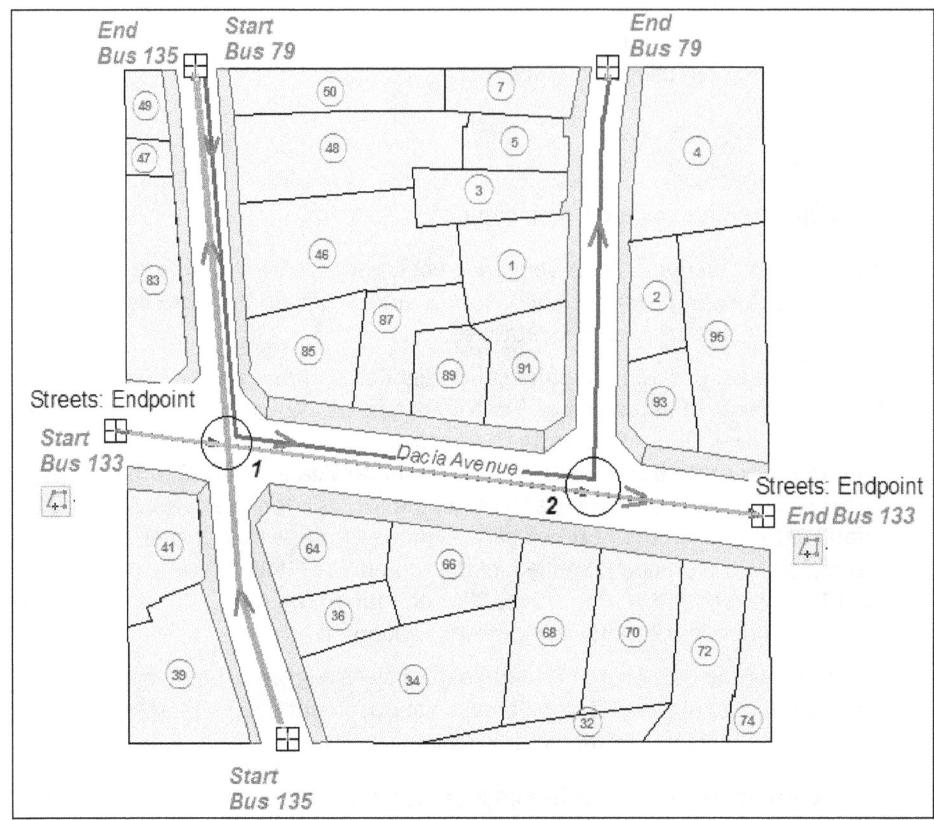

4. Start the edit session from the **Editor** toolbar. Set the editing environment: on the **Editor** toolbar by navigating to **Editor** | **Snapping** | **Snapping Toolbar**; select **Point**, **End**, **Vertex**, and **Edge**. From **Snapping** | **Options**, set **Tolerance**: `5` pixels and **Show tips**. In the **Table Of Contents** section, select **List by Selection**, and make `BusRoute_m` the only selectable layer.

5. Open the **Create Feature** window to see the feature templates. Select the `BusRoute_m` layer. Under **Construction Tools**, select **Line**. On the **Editor** toolbar, click on the **Trace** tool. You will digitize the new routes based on the `Street` reference layer. First click on the *Start Bus 133* point and digitize the route by tracing along the `Street` layer features until the *End Bus 133* point. Press the *F2* key to finish the sketch. Click on the **Attribute** button, and for **RouteID**, type `133`. Notice that your **SHAPE** geometry is `Polyline M`. This `M` means that your linear feature coordinates include measure values.

6. Repeat the previous step to create the routes `135` and `79`.
7. Save the edits and clear the selected features. Turn off the `Street` layer to better see the routes from the `BusRoute_m` layer. Symbolize the routes based on the `RouteID` field: **Layer Properties** | **Symbology** | **Categories** | **Unique values**.

   At step 3, you already decided to use distance for your measurement system. You will assign measures to your new routes:

8. On the **Editor** toolbar, click on the **Sketch Properties** button. To see the vertices' coordinates, select route number `133` with the **Select Features** tool, and click on **Edit Vertices**.
9. Your `M` values are unknown (`NaN`—**Not a Number**). In the **Attribute** window, you already noticed the value of **SHAPE_Length**: `132.668` for **RouteID**: `133`. You have the following two options:
   - Enter manually in the **Edit Sketch Properties** window, the start and end measures: `Point 0: 0.00` and `Point 4:132.668`. In the map, right-click on the selected route and go to **Route Measure Editing** | **Calculate NaN**. Notice the values in the **Edit Sketch Properties** window. Close the window. If you want to return to the NaN values, choose **Drop Measures**.
   - In the map, right-click on the selected route, and go to **Route Measure Editing** | **Set From/To**. In the second cell, type the value `132.668`. Press the *Enter* key to close and calculate the vertices' measure.

   For the bus route number `133`, you measured with a total station (surveying instrument) the `75` meters between the street intersections **1** and **2** (underlined in the previous screenshot by two circles). You will use this more accurate distance to update the segment measures of the route. This method is known as **calibration**. Firstly, add the **Route Editing** toolbar from the **Editor** toolbar by navigating to **Editor** | **More Editing Toolbars**. On the **Route Editing** toolbar, add the **Calibrate Route** tool from **Customize** | **Commands** | **Categories** | **Linear Referencing**.

10. Select route number `133`. On the **Route Editing** toolbar, click on the **Calibrate** tool. Move the dialog box to see the map. Select **Add Calibration Points Tool**. In the map, snap the mouse pointer to the first intersection, and click on it once. Snap the mouse pointer again to the second intersection. Notice that two records are added in your **Calibrate Route** window. For the first record, type `New M: 20.944`. For the second record, type `New M: 95.944`. In the **Options** panel, check **Interpolate between points**, **Extrapolate at the end**, and **Use existing measures**. Click on **Calibrate Route**.
11. Open the **Edit Sketch Properties** window to check the results. Notice that the last measure value has been changed to `136.783`. Now your measures reflect the reality from the field.

12. Assign measures to the last two routes: `135` and `79`. Try to calibrate route number `79` using the same segment value between the two street intersections. Save the edits and stop the editing session.

    Let's find a specific section along the route using the **Find** tool from the **Standard** toolbar:

13. Click on **Find** | **Linear Referencing**. Choose `BusRoute_m` and click on **Load Routes**. Choose route number `133`. For **Type**, check the **Line** option. You want to find the section of route `133` from `25` meters to `75` meters from the start of the route. Set for **From**: `25` and for **To**: `75`. Click on **Find**.
14. Select the record and right-click to explore the options. You can flash the segment. You can also **Draw Route Location**. You should turn off the `BusRoute_m` layer to see the segment that is a graphic element stored only in your map documents. To work with it, use the **Select Elements** tool from the **Drawing** toolbar or the **Standard** toolbar.

    Let's find the measure value for a point along the route using the **Identify Route Location Results** tool from the **Route Editing** toolbar:

15. Click on the **Identify Route Location Results** tool. In the map, click on the first street intersection until you see all three routes in the **Identify Route Location Results** window. Inspect all of them. The **Measure** field displays the measure units (*m* value) calculated from the start of the route.
16. Save your map document as `MyRoutes.mxd` at `...\Data\LinearReferencing`.

You can find the results of this recipe at `...\Data \LinearReferencing\Routes`.

How it works...

A route can be edited or updated in an ArcMap **Edit** session. At step 9, the **Use existing measures** option will constrain the interpolation process to use your measured value for the vertices of the calibrated segment (`75` meters) instead of the distance value calculated from the vertices' *x* and *y* coordinates (`72.327` meters).

There's more...

If you want to store routes with different measure units, you should separate them into different route feature classes. You can also update a route feature by following two steps: edit the route geometry (move, add and delete vertices, merge, and add new features) and re-measure the route (entire route or a portion of the route) using the **Define Line Portion** tool from the **Route Editing** toolbar. If the tool is not on your toolbar, you can add it by navigating to **Customize** | **Customize mode** | **Commands** | **Categories** | **Linear Referencing**.

Creating and editing events

A route event is a point or linear feature that occurs along a route feature. Route events are stored in tables called route event tables. A route table contains a route identifier field (for example, `RouteID`) that associates a specific route to the route point or linear event. Based on this route identifier field, a route table can store multiple events associated with a single route or can reference multiple routes. Also, multiple route events can reference a single route.

In ArcMap, you can create a layer of point or line features based on the information from the route event table. When you change a value in the route event table, the shape of the event (point or line) is automatically moved.

The features from the event layer are stored in a table and not in a feature class. You can export those event features to a feature class, but the dynamic connection between the route event table and routes will be lost.

Getting ready

In this recipe, you will create two route event tables, as shown in the following screenshot:

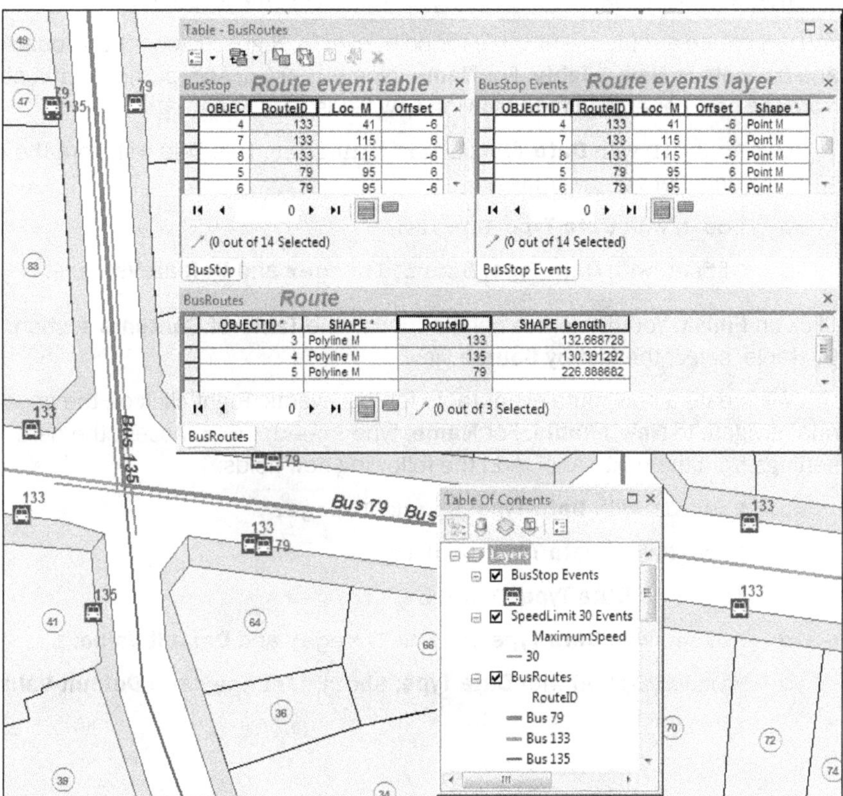

Firstly, the BusStop point route event table will have the following special fields: RouteID, Loc_M, and Offset. BusStop refers to the bus stops along the bus routes.

Secondly, the MaximumSpeed line route event table will have the following special fields: RouteID, From_M, To_M, and Offset. MaximumSpeed refers to the speed restriction segments along the bus routes.

The field RouteID (route identifier field) defines the connection between the route and the route events. Lo_M, From_M, and To_M are measure location field(s). Offset represents the symbol translation from the route in map units.

You can continue to work in your map document MyRoutes.mxd and skip step 1.

If you didn't succeed in finishing the previous exercise, start with step 1.

How to do it...

Follow these steps to create events using ArcMap:

1. Start ArcMap and open the existing map document StartEvents.mxd from <drive>:\PacktPublishing\Data\LinearReferencing\Routes. Open the **Catalog** window. Expand VeloGIS.gdb from the same location.
2. You will create a route event table for point events. Right-click on the geodatabase and navigate to **New** | **Table**. For **Name**, type BusStop. Accept the default settings from the next panel. Add the following new fields:
   - RouteID with **Data Type**: Short Integer. This field will store the unique identifiers of the routes.
   - Loc_M with **Data Type**: Double.
   - Offset with **Data Type**: Short Integer and **Default Value**: 6.
3. Click on **Finish**. Your table has been added in the **Table Of Contents** section. To see the table, select the **List by Source** view.
4. You will create a new route event table for *line* events. Right-click on the geodatabase and navigate to **New** | **Table**. For **Name**, type SpeedLimit. Accept the default settings from the next panel. Add the following new fields:
   - RouteID with **Data Type**: Short Integer
   - From_M with **Data Type**: Double
   - To_M with **Data Type**: Double
   - Offset with **Data Type**: Short Integer and **Default Value**: 1
   - MaximumSpeed with **Data Type**: Short Integer and **Default Value**: 50

5. Click on **Finish**.

   You now have two empty tables. Let's populate those tables. Open the `YourEvents.pdf` file from `...\Data\LinearReferencing`. The file contains all information necessary to complete the route event tables:

6. Open the table `BusStop`. Start an edit session. Click inside the first empty cell under the `RouteID` field and type `133`. For `Loc_M`, type `7`. The `Offset` field is already populated with the default value `6`. Notice also that the `ObjectID` field value is automatically generated. Continue to complete the table with the next 13 records using the information from the `YourEvents.pdf` file. Save the edits.
7. Open the `SpeedLimits` table. Complete the table with nine records. Save the edits and stop the edit session.

   Let's create the point event layer based on the `BusStop` route event table:

8. In the **Table Of Contents** section, right-click on the `BusStop` table, and select **Display Route Events**. In the dialog box, set the following parameters:
   - **Route Reference**: `BusRoutes` and **Route Identifier**: `RouteID`
   - For **Event Table** | **BusStop**, select **Route Identifier**: `RouteID`
   - Check **Point Events: Occur at a precise location along the route**
   - **Measure**: `Loc_M` and **Offset**: `Offset`; click on **OK**
9. The new layer named `BusStop Events` has been added in the **Table Of Contents** section. Symbolize the point route event layer: **Layer Properties** | **Symbology** | **Symbol** | **Style References** | **Public Sign**: `Bus`.
10. Add labels to the event layer: **Layer Properties** | **Labels** | **Label fields**: `RouteID`. Don't forget to check **Label features in this layer**. Click on **OK**. Explore the attribute table of `BusStop Events`.

Let's create the line event layer based on the `SpeedLimits` table. On the **Route Editing** toolbar, click on the **Add Route Events** tool, and set the parameters as shown in the following screenshot:

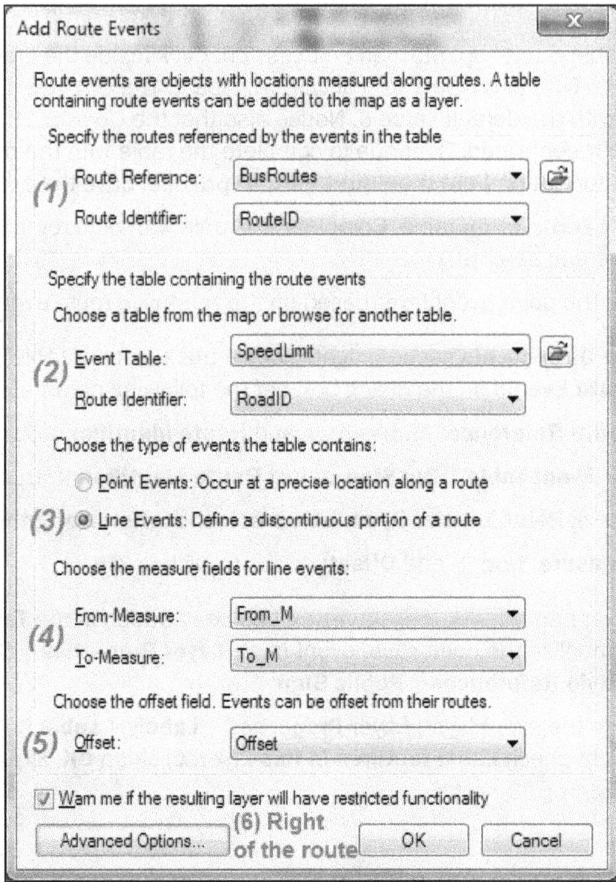

11. Click on **Advanced Options** and check **Right of the route. The side is determined by the digitized direction**. Once more, you can see how important the direction of the sketch is. Click on **OK**.
12. The `SpeedLimit Events` layer has been added. Symbolize the line route event layer based on the `MaximumSpeed` supplementary field: **Layer Properties** | **Symbology** | `Category: Unique values`.

    The city speed limit for public transport is **50 km per hour** and only a speed limit of **30 km per hour** is a critical restriction. For this reason, let's update the `SpeedLimits` route event table by erasing the records with a maximum speed of **50 km per hour**:

13. Start an edit session. A warning message tells you that event layers are not editable. Point and line events are temporary layers and are stored only in your map document. Click on **Continue**.
14. Open the attribute table of the `SpeedLimits` event table. With **Select by Attributes**, select all line events with `MaximumSpeed` as `50`. Click on **Delete Selected** to delete the selected records. Notice that line events with `MaximumSpeed` of `50` were deleted from the map. Use the *F5* key to refresh the map. Explore the attribute table of the `SpeedLimits Events` layer to see changes. You can use the **Undo Database Row(s) Deleted** and **Redo** tools to better see again how your line events are updating. Rename your layer: `SpeedLimits30 Events`.
15. Save your map document as `MyEevents.mxd`.

You can find the results of this recipe at `...\Data \LinearReferencing\Events`.

How it works...

You have stored the point and line events in separate tables. Because an event represents an attribute of the route, you can add additional information in your event table. The `SpeedLimits` event table contains the `MaximumSpeed` field that stores the maximum allowed speed for public transportation.

You have already noticed that your event table references multiple routes (for example, 79 bus line, 133 bus line) from a single route feature class.

Events with a negative `Offset` value will be placed on the right-hand side of the route. In this way, your bus stops that have the same location along the same route (for example, bus line 133) will be displayed on the left-hand side and the right-hand side of the route.

At step 13, you have seen that ArcGIS cannot edit the event layers. If you want to make it editable, export it as a feature class or a shape file. By exporting the event layer, the dynamic association between the event layer and the route event table will be lost.

Working with complex routes

Complex routes obey the simple route rules. Complex routes contain the *close loop*, *self-intersecting*, *double-back (overlap)*, and *branching* route features.

Getting ready

In this section, you will create a delivery route for customers of the `Farm Food Delivery` service. Your complex route will be used for:

- Measurement system: time
- Measure unit: minutes

Your complex route will use the street centerlines as reference lines from the `Street` feature class. The complex route will have eight simple routes as shown in the following screenshot:

In the end, your complex route will form a closed loop by double backing on itself and self-intersecting.

You can continue to work in your map document `MyEvents.mxd`, add the `FarmFoodDelivery` layer, and skip step 1. Open the `YourEvents.pdf` file from `...\Data\LinearReferencing`. The file contains all necessary information to complete the complex route.

How to do it...

Follow these steps to create a complex route in ArcMap:

1. Start ArcMap and open the existing map document `StartComplexRoute.mxd` from `<drive>:\PacktPublishing\Data \LinearReferencing\Events`.
2. Open the **Catalog** window and expand `VeloGIS.gdb` from the same location. Create a new route feature class named `FarmFoodDelivery_h`.

   The letter `h` should indicate the measure unit that will be used by your complex route (hours, minutes, seconds). Follow the steps from the *Creating and editing routes* recipe. Document the measurement system and measure units in your feature class metadata: `Delivery service`. The measurement system is time. The measure units are minutes.

3. Start the edit session. Prepare the editing environment as follows:
   - For **List by Selection** | **Selectable**, set `Streets` as the only selectable layer
   - **Snapping toolbar** | **Options** | **Tolerance**: `5` pixels
   - **Snapping toolbar** | **Snapping**: `End`, `Vertex`, and `Edge`

   The `YourEvents.pdf` file contains the measure values for all your simple routes. You will create eight simple routes, as shown in the previous screenshot.

4. Let's create the first simple route. Click on the **Edit** tool and select the street feature with **OBJECTID** as `3` that will represent **Route 1**.
5. On the **Route Editing** toolbar, select the **Make Route** tool. Move the dialog box to see the selected feature.
6. From **Click on the start point**, select the **Start Point** button.
7. On the map, click and snap on the start point: `Farm Food Delivery` (point **1** on the preceding screenshot).
8. In the **Make Route** dialog box, select **From/To measure**, and type `0` and `7`. Click on **Make Route**. Your first simple route starts from `0` minutes and ends at `7` minutes. Clear the selected feature with the **Clear Selected Feature** tool.
9. Create the second simple route. Select the second street with **OBJECTID** as `1` that will represent **Route 2**. Click on **Make Route** and click on *intersection 1* (or point **2** on the screenshot).

10. Type the values for **From/To measure**: `7` and `27`. Click on **Make Route**. Your second simple route starts from `7` minutes and ends at `27` minutes. Continue to create the remaining routes as shown in `YourEvents.pdf`.
11. Save the edits and clear all selected features.

    Let's merge all eight simple routes into a complex route:

12. In the **Table Of Contents** section, right-click on `FarmFoodDelivery_h`, and navigate to **Selection** | **Select All**. From the **Editor** menu, select **Merge**, and select the first record. Check **Preserve Overlapping segments** and click on **OK**.
13. On the **Editor** toolbar, select the **Attributes** tool, and for the `RouteID` field, enter the value `1`. Save the edits.
14. Set `FarmFoodDelivery_h` as the selectable layer. Explore your route using the **Edit Sketch Properties** window and the **Identify Route Location Results** tool.
15. Stop the edit session. Save your map document as `MyComplexRoute.mxd`.

You can find the results of this recipe at `...\Data\LinearReferencing\ComplexRoutes`.

How it works...

Generally, the digitized direction of the reference feature (for example, street centerline) determines the direction of the simple route. In this recipe, you saw that the direction of the increasing measure does not have to coincide with the digitized direction of the reference feature. Assign carefully the correct measures to the simple routes that will merge. With the **Identify Route Location Results** tool, click on the double-back segment of the complex route (**Routes 6** and **Route 7** from the screenshot). You will find two route identifiers with the same **RouteID** as `1`. Even if it is a single route, **Measure** will report a value for the going path and another value for the return path.

Analyzing events

Events support two overlay operations: union and intersection. The event tables or event layers can be used as input in an overlay analysis operation.

To avoid errors, respect the following conditions:

- Use the same measure system
- Measures should be incremented in the same direction and reference the same route
- Event layers should cover the same area

The result of an overlay operation is a new event table that contains the attributes from the input tables.

Getting ready

In this section, you want to identify the possible traffic jam sections generated by the speed restriction and frequent stops of the buses as shown in the following screenshot:

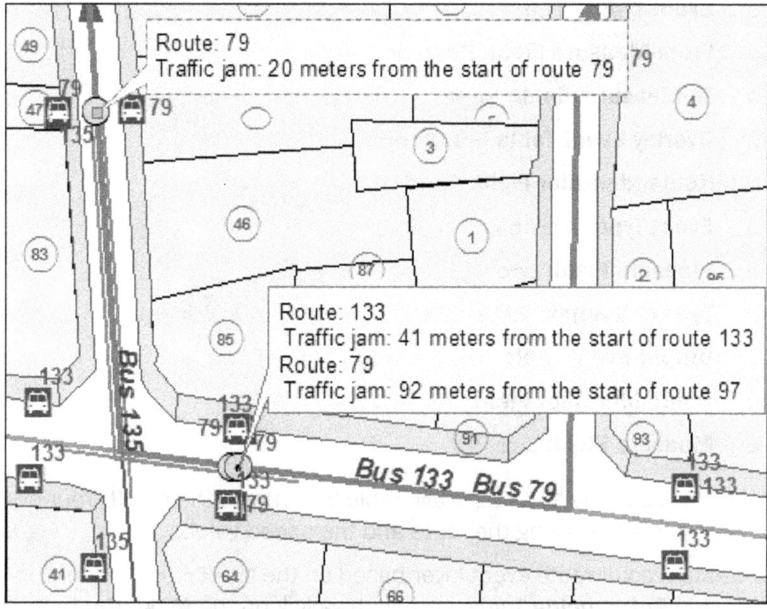

You can use this information to relocate some bus stops or bus lines in order to reduce traffic congestion.

You will perform a line-on-point intersection of two event tables: `BusStop` and `SpeedLimits`. Your `SpeedLimits` line event table contains only the section with the speed limit of **30 km per hour**. Both event tables reference the `BusRoute_m` route and have the same measurement system and measure unit: *distance* and *meters*.

You can continue to work in your map document `MyEvents.mxd` and skip step 1.

How to do it...

Follow these steps to analyze the traffic along a bus route using ArcMap:

1. Start ArcMap and open the existing map document `Analyze.mxd` from `<drive>:\PacktPublishing\Data\LinearReferencing\Events`.

2. Open **ArcToolbox**, expand **Linear Referencing Tools**, and double-click on the **Overlay Route Events** tool to open the dialog box. Set the following parameters:
   - **Input Event Table**: `SpeedLimit_30`
   - **Route Identifier Field**: `RoadID`
   - **Event Type**: `Line`
   - **From-Measure Field**: `From_M`
   - **To-Measure Field**: `To_M`
   - **Overlay Event Table**: `BusStop`
   - **Route Identifier Field**: `RoadID`
   - **Event Type**: `Point`
   - **Measure Field**: `Loc_M`
   - **Type of Overlay**: `INTERSECT`
   - **Output Event Table**: `VeloGIS.gdb\TrafficJam`
   - **Route Identifier Field**: `RoadID`
   - **Measure Field**: `Loc_M`
3. Click on **OK**. Open the `TrafficJam` table to examine the result. Your event table now has the location along the route and the speed value.
4. Let's create a point route event layer based on the `TrafficJam` table. In **ArcToolbox**, go to **Linear Referencing Tools**, and double-click on the **Make Route Event Layer** tool to open the dialog box. Set the following parameters:
   - **Input Route Feature**: `BusRoutes`
   - **Route Identifier Field**: `RoadID`
   - **Input Event Table**: `VeloGIS.gdb\TrafficJam`
   - **Route Identifier Field**: `RoadID`
   - **Event Type**: `Point`
   - **Measure Field**: `Loc_M`
   - **Layer Name or Table View**: `TrafficJam Events`
   - **Offset Field**: `none`
5. Leave all other options unchecked. Click on **OK**. The `TrafficJam Events` layer has been added in the **Table Of Contents** section.

6. Open the `TrafficJam Events` layer table to examine the result. Right-click on the `RouteID` field and select **Summarize**. For **Select a field to summarize**, select `RouteID`. For Specify output table, go to your geodatabase, and save the table as `Sum_TrafficJam`. Click on **Yes** to add the table in the **Table Of Contents** section. Open the `Sum_TrafficJam` table to examine the results.

   It seems that route `79` has the highest bus stop count in the speed restriction zone. The locations of `Route 133: 41` and `Route 79:92` refer to the same location on Dacia Avenue. There are double stops for bus 79 and 133 and this could generate a traffic jam in the area.

7. Save your map document as `MyAnalyze.mxd`.

You can find the results of this recipe at `...\Data \LinearReferencing\Analyze`.

How it works...

The line-on-point intersection returned an output table with attributes from both the input and overlay tables, as follows:

- The `RouteID` field
- The `Loc_M` field-point locations along the route from the `BusStop` event table
- The `MaximumSpeed` optional field from the `SpeedLimit_30` event table

Based on the output table, you created a new point event layer. If you want to save it as a feature class in your geodatabase, right-click on the `TrafficJam Events` layer, and go to **Data | Export Data**.

9
Working with Spatial Analyst

In this chapter, we will cover the following recipes:

- Analyzing surfaces
- Interpolating data
- Reclassifying a raster
- Working with Map Algebra
- Working with Cell Statistics
- Generalizing a raster
- Creating density surfaces
- Analyzing the least-cost path

Introduction

The **ArcGIS Spatial Analyst** extension offers a lot of great tools for geoprocessing raster data. Most of the *Spatial Analyst* tools generate a new raster output. Before starting a raster analysis session, it's best practice to set the main analysis environment parameters settings (for example, scratch the workspace, extent, and cell size of the output raster). In this chapter, you will store all raster datasets in file geodatabase as **file geodatabase raster datasets**.

For theoretical aspects about spatial data analysis, please refer to *Geographic Information Analysis, David O'Sullivan* and *David Unwin, John Wiley & Sons, Inc., 2003.*

For more details about the Spatial Analyst toolbox, please refer to *ArcGIS Help (10.2)* at `http://resources.arcgis.com/en/help/main/10.2 online.`

Analyzing surfaces

In this recipe, you will represent 3D surface data in a two-dimensional environment. To represent 3D surface data in the ArcMap 2D environment, you will use hillshades and contours. You can use the `hillshade` raster as a background for other raster or vector data in ArcMap.

Using the surface analysis tools, you can derive new surface data, such as slope and aspect or locations visibility.

Getting ready

In the surface analysis context:

- The term slope refers to the steepness of raster cells
- Aspect defines the orientation or compass direction of a cell
- Visibility identifies which raster cells are visible from a surface location

In this recipe, you will prepare your data for analysis by creating an elevation surface named `Elevation` from vector data. The two feature classes involved are the `PointElevation` point feature class and the `ContourLine` polyline feature class. All other output raster datasets will derive from the `Elevation` raster.

How to do it...

Follow these steps to prepare your data for spatial analysis:

1. Start ArcMap and open the existing map document, `SurfaceAnalysis.mxd`, from `<drive>:\PacktPublishing\Data\SpatialAnalyst`. Go to **Customize** | **Extensions** and check the **Spatial Analyst** extension.

2. Open **ArcToolbox**, right-click on the **ArcToolbox** toolbox, and select **Environments**. Set the geoprocessing environment as follows:
   - **Workspace | Current Workspace**: `Data\SpatialAnalyst\TOPO5000.gdb` and **Scratch Workspace**: `Data\SpatialAnalyst\ScratchTOPO5000.gdb`.
   - **Output Coordinates**: Same as **Input**.
   - **Raster Analysis | Cell Size**: **As Specified below**: type `0.5` with unit as **m**.
   - **Mask**: `SpatialAnalyst\TOPO5000.gdb\Trapezoid5k`.
   - **Raster Storage | Pyramid**: check **Build pyramids** and **Pyramid levels**: type `3`. Click on **OK**.
3. In **ArcToolbox**, expand **Spatial Analyst Tools | Interpolation**, and double-click on the **Topo to Raster** tool to open the dialog box. Click on **Show Help** to see the meaning of every parameter. Set the following parameters:
   - **Input feature data**:

     PointElevation Field: `Elevation` and **Type**: `PointElevation`

     ContourLine Field: `Elevation` and **Type**: `Contour`

     WatercourseA Type: `Lake`
   - **Output surface raster**: `...ScratchTOPO5000.gdb\Elevation`
   - **Output extent (optional)**: `ContourLine`
   - **Drainage enforcement (optional)**: `NO_ENFORCE`
4. Accept the default values for all other parameters. Click on **OK**. The `Elevation` raster is a *continuous thematic raster*. The raster cells are arranged in 4,967 rows and 4,656 columns.
5. Open **Layer Properties | Source of the raster** and explore the following properties: **Data Type** (`File Geodatabase Raster Dataset`), **Cell Size** (`0.5` meters) or **Spatial Reference** (EPSG: `3844`).
6. In the **Layer Properties** window, click on the **Symbology** tab. Select the **Stretched** display method for the continuous raster cell values as follows: **Show**: `Stretched` and **Color Ramp**: `Surface`. Click on **OK**.

7. Explore the cell values using the following two options:
   - Go to **Layer Properties** | **Display** and check **Show MapTips**
   - Add the **Spatial Analyst** toolbar, and from **Customize** | **Commands**, add the **Pixel Inspector** tool

   Let's create a hillshade raster using the `Elevation` layer:

8. Expand **Spatial Analyst Tools** | **Interpolation** and double-click on the **Hillshade** tool to open the dialog box. Set the following parameters:
   - **Input raster**: `ScratchTOPO5000.gdb\Elevation`
   - **Output raster**: `ScratchTOPO5000.gdb\Hillshade`
   - **Azimuth (optional)**: `315` and **Altitude (optional)**: `45`

9. Accept the default value for **Z factor** and leave the **Model shadows** option unchecked. Click on **OK**.

   From time to time, please ensure to save the map document as `MySurfaceAnalysis.mxd` at `...\Data\SpatialAnalyst`.

10. The `Hillshade` raster is a *discrete thematic raster* that has an associated attribute table known as **Value Attribute Table** (**VAT**). Right-click on the `Hillshade` raster layer and select **Open Attribute Table**.
11. The `Value` field stores the illumination values of the raster cells based on the position of the light source. The `0` value (black) means that `25406` cells are not illuminated by the sun, and `254` value (white) means that `992` cells are entirely illuminated. Close the table.
12. In the **Table Of Contents** section, drag the `Hillshade` layer below the `Elevation` layer, and use the **Effects** | **Transparency** tool to add a transparency effect for the `Elevation` raster layer, as shown in the following screenshot:

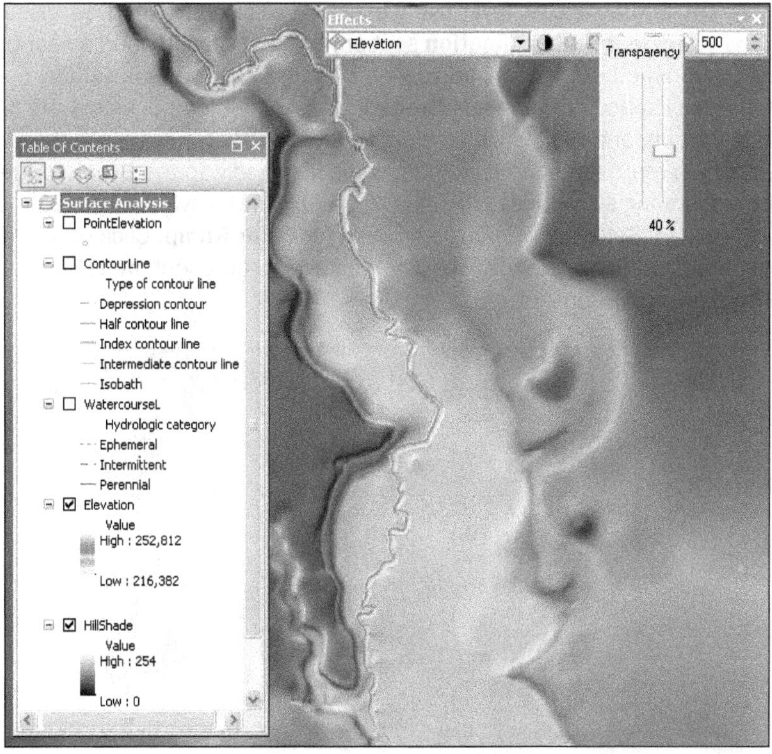

In the next step, you will derive a raster of slope and aspect from the `Elevation` layer.

13. Expand **Spatial Analyst Tools | Interpolation** and double-click on the **Slope** tool to open the dialog box. Set the following parameters:
    - **Input raster**: `Elevation`
    - **Output raster**: `ScratchTOPO5000.gdb\SlopePercent`
    - **Output measurement (optional)**: `PERCENT_RISE`

14. Click on **OK**. Symbolize the layer using the `Classified` method, as follows: **Show**: `Classified`. In the **Classification** section, click on **Classify** and select the **Manual** classification method. You will add *seven classes*. To add break values, right-click on the empty space of the **Insert Break** graph. To delete one, select the break value from the graph, and right-click to select **Delete Break**. Do not erase the last break value, which represents the maximum value. Secondly, in the **Break Values** section, edit the following six values: `5`; `7`; `15`; `20`; `60`; `90`, and leave unchanged the seventh value (`496,6`). Select `Slope` (green to red) for **Color Ramp**. Click on **OK**. The green areas represent flatter slopes, while the red areas represent steep slopes, as shown in the following screenshot:

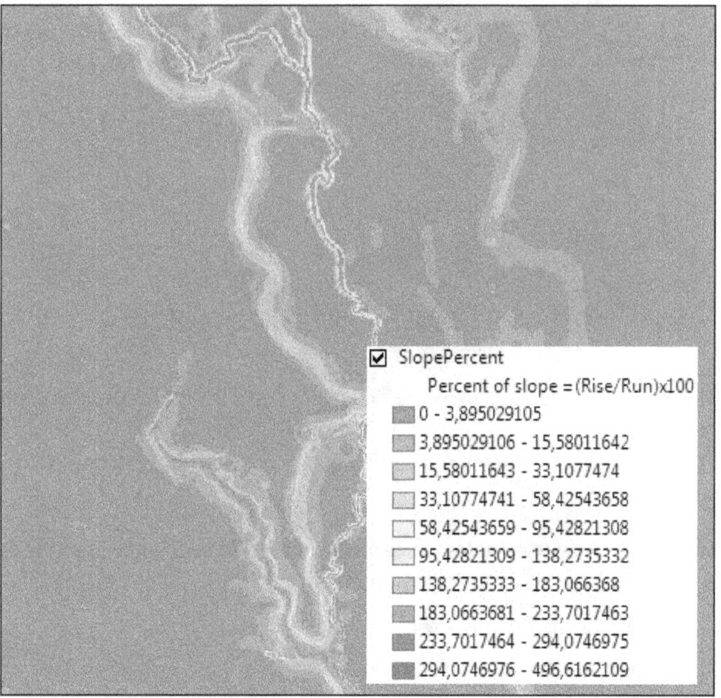

15. Expand **Spatial Analyst Tools** | **Interpolation** and double click on the **Aspect** tool to open the dialog box. Set the following parameters:
    - **Input raster**: `Elevation`
    - **Output raster**: `ScratchTOPO5000.gdb\Aspect`

16. Click on **OK**. Symbolize the `Aspect` layer. For **Classify**, click on the **Manual** classification method. You will add *five classes*. To add or delete break values, right-click on the empty space of the graph, and select **Insert / Delete Break**. Secondly, edit the following four values: `0`; `90`; `180`; `270`, leaving unchanged the fifth value in the **Break Values** section. Click on **OK**.

17. In the **Symbology** window, edit the labels of the five classes as shown in the following picture. Click on **OK**. In the **Table Of Contents** section, select the **<VALUE>** label, and type `Slope Direction`. The following screenshot is the result of this action:

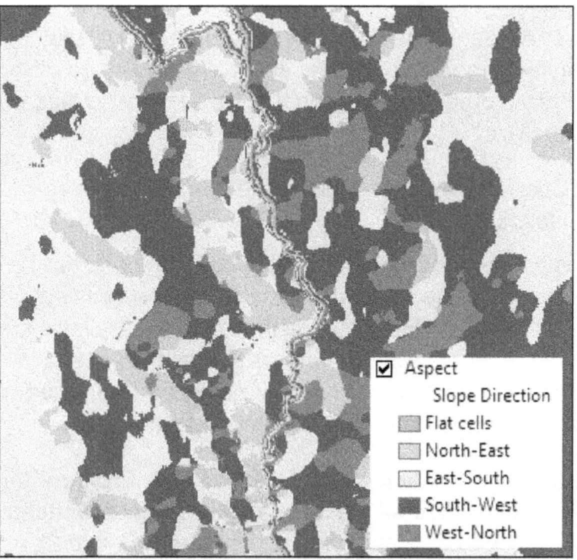

In the next step, you will create a raster of visibility between two geodetic points in order to plan some topographic measurements using an electronic theodolite. You will use the `TriangulationPoint` and `Elevation` layers:

18. In the **Table Of Contents** section, turn on the `TriangulationPoint` layer, and open its attribute table to examine the fields.

    There are two geodetic points with the following supplementary fields: `OffsetA` and `OffsetB`. `OffsetA` is the proposed height of the instrument mounted on its tripod above stations `8` and `72`. `OffsetB` is the proposed height of the reflector (or target) above the same points.

19. Close the table. Expand **Spatial Analyst Tools** | **Interpolation** and double-click on the **Visibility** tool to open the dialog box. Click on **Show Help** to see the meaning of every parameter. Set the following parameters:
    - **Input raster**: `Elevation`
    - **Input point or polyline observer features**: `TOPO5000.gdb\GeodeticPoints\TriangulationPoint`
    - **Output raster**: `ScratchTOPO5000.gdb\Visibility`
    - **Analysis type (optional)**: `OBSERVERS`

- **Observer parameters** | **Surface offset (optional)**: `OffsetB`
- **Observer offset (optional)**: `OffsetA`
- **Outer radius (optional)**: For this, type `1600`

Notice that `OffsetA` and `OffsetB` were automatically assigned. The **Outer radius** parameter limits the search distance, and it is the rounded distance between the two geodetic points. All other cells beyond the 1,600-meter radius will be excluded from the visibility analysis.

20. Click on **OK**. Open the attribute table of the `Visibility` layer to inspect the fields and values.

    The `Value` field stores the value of cells. Value `0` means that cells are not visible from the two points. Value `1` means that `6,608,948` cells are visible only from point 8 (first observer `OBS1`). Value `2` means that `1,813,578` cells are visible only from point 72 (second observer `OBS2`). Value `3` means that `4,351,861` cells are visible from both points. In conclusion, there is visibility between the two points if the height of the instrument and reflector is 1.5 meters. Close the table.

21. Symbolize the `Visibility` layer, as follows: **Show**: `Unique Values` and **Value Field**: `Value`. Click on **Add All Values** and choose **Color Scheme**: `Yellow-Green Bright`. Select **<Heading>** and change the **Label** value to `Height 1.5 meters`. Double-click on the symbol for **Value** as `0`, and select **No Color**. Click on **OK**. The `Visibility` layer is symbolized as shown in the following screenshot:

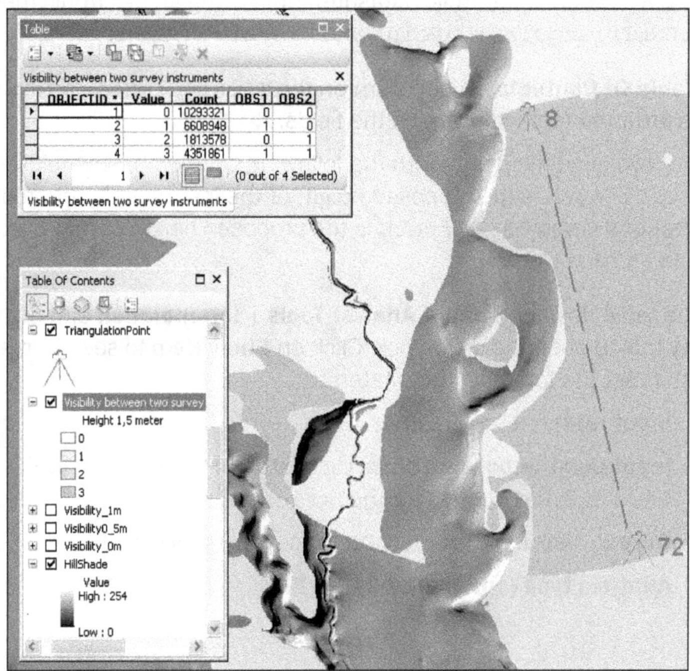

22. Turn off all layers except the `Visibility`, `TriangulationPoint`, and `Hillshade` layers. Save your map as `MySurfaceAnalysis.mxd` and close ArcMap.

You can find the final results at `<drive>:\PacktPublishing\Data\SpatialAnalyst\SurfaceAnalysis`.

How it works...

You have started the exercise by setting the geoprocessing environment. You will override those settings in the next recipes. At the application level, you chose to build pyramids. By creating pyramids, your raster will be displayed faster when you zoom out. The pyramid levels contain the copy of the original raster at a low resolution. The original raster will have a cell size of `0.5` meters. The pixel size will double at each level of the pyramid, so the first level will have a cell size of `1` meter; the second level will have a cell size of `2` meters; and the third level will have a cell size of `4` meters.

Even if the values of cells refer to heights measured above the local mean sea level (zero-level surface), you should consider the planimetric accuracy of the dataset. Please remember that `TOPO5000.gdb` refers to a product at the scale `1:5,000`. This is the reason why you have chosen `0.5` meters for the raster cell size. At step 4, you used the `PointElevation` layer as supplementary data when you created the `Elevation` raster.

If one of your **ArcToolbox** tools fails to execute or you have obtained an empty raster output, you have some options here:

- Open the **Results** dialog from the **Geoprocessing** menu to explore the error report. This will help you to identify the parameter errors. Right-click on the previous execution of the tool and choose **Open** (step **1**). Change the parameters and click on **OK** to run the tool. Choose **Re Run** if you want to run the tool with the parameters unchanged (step **2**) as shown in the following screenshot:

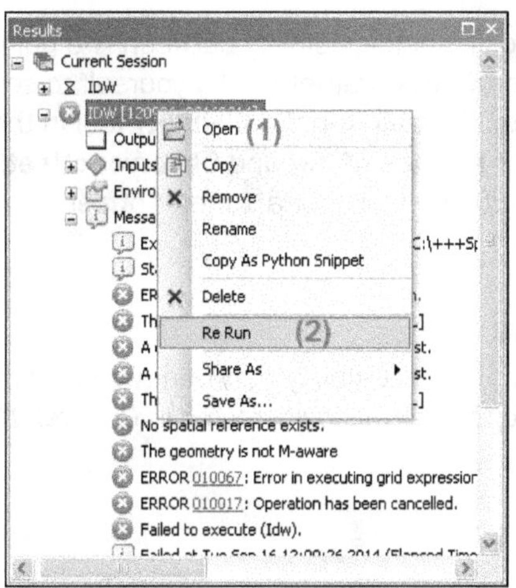

- Run the **ArcToolbox** tool from the ArcCatalog application. Before running the tool, check the geoprocessing environment in ArcCatalog by navigating to **Geoprocessing** | **Environments**.

There's more...

In *Chapter 4, Geoprocessing*, you learned to work with **Model Builder**. What if you have a model with all previous steps?

Open ArcCatalog, and go to `...Data\SpatialAnalyst\ModelBuilder`. In the `ModelBuilder` folder, you have a toolbox named `MyToolbox`, which contains the **Surface Analysis** model. Right-click on the model and select **Properties**. Take your time to study the information from the **General**, **Parameters**, and **Environments** tabs. The output (derived data) will be saved in **Scratch Workspace**: `ModelBuilder\ ScratchTopo5000.gdb`. Click on **OK** to close the **Surface Analysis Properties** window. Running the entire model will take you around 25 minutes. You have two options:

- **Tool dialog option**: Right-click on the **Surface Analysis** model and select **Open**. Notice the model parameters that you can modify and read the **Help** information. Click on **OK** to run the model.
- **Edit mode**: Right-click on the **Surface Analysis** model and select **Edit**. The colored model elements are in the second state—they are ready to run the **Surface Analysis** model by using one of those two options:
    - To run the entire model at the same time, select **Run Entire Model** from the **Model** menu.
    - To run the tools (yellow rounded rectangle) one by one, select the **Topo to Raster** tool with the **Select** tool, and click on the **Run** tool from the **Standard** toolbar. Please remember that a shadow behind a tool means that the model element has already been run.

You used the **Visibility** tool to check the visibility between two points with `1.5` meters for the **Observer offset** and **Surface offset** parameters. Try yourself to see what happens if the offset value is less than `1.5` meters. To again run the **Visibility** tool in the **Edit** mode, right-click on the tool, and select **Open**. For **Surface offset** and **Observer offset**, type `0.5` meters and click on **OK** to run the tool. Repeat these steps for a `1` meter offset.

See also

- If you want to determine the visibility between two or more points using other great tools, please refer to the *Intervisibility* recipe in *Chapter 10, Working with 3D Analyst*

Interpolating data

Spatial interpolation is the process of estimating an unknown value between two known values taking into account Tobler's First Law:

> *"Everything is related to everything else, but near things are more related than distant things."*

This recipe does not undertake to teach you the advanced concept of interpolation because it is too complex for this book. Instead, this recipe will guide you to create a terrain surface using the following:

- A feature class with sample elevation points
- Two interpolation methods: **Inverse Distance Weighted** (**IDW**) and **Spline**

For further research, please refer to: *Geographic Information Analysis, David O'Sullivan* and *David Unwin, John Wiley & Sons, Inc., 2003,* specifically the *8.3 Spatial interpolation* recipe of *Chapter 8, Describing and Analyzing Fields, pp.220-234.*

Getting ready

In this recipe, you will create a terrain surface stored as a raster using the `PointElevation` sample points. Your sample data has the following characteristics:

- The average distance between points is 150 meters
- The density of sample points is not the same on the entire area of interest
- There are not enough points to define the cliffs and the depressions
- There are not extreme differences in elevation values

How to do it...

Follow these steps to create a terrain surface using the **IDW** tool:

1. Start ArcMap and open an existing map document `Interpolation.mxd` from `<drive>:\PacktPublishing\Data\SpatialAnalyst`.
2. Set the geoprocessing environment, as follows:
   - **Workspace** | **Current Workspace**: `Data\SpatialAnalyst\TOPO5000.gdb` and **Scratch Workspace**: `Data\SpatialAnalyst\ScratchTOPO5000.gdb`

- **Output Coordinates**: Same as the `PointElevation` layer
- **Raster Analysis** | **Cell Size**: **As Specified Below**: `1`
- **Mask**: `Data\SpatialAnalyst\TOPO5000.gdb\Trapezoid5k`

In the next two steps, you will use the **IDW** tool. Running **IDW** with barrier polyline features will take you around 15 minutes:

3. In **ArcToolbox**, go to **Spatial Analyst Tools** | **Interpolation**, and double-click on the **IDW** tool. Click on **Show Help** to see the meaning of every parameter. Set the following parameters:
   - **Input point features**: `PointElevation`
   - **Z value field**: `Elevation`
   - **Output raster**: `ScratchTOPO5000.gdb\IDW_1`
   - **Power (optional)**: `0.5`
   - **Search radius (optional)**: `Variable`
   - **Search Radius Settings** | `Number of points`: `6`
   - **Maximum distance**: `500`
   - **Input barrier polyline features (optional)**: `TOPO5000.gdb\Hydrography\WatercourseL`
4. Accept the default value of **Output cell size (optional)**. Click on **OK**.
5. Repeat step 3 by setting the following parameters:
   - **Input point features**: `PointElevation`
   - **Z value field**: `Elevation`
   - **Output raster**: `ScratchTOPO5000.gdb\IDW_2`
   - **Power (optional)**: `2`

6. The rest of the parameters are the same as in step 3. Click on **OK**. Symbolize the `IDW_1` and `IDW_2` layers as follows: **Show**: `Classified`; **Classification**: **Equal Interval**: `10` classes; **Color Scheme**: `Surface`. Click on **OK**. You should obtain the following results:

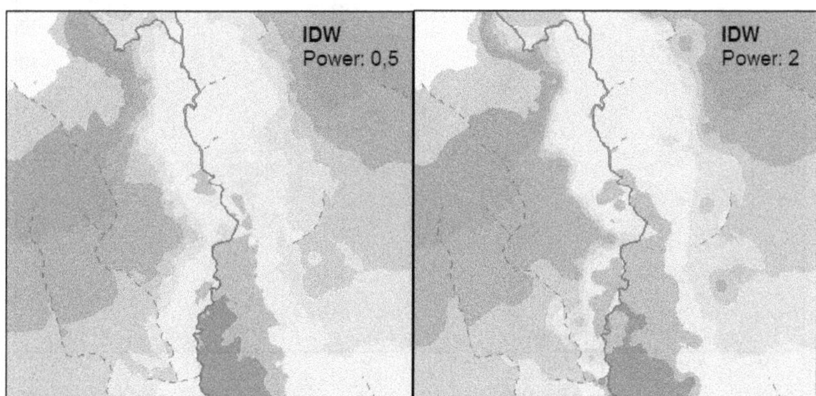

In the following steps, you will use the **Spline** tool to generate the terrain surface:

7. In **ArcToolbox**, go to **Spatial Analyst Tools** | **Interpolation**, and double-click on the **Spline** tool. Set the following parameters:
   - **Input point features**: `PointElevation`
   - **Z value field**: `Elevation`
   - **Output raster**: `ScratchTOPO5000.gdb\Spline_Regular`
   - **Spline type (optional)**: `REGULARIZED`
   - **Weight (optional)**: `0.1` and **Number of points (optional)**: `6`
8. Accept the default value of **Output cell size (optional)**. Click on **OK**.
9. Run again the **Spline** tool with the following parameters:
   - **Input point features**: `PointElevation`
   - **Z value field**: `Elevation`
   - **Output raster**: `ScratchTOPO5000.gdb\Spline_Tension`
   - **Spline type (optional)**: `TENSION`
   - **Weight (optional)**: `0.1` and **Number of points (optional)**: `6`

10. Accept the default value of **Output cell size (optional)**. Click on **OK**. Symbolize the `Spline_Regular` and `Spline_Tension` raster layers using the **Equal Interval** method classification with `10` classes and the `Surface` color ramp:

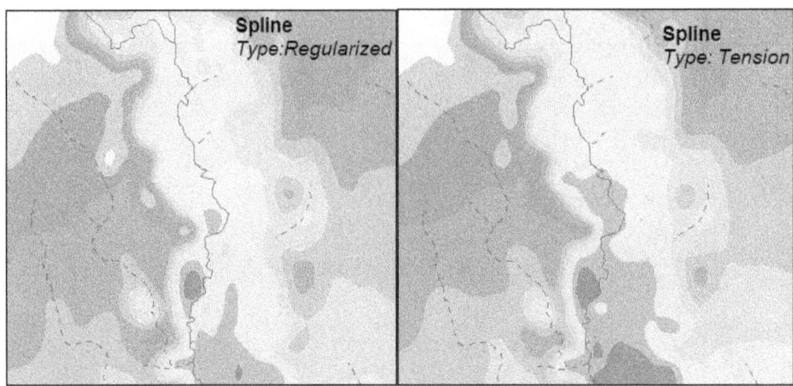

In the next steps, you will use the **Spline** with **Barriers** tool to generate a terrain surface using an increased number of sample points. You will transform the `ContourLine` layer in a point feature class. You will combine those new points with features from the `PointElevation` layer:

11. In **ArcToolbox**, go to **Data Management Tools** | **Features**, and double-click on the **Feature vertices to Points** tool. Set the following parameters:
    - **Input features**: `ContourLine`
    - **Output Feature Class**: `TOPO5000.gdb\Relief\ ContourLine_FeatureVertices`
    - **Point type (optional)**: `ALL`
12. Click on **OK**. Inspect the attribute table of the newly created layer. In the **Catalog** window, go to `...TOPO5000.gdb\Relief` and create a copy of the `PointElevation` feature class. Rename the new feature class as `ContourAndPoint`. Right-click on `ContourAndPoint` and select **Load** | **Load Data**. Set the following parameters from the second and fourth panels:
    - **Input data**: `ContourLine_FeatureVertices`
    - **Target Field**: `Elevation`
    - **Matching Source Field**: `Elevation`
13. Accept the default values for the rest of the parameters and click on **Finish**.

14. In **ArcToolbox**, go to **Spatial Analyst Tools** | **Interpolation**, and double-click on the **Spline with Barriers** tool. Set the following parameters:
    - **Input point features**: `ContourAndPoint`
    - **Z value field**: `Elevation`
    - **Input barrier features (optional)**: `TOPO5000.gdb\Hydrography\WatercourseA`
    - **Output raster**: `ScratchTOPO5000.gdb\Spline_WaterA`
    - **Smoothing Factor (optional)**: `0`
15. Accept the default value of **Output cell size (optional)**. Click on **OK**. You should obtain a similar terrain surface to what's shown here:

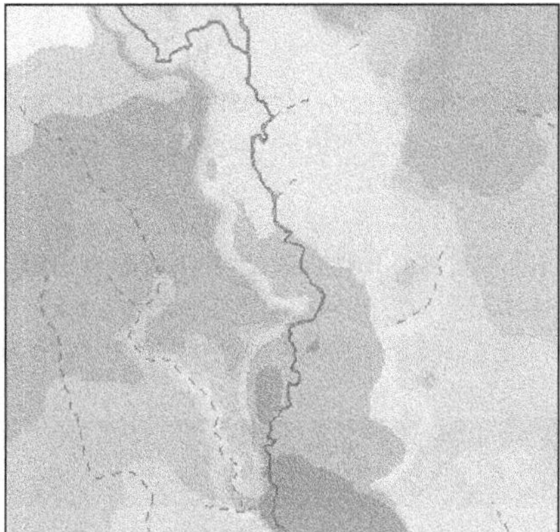

    Explore the results by comparing the similarities or differences of the terrain surface between interpolated raster layers and the `ContourLine` vector layer. The IDW method works well with a proper density of sample points. Try to create a new surface using the **IDW** tool and the `ContourAndPoint` layer as sample points.

16. Save your map as `MyInterpolation.mxd` and close ArcMap.

You can find the final results at `<drive>:\PacktPublishing\Data\SpatialAnalyst \Interpolation`.

How it works...

The IDW method generated an average surface that will not cross through the known point elevation values and will not estimate the values below the minimum or above the maximum given point values. The **IDW** tool allows you to define polyline barriers or limits in searching sample points for interpolation. Even if the `WatercourseL` polyline feature classes do not have elevation values, river features can be used to interrupt the continuity of interpolated surfaces.

To obtain fewer averaged estimated values (reduce the IDW smoother effect) you have to:

- Reduce the sample size to `6` points
- Choose a variable search radius
- Increase the power to `2`

The **Power** option defines the influence of sample point values. This value increases with the distance. There is a disadvantage because around a few sample points, there are small areas raised above the surrounding surface or small hollows below the surrounding surface.

The **Spline** method has generated a surface that crosses through all the known point elevation values and estimates the values below the minimum or above the maximum sample point values. Because the density of points is quite low, we reduced the sample size to **6** points and defined a variable search radius of 500 meters in order to reduce the smoothening effect.

The **Regularized** option estimates the hills or depressions that are not cached by the sample point values. The **Tension** option will force the interpolated values to stay closer to the sample point values.

Starting from step 12, we increased the number of sample points in order to better estimate the surface. At step 14, notice that the **Spline with Barriers** tool allows you to use the polygon feature class as breaks or barriers in searching sample points for interpolation.

Reclassifying a raster

When you reclassify a raster, you reassign one or more cell values or a range of values to new output cell values. You may need to reclassify a raster to:

- Update the cell values with new information from the field
- Simplify raster data
- Add cell values to a common scale
- Exclude values from analysis by setting those values to `NoData`
- Include `NoData` cells in analysis by setting values to those cells

The classification method and the number of classes you define will affect the raster displaying and interpreting process.

Getting ready

In this recipe, you will add the cell values to a common scale for three thematic raster layers: `LandUse`, `SlopeProcent`, and `Aspect`.

How to do it...

Follow these steps to reclassify three raster layers:

1. Start ArcMap and open a new map document. Add the following datasets to the map:
   - The `LandUse` feature class from `...Data\SpatialAnalyst\TOPO5000.gdb\LandUse`
   - The `Elevation`, `SlopeProcent`, and `Aspect` rasters from `SpatialAnalyst\ScratchTOPO5000.gdb`
2. Save your map as `ReclassifyRaster.mxd`. Load the **Spatial Analyst** extension and set the analysis environment, as you learned in the previous recipe.

   Firstly, you will convert the `LandUse` vector layer in a discrete thematic raster with a cell size of `0.5` meters:
3. In **ArcToolbox**, expand **Conversion Tools** | **To Raster**, and double-click on the **Polygon to Raster** tool to open the dialog box. Set the following parameters:
   - **Input Features**: `LandUse`
   - **Value Field**: `CAT`
   - **Output Raster Dataset**: `...ScratchTOPO5000.gdb\LandUse`
   - **Cell assignment type (optional)**: `MAX_AREA`
4. Accept the default values for all other parameters. Click on **OK**. Add the `LandUse` raster to your map.
5. Right-click on the `LandUse` raster layer and select **Properties** | **Symbology**. For the `Unique Values` display, change **Color Schema** to `Terra Tones`. Select the `Black` color for **Display NoData as**. Click on **OK**.
6. A discrete raster has associated an attribute table named the Value Attribute Table (VAT). Open its Value Attribute Table. The **Value** field stores the subtype codes from the input `LandUse` vector layer. You worked with that code in the *Creating subtypes recipe* of *Chapter 1, Designing Geodatabase*.

In the next steps, you will manually reclassify the cell values of the `LandUse` raster layer, as shown in the following table:

CAT Description: Category	Code	New cell values	CAT Description: Category	Code	New cell values
Hydrography Transportation	`31` `41`	`NoData`	Forest	`21`	`4`
Arable Fruit orchard	`11` `15`	`3`	Vineyard Other terrains	`14` `42`	`5`
Pasture Meadow	`12` `13`	`1`	Unproductive	`51`	`2`

7. In **ArcToolbox**, expand **Spatial Analyst Tools** | **Reclass**, and double-click on the **Reclassify** tool to open the dialog box. Set the following parameters:
    - **Input Raster**: `LandUse`
    - **Reclass field**: `Value`
    - **Reclassification**: For this, click on the **Unique** button to see all values
8. You will delete the `Hydrography` and `Transportation` values. Press the *Ctrl* key, and select the small box to the left of the rows with old values as `31` and `41`. The rows will become yellow. Click on **Delete Entries**. You will modify the values of the **New cell values** column as shown in the preceding table; for example, the old value `11` will have the new value as `4`.
9. After you finish adding all new values, click on the **Save** button to save the values in a *remap table* (information table) named `ReclassLU`. You can reuse this information table if you want to repeat the step.
10. Continue to set the last two parameters:
    - **Output Raster**: `...ScratchTOPO5000.gdb\ReclassLU`
    - Check **Change missing values to NoData (optional)**
11. Click on **OK**. Add the `ReclassLU` raster to your map. Right-click on the raster layer and select **Properties** | **Symbology**. For the **Unique Values** display, change **Color Schema** to `Yellow-Green bright`. Click on the **Symbols** column to select **Flip Colors**. For **Display NoData as**, select `Black`. Click on **OK**.
12. Open the `ReclassLU` raster attribute table to inspect the result.

13. Turn off all layers except the `SlopeProcent` raster layer. In **ArcToolbox**, expand **Spatial Analyst Tools | Reclass**, and double-click on the **Reclassify by Table** tool to open the dialog box. Set the following parameters:
    - **Input Raster**: `SlopePercent`
    - **Input remap table**: `Data\SpatialAnalyst\ReclassifySlopePercent`
    - **From value field**: `FROM_`
    - **To value field**: `TO`
    - **Output value field**: `NewValue`
    - **Output Raster**: `...ScratchTOPO5000.gdb\ReclassSlope`
    - Check **Change missing values to NoData (optional)**
14. Click on **OK**. Add the `ReclassSlope` raster to your map. Symbolize the raster as follows: **Unique Values** display method, **Color Schema**: `Slope`, and **Display NoData as**: `Black`.

    In the next step you will reclassify the cell values of the `Aspect` raster layer:
15. Turn off all layers except the `Aspect` raster layer. In **ArcToolbox**, expand **Spatial Analyst Tools | Reclass**, and double-click on the **Reclassify** tool to open the dialog box. Set the following new values:

Old values	New values
-1 - 0	3
0 - 90	2
90 - 180	1
180 - 270	1
270 - 359,999237	5
NoData	NoData

16. Add the `ReclassAspect` raster to your map. Symbolize the raster as follows: **Unique Values** display method, **Color Schema**: `Red to Green`, **Flip** colors, and **Display NoData as**: `Black`.
17. Save your map and close ArcMap.

You can find the final results at `<drive>:\PacktPublishing\Data\SpatialAnalyst\Reclassify`.

How it works...

This is how these steps work:

- At step 5, you saw `NoData` for the `LandUse` raster layer. The surface from the `LandUse` feature class follows the trapezoid shape of the scanned map. Your output `LandUse` raster is a grid of square cells (4,967 rows and 4,656 columns).
- At step 8, you set `NoData` for the `Hydrography` and `Transportation` cell values in order to exclude them from future analysis.

The *reclassification range* is common for all three raster layers: *from 1 to 5*. The value `1` means the most suitable value, while the value `5` means the least suitable value.

There's more...

Let's suppose we want to find the best places for fruit orchards by overlaying the following rasters: `ReclassLU`, `ReclassSlope`, and `ReclassAspect`. In our scenarios, we are searching the `Pasture`, `Meadow`, or `Unproductive` surface with slope range value from 7 percent, to 15 percent and direction range from southeast to southwest.

In the next steps, you will find suitable places for the fruit orchards using weight raster layers:

1. Open the last saved map document, `ReclassifyRaster.mxd`. In **ArcToolbox**, expand **Spatial Analyst Tools** | **Overlay** and double-click on the **Weighted Overlay** tool.
2. For the **Evaluation** scale, select `1 to 5 by 1` from the drop-down list.
3. Click on the **Add raster row** button to add `ReclassLU`, `ReclassSlope`, and `ReclassAspect`. **Field** contains the reclassified values of the `LandUse` raster, and **Scale Value** refers to the ranking of suitability. Be sure those columns have the same values.

   Define the percent of influence for all raster layers.
4. Assign to `ReclassLU` an influence of 50 percent, to `ReclassSlope` an influence of 30 percent, and to `ReclassAspect` an influence of 20 percent. Set **Output raster**: `...ScratchTOPO5000.gdb\Fruit_Orchards`. Click on `OK`. Open its `Fruit_Orchards` attribute table to inspect the result. There are only three cells with value `1`, which represent the most suitable surface.
5. Save your map and close ArcMap.

Thus, we conclude this:

- We didn't find the most suitable surface with value `1`—appropriate land use, slope, and aspect for a fruit orchard
- We obtained a reasonable alternative value `2`

Working with Map Algebra

Map Algebra is the math for raster datasets. The math-like expressions contain operators (for example, `Divide`, `Equal To`, `And`, `Over`, and `InList`) and functions (for example, `Mean`, `Con`, `Slope`, and `Aspect`).

Some Map Algebra operators and functions return logical values: `True` (logical value `1`) and `False` (logical value `0`). Be aware of `NoData` cell values because Map Algebra will not interpret this nonzero value as a true condition.

Getting ready

In this recipe, you will use Map Algebra expressions to identify the suitable sites for fruit orchards and reforestation. You will update the `LandUse` raster dataset with the proposed forest and fruit orchard sites using a Map Algebra operator called `Over`.

How to do it...

Follow these steps to learn how to work with Map Algebra:

1. Start ArcMap and open the existing map document `MapAlgebra.mxd` from: `<drive>:\PacktPublishing\Data\SpatialAnalyst`.
2. Add the following rasters created in the *Reclassifying a raster* recipe: `ReclassAspect` and `LandUse`. Load the **Spatial Analyst** extension and set the analysis environment. In the **Table Of Contents** section, turn off all layers except `SlopePercent`.

   In the next step, you will convert the floating point slope value to an integer value:

3. In **ArcToolbox**, expand **Spatial Analyst Tools** | **Map Algebra** and double-click, the **Raster Calculator** tool to open the dialog box, as shown in the following screenshot:

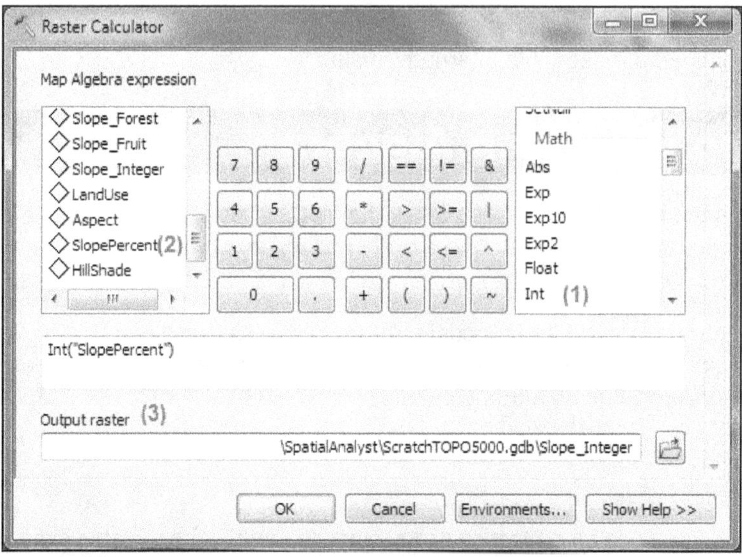

4. Double-click on the `Int` operator to start building the expression. Move the cursor into the brackets and double-click on the `SlopePercent` layer. Rename the **Output raster**: `ScratchTOPO5000.gdb\Slope_Integer`. Click on **OK**.
5. Explore the value attribute table of the `Slope_Integer` raster layer.

   You will create a new raster with slope percent equal to or greater than 15 percent:

6. Reopen the **Raster Calculator** tool. Set the following parameters:
   - Build the following expression: `"Slope_Integer">= 15`
   - **Output raster**: `ScratchTOPO5000.gdb\Slope_Forest`
7. Click on **OK**. Open its value attribute table to inspect the fields and values. The cells with a slope value equal to or greater than 15 percent have the logical value `1` or `true`. Cells with other values have the logical value `0` or `false`.
8. You will create a new raster with slope percent from 7 to 14. Open the **Raster Calculator** tool. Set the following:
   - Build the following expression:

     `InList("Slope_Integer",[7,8,9,10,11,12,13,14])`
   - **Output raster**: `ScratchTOPO5000.gdb\Slope_Fruit`

9. Click on **OK**. Open its value attribute table to inspect the fields and values. The raster cells have values from 7 to 14. All other cells are assigned `NoData`, which means there is no assigned (or recorded) value. Please remember that `NoData` does not mean zero value.
10. For **Display NoData as**, select the `Black` color to see the cells with `NoData`.
11. You will extract from the `LandUse` raster only `Arable`, `Pasture`, and `Meadow` land use types. Open the **Raster Calculator** tool. Set the following:
    - Build the following expression:

      ```
      InList("LandUse",[11, 12,13])
      ```

    - **Output raster**: `ScratchTOPO5000.gdb\LandUse_Good`
12. Click on **OK**. Open the value attribute table to inspect the fields and values. For **Display NoData as**, select the `Black` color to see the cells with `NoData`.

    You will identify areas suitable for fruit orchards by putting together `LandUse_Good` and `Slope_Fruit` using the **Boolean And** tool from **Spatial Analyst Tools** | **Math** | **Logical Toolset**.
13. Open the **Boolean And** tool and set the parameters, as shown in the following screenshot:

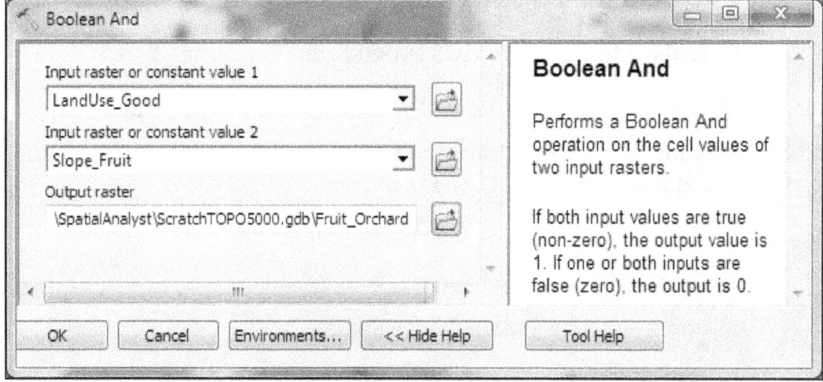

14. Click on **OK**. Open the value attribute table to inspect the fields and values. The cells with the logical true value `1` refer to the area suitable for a fruit orchard. All other cells are assigned `NoData`.

    In the previous recipe, we have identified other important criteria: direction of slope and aspect of cells. The `ReclassAspect` cells already have integer values because they were reclassified. Let's use the **Combinatorial And** tool to add the corresponding cell values from the `ReclassAspect` raster to the value attribute table of `Fruit_Orchard`. This will be equivalent to geoprocessing the overlay tool for vector datasets.

15. In **ArcToolbox**, expand **Spatial Analyst Tools** | **Math** | **Logical**, and double-click on the **Combinatorial And** tool to open the dialog box. Set the following parameters:

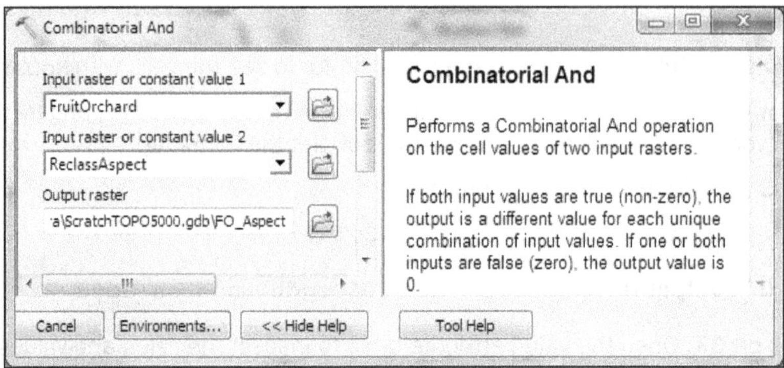

16. Click on **OK**. Open the `FO_Aspect` value attribute table to inspect the fields and values.
17. You will update the `LandUse` raster with the proposed fruit orchard sites using a conditional `if-else` function. Open the **Raster Calculator** tool. Set the following:
    - Build the following expression:

      `Con(IsNull("FO_Aspect"),"LandUse", 100)`
    - **Output raster**: `ScratchTOPO5000.gdb\LandUse_1`
18. Click on **OK**. Open the value attribute table to inspect the fields and values.

    In the next steps, you will identify the suitable sites for reforestation by putting together `LandUse_Good` and `Slope_Forest`:
19. Open the **Raster Calculator** tool. Set the following:
    - Build the following expression:

      `"LandUse_Good" & "Slope_Forest"`
    - **Output raster**: `ScratchTOPO5000.gdb\Reforestation`
20. Click on **OK**. Open the value attribute table to inspect the fields and values. The cells with the logical true value `1` refer to the area that should be reforested.
21. You should also update the `LandUse_1` raster with the proposed forest sites using the `Over` operator. This operator will paste the `Reforestation` value `1` over the corresponding cells in the `LandUse_1` raster. The value `0` will be replaced by the corresponding cell values in the `LandUse_1` raster. The `NoData` cell from `Reforestation` will be transferred to the `LandUse_1` raster. For this reason, we don't want to have `NoData` cells in the `Reforestation` raster.

22. Firstly, we have to change the `NoData` cells to `0`. You have two main options:
    - **A shortcut**: Open the **Reclassify** tool and set the following new values:

      Input Raster: `Reforestation`

      Reclass field: `Value`

Old values	New values
0	0
1	101
NoData	0

      Output raster: `ScratchTOPO5000.gdb\Reclass_Forest`

      Don't check **Change missing values to NoData (optional)**
    - Follow the next steps using the **Raster Calculator** dialog::

      Step 1: `IsNull("Reforestration")` and **Output raster**: `ForestNoData`

      Step 2: `Con("ForestNoData"= =0,"Reforestation",0)` and **Output raster**: `Forest_Final`

      Step 3: `Con("Forest_Final"= =1,101,0);` and **Output raster**: `ForestReclass`
23. Open the **Raster Calculator** tool and build the following expression: `Over("Reclass_Forest","LandUse_1")`.
24. Change the **Output raster**: `ScratchTOPO5000.gdb\LandUse_Update`. Click on **OK**. You should obtain a raster similar to the one shown in the following screenshot:

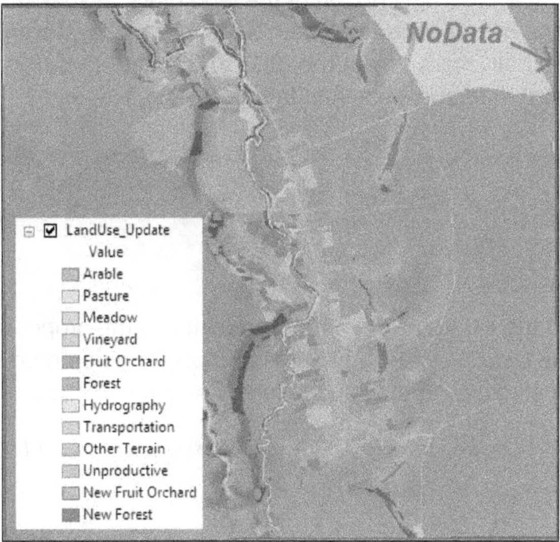

25. Symbolize the layer as shown in the following screenshot:

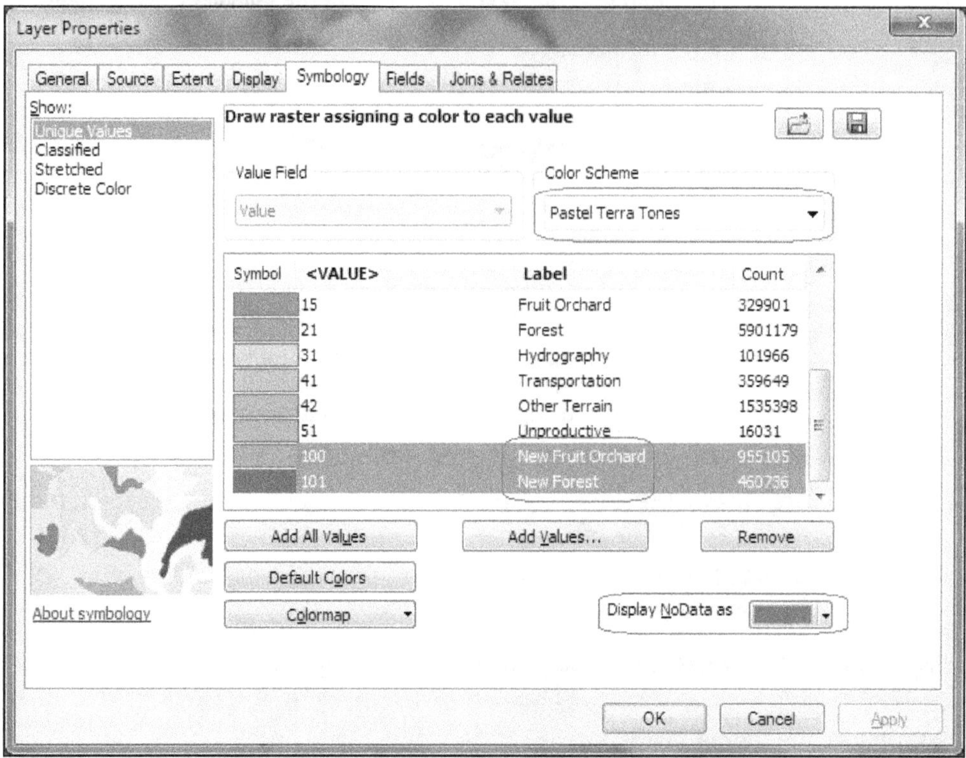

26. For **Label**, use the code description of the land use subtypes from the table presented in the *Reclassifying a raster* recipe. Set the layer transparency to 40 percent. Turn off all layers except the Hillshade and LandUse_Update raster layers. Save your map and close ArcMap.

You can find the results at <drive>:\PacktPublishing\Data\SpatialAnalyst\MapAlgebra.

How it works...

At step 15, you created the FO_Aspect raster that contains the proposed areas for fruit orchards. To update the LandUse raster with the proposed fruit orchard sites, you built the following expression at step 17: Con(IsNull("FO_Aspect"),"LandUse", 100).

The Con syntax is Con (<condition>, <true_expression>, {false_expression}).

The expression means *if* (a cell from `FO_Aspect` contains `NoData`), *then* return the `LandUse` cell value, *else* return a value of `100`.

You used the `IsNull` function with the `Con` function to replace the `NoData` value with the value `100`.

Starting with step 22, you want to update once again the `LandUse` raster with the proposed forest sites using the `Over` operator. But first you have to assign the `NoData` cells a value. The first option is quite intuitive if you use the **Reclassify** tool.

The second option describes the steps when you use the **Raster Calculator** dialog:

- `IsNull("Reforestration")`

  The `IsNull` function tests the `Reforestration` raster cells for the `NoData` value and returns `true` (`1`). All other values from `Reforestration` will have the `false` value (`0`).

- `Con("ForestNoData"= =0,"Reforestation",0)`

  This means that *if* a cell from `ForestNoData` is equal to `0`, *then* return the `Reforestation` cell value, *else* return a value of `0`.

- `Con("Forest_Final"= =1,101,0)`

  This means that *if* a cell from `Forest_Final` is equal to `1`, *then* return the `101` value, *else* return a value of `0`. The `Con` function will return `NoData` when no value is given for the false argument.

You have used the `IsNull` function with the `Con` function to replace the `NoData` value with the `LandUse` cell values.

Working with Cell Statistics

ArcGIS Spatial Analyst classifies the statistical functions in three basic groups: local, neighborhood, and zonal. In this recipe, you will work with the **Local Cell Statistics** tool. The `Cell Statistics` function compares the corresponding cells in two or more raster datasets and calculates descriptive statistics, such as: `Majority`, `Mean`, `Median`, or `Standard Deviation`.

For theoretical aspects about statistical methods, please refer to *Statistical Methods For Geography, Peter Rogerson, SAGE Publications Ltd., Inc., 2001.*

Getting ready

In the previous recipe, *Working with Map Algebra*, you have updated the `LandUse` raster dataset with new forest and fruit orchard sites. In this recipe, you will compare the `LandUse` raster created in 2010 with the `LandUse_2015` raster created in 2015.

How to do it...

Follow these steps to learn how to work with the **Cell Statistics** tool:

1. Start ArcMap and open the existing map document `Statistics.mxd` from `<drive>:\PacktPublishing\Data\SpatialAnalyst\CellStatistics`.
2. Open **ArcToolbox**. Go to **Geoprocessing** | **Environments** and set the geoprocessing environment as follows:
    - **Workspace** | **Current Workspace**: `...\SpatialAnalyst\CellStatistics\TOPO.gdb` and **Scratch Workspace**: `...\SpatialAnalyst\CellStatistics\ScratchTOPO.gdb`
    - **Output Coordinates**: Same as the `LandUse` layer
    - **Raster Analysis** | **Cell Size**: Same as the `LandUse` layer and **Mask**: `LandUse`
3. To find the logical difference between the `LandUse` and `LandUse_2015` raster layers, you will use the `Cell Statistics` functions.
4. In **ArcToolbox**, expand **Spatial Analyst Tools** | **Local**, and double-click on the **Cell Statistics** tool. Set the following parameters:
    - **Input raster or constant value 1**: `ScratchTOPO.gdb\LandUse`
    - **Input raster or constant value 2**: `ScratchTOPO.gdb\LandUse_2015`
    - **Output raster**: `ScratchTOPO.gdb\LandUse_Diff`
    - **Overlay statistic (optional)**: `VARIETY`
    - **Output raster**: `ScratchTOPO.gdb\LandUse_Diff`
    - Check **Ignore NoData in calculation (optional)**
5. Click on **OK**. The cells with **value 2** are representing areas that have changed. Turn off all raster layers except the `LandUse_Diff` layer.

    Let's find the land use types that have been changed:

6. In **ArcToolbox**, expand **Spatial Analyst Tools** | **Conditional**, and double-click on the **Con** tool. Set the parameters as shown in the following screenshot:

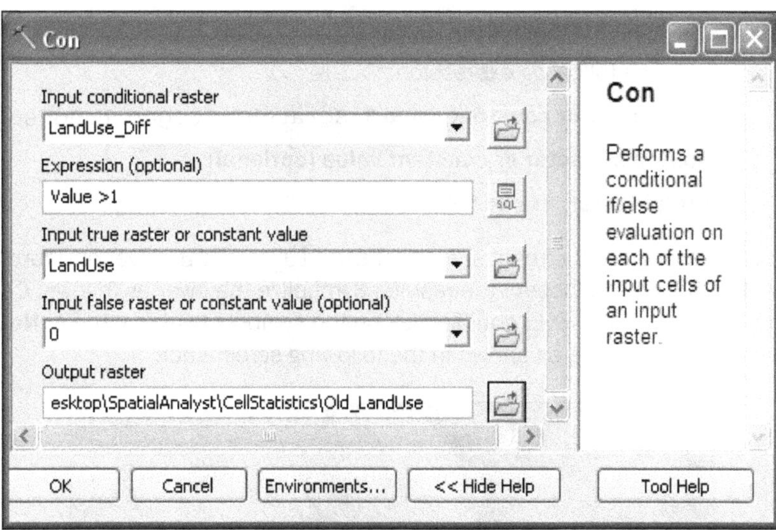

7. Click on **OK**. Turn off the `LandUse_Diff` layer, and turn on **LandUse** and `Old_LandUse`. Inspect the results. Symbolize the layer: **Color Schema**: `Terra Tones`; double-click on the symbol of value `0` to set **NoColor**; and edit the **Label** column, as shown in the following screenshot:

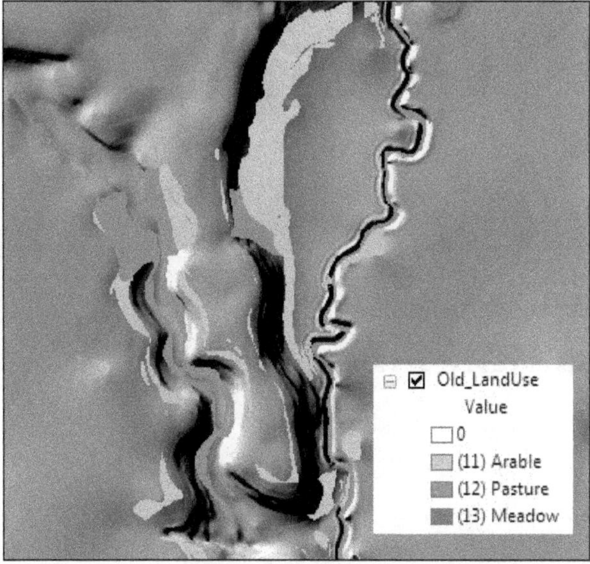

Now, we want to know the new land use types:

8. Reopen the **Con** tool. Set the following:
    - **Input conditional raster**: `ScratchTOPO.gdb\LandUse_Diff`
    - Build the following expression: `Value >1`
    - **Input raster or constant value 1**: `ScratchTOPO.gdb\LandUse_2015`
    - **Input false raster or constant value (optional)**: `0`
    - **Output raster**: `ScratchTOPO5000.gdb\New_LandUse`
9. Click on **OK**. Inspect the results. Turn off the `Old_LandUse` layer and turn on the **New_LandUse** layer. Inspect the results. Symbolize the layer as follows: **Color Schema**: `Green Bright`; double-click on the symbol of value `0` to set **NoColor**; and edit the **Label** column, as shown in the following screenshot:

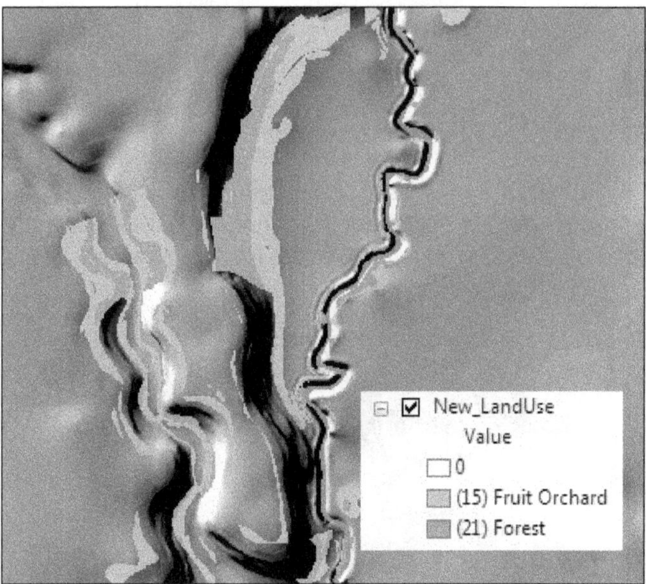

10. Save your map and close ArcMap.

You can find the results at `<drive>:\PacktPublishing\Data\SpatialAnalyst\CellStatistics`.

How it works...

Using the **Cell Statistics** tool is quite simple, but the output raster contains only two values, `1`, which means no change in land use, and `2`, which indicates land use change.

In the previous steps, you answered the following questions:

- **Question 1**: What are the old land use types that have been changed?

  You built the following expression: *if* a cell from `LandUse_Diff` is greater than `1`, *then* return the `LandUse` value, *else* return a value of `0`.

  Answer: `Arable`, `Pasture`, and `Meadow` land use.

- **Question 2**: What are the new land use types for the changed areas?

  You built the following expression: *if* a cell from `LandUse_Diff` is greater than `1`, *then* return the `LandUse_2015` value, *else* return a value of `0`.

  Answer: `Fruit orchard` and `Forest` land use.

There's more...

There is a shortcut to obtain the old land use types that have been changed (`Old_LandUse` results):

1. Start ArcMap and open the previous map document `Statistics.mxd`.
2. In **ArcToolbox**, expand **Spatial Analyst Tools** | **Math** | **Logical** and double-click on the **Diff** tool. Set the following parameters:
   - **Input raster or constant value 1**: `ScratchTOPO5000.gdb\LandUse`
   - **Input raster or constant value 2**: `ScratchTOPO5000.gdb\LandUse_Update`
   - **Output raster**: `ScratchTOPO5000.gdb\Difference`
3. Click on **OK**. Save your map and close ArcMap.

You have obtained the *old land use* for those areas planned for plantation. The cells with value `0` represent the cells that have the same values in both rasters. Be aware of the order of the input raster because the first raster cell values will be returned if the cell values are different in the first raster and the second raster.

Generalizing a raster

In the raster dataset, a **zone** is a class of elements represented by one or more cells that are connected or disconnected and that have the same value. The `LandUse` layer has zones or types of parcels, such as `Arable`, `Pasture`, `Forest`, and so on.

The `Arable` zone with the cell value `11` has a lot of disconnected areas or regions. A **region** is a group of connected cells having the same value. A zone is composed of one or more regions.

Getting ready

In this recipe, you will generalize the `LandUse_2015` raster. The generalization process will change the values of the small-region areas to the values of the larger neighbor regions. The parcels with area below `625` meters square are not represented in a classic topographic map at scale `1:5,000`. On the sheet map, this polygon area means a square of `5` millimeters. The `LandUse_2015` raster has a cell size of `0.5` meters. For this reason, you will dissolve the regions smaller than *50 x 50 = 2500* cells.

The roads, rivers, and build-up areas should not be affected by the generalization process (for example, edge changed or dissolved). Surface types such as arable, pasture, and vineyard will be affected by generalization. Here's how the resultant map looks:

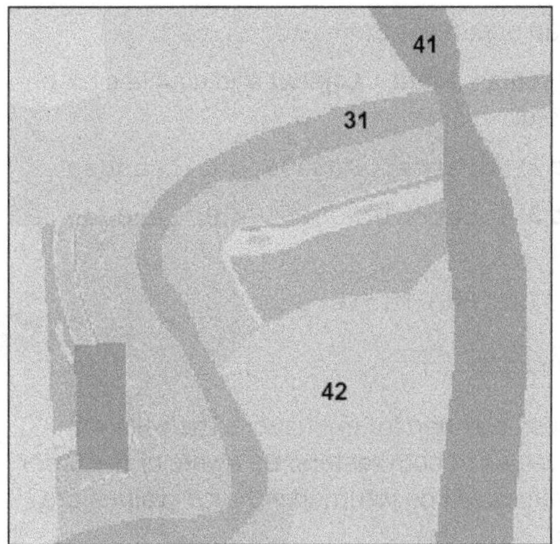

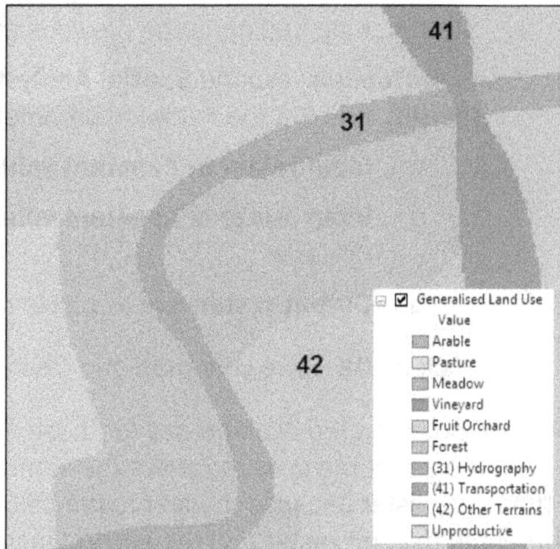

How to do it...

Follow these steps to generalize the `LandUse_2015` raster:

1. Start ArcMap and open an existing map document `Generalised.mxd` from `<drive>:\PacktPublishing\Data\SpatialAnalyst\Generalised`.
2. Open the **ArcToolbox**. Go to **Geoprocessing** | **Environments** and set the geoprocessing environment, as follows:
    - **Workspace** | **Current Workspace**: `Data\SpatialAnalyst\Generalised\TOPO.gdb` and **Scratch Workspace**: `Data\SpatialAnalyst\Generalised\ScratchTOPO.gdb`
    - **Output Coordinates**: Same as the `LandUse_2015` layer
    - **Raster Analysis** | **Cell Size**: Same as the `LandUse_2015` layer and **Mask**: `<none>`

    In the next step, you will exclude from spatial analysis the following zones: `Hydrography`, `Transportation`, and `Other Terrains` (for example, built-up area):

3. In **ArcToolbox**, go to **Spatial Analyst Tools** | **Conditional**, and double-click on the **Con** tool. Set the following parameters:
    - **Input conditional raster**: `LandUse_2015`
    - **Expression**: Click on the **SQL** button and select **Load** to add `Exclude.exp` from the `Generalised` folder
    - **Input true raster or constant value**: `LandUse_2015`
    - **Input false raster or constant value (optional)**: `<none>`
    - **Output raster**: `ScratchTOPO.gdb\Con_LandUse`
4. Click on **OK**. In **ArcToolbox**, go to **Spatial Analyst Tools** | **Generalization**, and double-click on the **Region Group** tool. Set the following parameters:
    - **Input raster**: `Con_LandUse`
    - **Output raster**: `ScratchTOPO.gdb\Regions`
5. Accept the default values for all other parameters. Click on **OK**. In **ArcToolbox**, go to **Spatial Analyst Tools** | **Zonal**, and double-click on the **Zonal Geometry** tool. Set the following parameters:
    - **Input raster or feature zone data**: `Regions`
    - **Zone field**: `Value`
    - **Output raster**: `ScratchTOPO.gdb\LandUse_Area`

- **Geometry type (optional)**: `AREA`
- **Output cell size (optional)**: `0.5`

6. Click on **OK**. In **ArcToolbox**, go to **Spatial Analyst Tools | Conditional**, and double-click on the **Set Null** tool.
7. You have two options: using `LandUse_Area` and area, or using `Regions` and number of cells. Set the following parameters:
   - **Input conditional raster**: `LandUse_Area`, or type `Regions`
   - **Expression**: `Value < 625` (second option: `Count < 2500`)
   - **Input false raster or constant value**: `1`
   - **Output raster**: `ScratchTOPO.gdb\BigArea`
8. Click on **OK**. In **ArcToolbox**, go to **Spatial Analyst Tools | Generalization**, and double-click on the **Nibble** tool. Set the following parameters:
   - **Input conditional raster**: `Con_LandUse`
   - **Input raster mask**: `BigArea`
   - **Output raster**: `ScratchTOPO.gdb\Generalization`
   - Uncheck **Use NoData value if they are the nearest neighbor**
9. Click on **OK**. In **ArcToolbox**, go to **Spatial Analyst Tools | Math | Logical**, and double-click on the **Is Null** tool. Set the following parameters:
   - **Input raster**: `Generalization`
   - **Output raster**: `ScratchTOPO.gdb\LandUseNoData`
10. Click on **OK**. In **ArcToolbox**, go to **Spatial Analyst Tools | Conditional**, and double-click on the **Con** tool. Set the following parameters:
    - **Input conditional raster**: `LandUseNoData`
    - **Expresion**: `Value = 1`
    - **Input true raster or constant value**: `ScratchTOPO.gdb\LandUse_2015`
    - **Input false raster or constant value (optional)**: `ScratchTOPO.gdb\Generalization`
    - **Output raster**: `ScratchTOPO.gdb\GeneralisedLU`
11. Click on **OK**. Symbolize the `GeneralisedLU` raster layer by importing `LandUse_2015.lyr` from `the Generalised` folder. Inspect the results with the **Pixel Inspector** tool.
12. Save your map and close ArcMap.

You can find the results at `<drive>:\PacktPublishing\Data\SpatialAnalyst \Generalised`.

How it works...

This is how these steps work:

- At step 3, you excluded three zones by assigning them a `NoData` value.
- At step 4, you delimitated the regions for all land use zones. Every region from the `Regions` layer has a unique value and its associated zone value. In this way, you can identify the regions with a small number of cells.
- At step 5, you calculated the area of each zone from the raster.

  In order to replace all regions smaller than `625` meters, you have to create two groups of regions: `1` = `regions larger` than `625` meters, and `NoData` = `smaller region`. When you use the `Set Null` function, you have two options: working with the area or the number of cells.

- At step 8, you used the `Nibble` clean-up function. You *replaced* the zone value from the `Con_LandUse` layer with the zone values of the *nearest neighbor regions* based on the *raster mask* built at step 7. The `NoData` cells from the mask raster indicate the cells that should change the zone value.

Your generalized raster does not contain the three zones excluded at step 3. At step 9, you assigned the true value `1` to `NoData` cells. At step 10, you updated the value `1` with the corresponding cell values from `LandUse_2015`.

There's more...

Let's work again with **Model Builder**. Open ArcCatalog and go to `...\SpatialAnalyst\ModelBuilder`. Expand **MyToolbox**, and open **Generalised a Thematic Raster** model in the **Edit** mode. Explore the model: it contains the entire workflow you already tested. Note that the output (derived raster) will be saved at `...\SpatialAnalyst\Generalised\TOPO_Model.gdb`. Running the entire model will take you around 15 minutes.

Creating density surfaces

Density surfaces estimate the concentrations of points or polylines per unit of area. ArcGIS allows you to create a density surface using two methods:

- Simple—calculate the cell values using a circular search area around the raster cell
- Kernel (**Kernel Density Estimation** or (**KDE**)—use a circular search area centered on the sample point or polyline feature; the cell value is calculated based on the overlapping kernel-density surface values

Getting ready

In this recipe, you will prepare data for analysis by representing the built-up area using the `Buildings` polygon feature class. You will use the **kernel** density function to estimate the number of buildings per square kilometer. Using the same function, you will create the density surfaces based on the selected features (for example, dwellings and dwellings made of clay) and based on an attribute field (for example, the number of families, population, and so on).

How to do it...

Follow these steps to create two density raster surfaces:

1. Start ArcMap and open the existing map document `DensitySurfaces.mxd` from `<drive>:\PacktPublishing\Data\SpatialAnalyst`.
2. Open **ArcToolbox** and set the geoprocessing environment, as follows:
    - **Workspace | Current Workspace**: `Data\SpatialAnalyst\TOPO5000.gdb` and **Scratch Workspace**: `Data\SpatialAnalyst\ScratchTOPO5000.gdb`
    - **Output Coordinates**: Same as **Input**
    - **Cartography | Reference Scale**: `5000`
    - **Raster Analysis | Cell Size**: `0.5`

    In the next step, you will delineate the built-up area that will define the density processing extent:

3. Open the `Buildings` attribute table. Notice the empty field named `BuiltUp`. You will use this field in the next step.
4. In **ArcToolbox**, go to **Cartography Tools | Generalization**, and double-click on the **Delineate Built-Up Areas** tool. Click on **Show Help** to see the meaning of every parameter. Set the following parameters:
    - **Input Building Layers:** `TOPO5000.gdb\Buildings\Buildings`
    - **Identifier Field (optional):** `BuiltUp`
    - **Output Feature Class:** `TOPO5000.gdb\Boundaries\BuiltUp_Area`
    - **Grouping Distance:** `1000` with unit as **Meters**
    - **Minimum Detail Size:** `5` with unit as **Meters**
    - **Edge Features (optional):** `TOPO5000.gdb\Transportation\Road`
    - **Minimum Building Count (optional):** `2`
5. Click on **OK.** Notice that the `BuiltUp` field from the `Buildings` value attribute table has been populated with the building status code.

6. Symbolize the `BuiltUp_Area` layer, as follows: **Hollow** style with **Outline**: `Boundary, Township` and **Outline Width**: `1.7`. In the **Table Of Contents** section, move the layer below the `Buildings` and `Road` layers.
7. Let's create a point feature class from the `Buildings` polygon feature class. In **ArcToolbox**, go to **Data Management Tools** | **Features**, and double-click on the **Feature to Point** tool. Set the following parameters:
    - **Input Feature**: `TOPO5000.gdb\Buildings\Buildings`
    - **Output Feature Class**: `TOPO5000.gdb\Buildings\BuildingsP`
    - Check **Inside (optional)**; click on **OK**
8. Turn off all layers except `BuildingsP` and `BuiltUp_Area`. Go to **Geoprocessing** | **Environments**, and update the following parameter: **Processing Extent**: `BuiltUp_Area`.

    Let's create a population density surface using a 200-meter search radius:
9. In **ArcToolbox**, go to **Spatial Analyst Tools** | **Density**, and double-click on the **Kernel Density** tool. Click on **Show Help** to see the meaning of every parameter. Set the following parameters:
    - **Input Building Layers**: `BuildingsP`
    - **Population field**: `NumberPersons`
    - **Output raster**: `ScratchTOPO5000.gdb\Population_200`
    - **Output cell size (optional)**: `0.5`
    - **Search radius (optional)**: `200`
    - **Area units (optional)**: `SQUARE_KILOMETERS`; click on **OK**

    Let's create a family's density surface using a 200-meter search radius:
10. In **ArcToolbox**, double-click on the **Kernel Density** tool. Set the following parameters:
    - **Input Building Layers**: `BuildingsP`
    - **Population field**: `NumberFamilies`
    - **Output raster**: `ScratchTOPO5000.gdb\Families_200`
    - Set the rest of the parameters the same as at step 9; click on **OK**

    Let's create a dwelling density surface using a 200-meters search radius:
11. Open the `BuildingsP` attribute table. Click on **Table Options** and select the **Select By Attributes** option. Build the following expression for the `Use Of Buildings` (`CDC`) field: `CDC = 1`. Click on **Apply**. The expression will select the `Dwelling` features. Based on the selected features, you will create a density surface.

12. In **ArcToolbox**, double-click on the **Kernel Density** tool. Set the following parameters:
    - **Input Building Layers**: `BuildingsP`
    - **Population field**: `<none>`
    - **Output raster**: `ScratchTOPO5000.gdb\Dwelling_200`
    - Set the rest of the parameters in the same way as in step 9; click on **OK**.

    Let's do an empirical analysis based on the *old dwellings made of clay* and *population density*. We assume that elderly people have, in general, old dwellings. We expect the population density to be quite low in areas with a high density of clay houses.

13. Firstly, select the clay dwelling with the following expression: `CMA = 4 AND CDC = 1`. Secondly, repeat step 12, and create the `OldDwellings_200` density surface based on the selected features. Explore the results by comparing the `Population_200` and `OldDwellings_200` density surfaces as shown in the following screenshot:

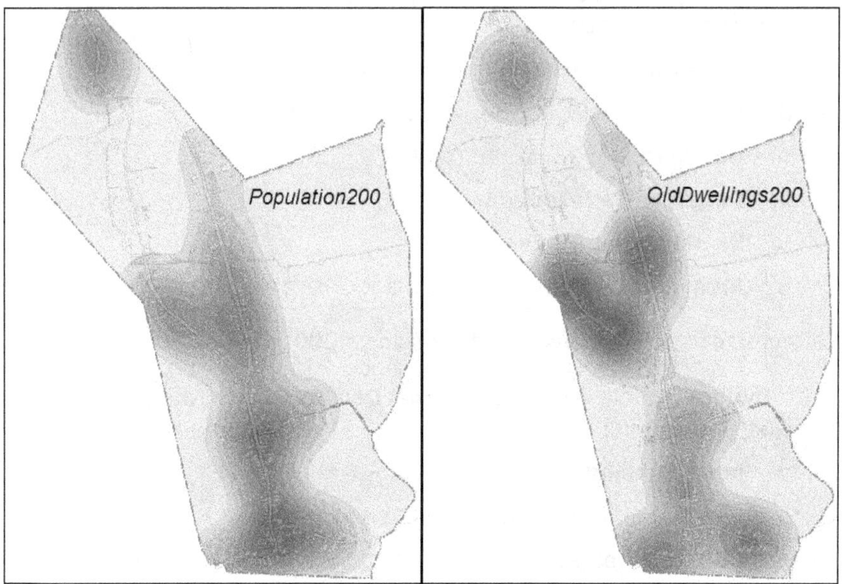

14. Save your map as `MyDensitySurfaces.mxd` and close ArcMap.

You can find the results at `<drive>:\PacktPublishing\Data\SpatialAnalyst \Density`.

How it works...

From steps 4 to 8, you prepared your data for density analysis. For the density surface, we have used a constant value for the radius of the circular search area around the buildings. If you increase the value of the search radius, you will obtain a more generalized surface of density. This might help you to identify particular areas of density or hotspots that are not so obvious at a 200-meter radius.

Analyzing the least-cost path

A least-cost path analysis uses a cost-weighted distance surface and a cost-weighted direction surface to create the smallest cost of travelling path from the source to the destination. The cost refers to one or more factors that could affect the travel. Cost factors, such as slopes or speed-restriction areas, define the cost surfaces. Those cost surfaces can be combined if the cost values use a common scale (for example, from value `1`—best to value `10`—worst). A least-cost path will follow the cells with the smallest accumulated cost.

Getting ready

In the *Analyzing surfaces* recipe, you studied the visibility between two geodetic points in order to plan some topographic surveys. Let's suppose you have two teams for this topographic survey campaign. Now, it is time to plan the trip from your office to the field. Your best routes will follow the roads with shallow slopes. The following model shows the workflow of the next spatial analysis:

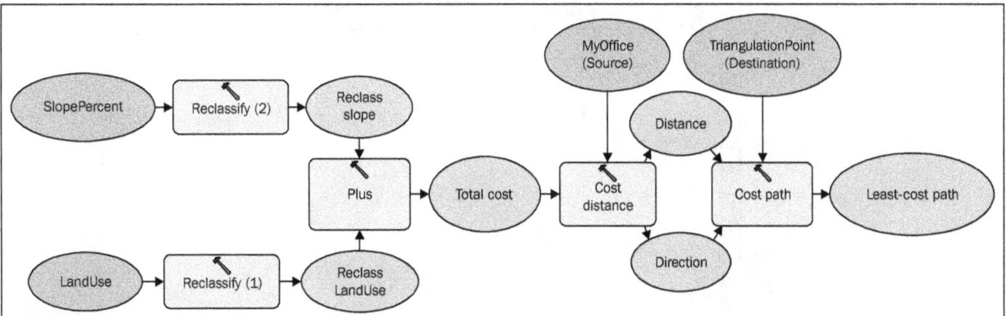

How to do it...

Follow these steps to create two least-cost paths:

1. Start ArcMap and open the existing map document `LeastCostPath.mxd` from `<drive>:\PacktPublishing\Data\SpatialAnalyst\ Least-CostPath`.
2. Go to **Geoprocessing | Environments** and set the geoprocessing environment as follows:
   - **Workspace | Current Workspace**: `Data\SpatialAnalyst\Least-CostPath\TOPO.gdb` and **Scratch Workspace**: `Data\SpatialAnalyst\Least-CostPath \ScratchTOPO.gdb`
   - **Output Coordinates**: Same as **Input**
   - **Processing Extent**: Same as the `Trapezoid5000` layer
   - **Raster | Analysis Cell Size**: Same as the `LandUse` layer

   In the next steps you will reclassify the cell values of the `LandUse` raster layer:

3. In **ArcToolbox**, expand **Spatial Analyst Tools | Reclass**, and double-click on the **Reclassify** tool to open the dialog box. Set the following parameters:
   - **Input Raster**: `LandUse`
   - **Reclass field**: `Value`
   - **Reclassification**: Modify the values of the **New values** column, as shown in the following table:

Old values	New values
11	3
12	4
13	5
14	6
15	7
21	8
31	10
41	1
42	9
51	2
NoData	NoData

4. Click on the **Save** button to save the values in the remap table (information table) named `MyReclassLand`. You can reuse this information table if you want to repeat the step. Continue to set the last two parameters:
   - **Output Raster**: `...ScratchTOPO.gdb\Reclass_Land`
   - Check **Change missing values to NoData (optional)**
5. Click on **OK**. Symbolize the raster layer as follows: **Color Schema** to `Red to Green`. Click on the **Symbols** column to select **Flip Colors**. For **Display NoData as**, select `Black`. Click on **OK**. Open the `Reclass_Land` raster attribute table to inspect the result.

   In the next steps, you will reclassify the cell values of the `SlopePercent` raster layer.
6. Open the **Reclassify** tool, and set the following parameters:
   - **Input Raster**: `SlopePercent`
   - **Reclass field**: `Value`
7. Click on the **Classify** button, and set the following parameters:

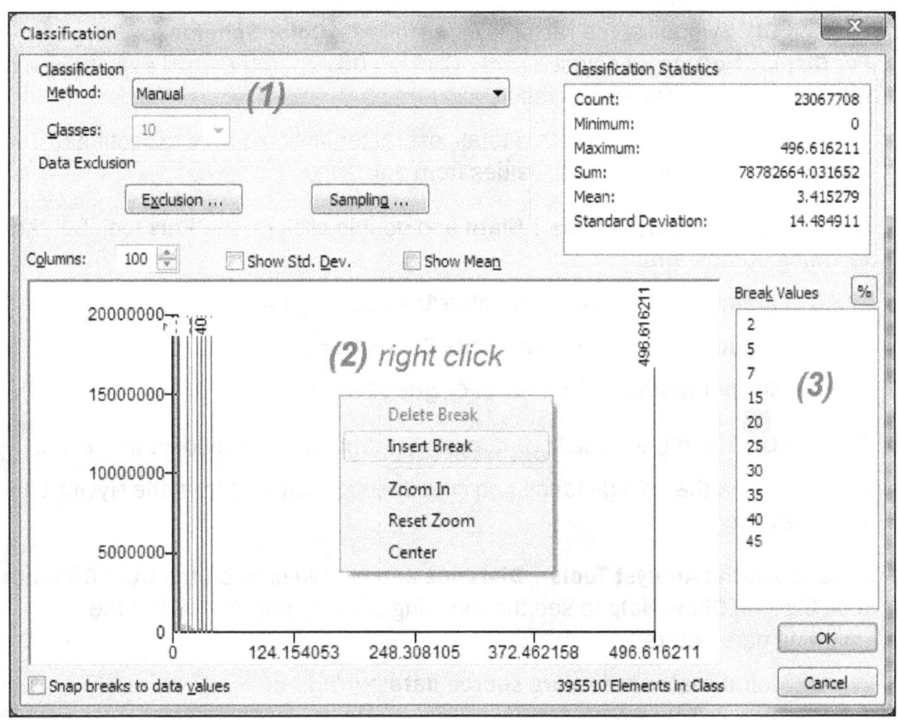

- **Reclassification**: Modify the values of the **New values** column, as shown in the following table:

Old values	New values
0 - 2	1
2 - 5	2
5 - 7	3
7 - 15	4
15 - 25	5
25 - 30	6
30 - 35	7
35 - 40	8
40 - 45	9
45 - 496.616211	10
NoData	NoData

- **Output Raster**: `...ScratchTOPO.gdb\Reclass_Slope`
- Check **Change missing values to NoData (optional)**

8. Click on **OK**. Symbolize the raster layer as follows: **Color Schema** to `Slope`. For **Display NoData as** select `Black`. Click on **OK**. Open the `Reclass_Slope` raster attribute table to inspect the result.

   In the next step, you will create a total-cost raster layer. You are expecting to obtain a total-cost raster layer with cell values from `2` to `20`:

9. Expand **Spatial Analyst Tools** | **Math** and double-click on the **Plus** tool. Set the following parameters:
   - **Input raster or constant value 1**: `Reclass_Land`
   - **Input raster or constant value 2**: `Reclass_Slope`
   - **Output raster**: `ScratchTOPO.gdb\TotalCost`

10. Click on **OK**. Open the `TotalCost` raster attribute table to inspect the result.

    Let's calculate the cost distance and cost direction starting from the `MyOffice` vector layer:

11. Expand **Spatial Analyst Tools** | **Distance** and double-click on the **Cost Distance** tool. Click on **Show Help** to see the meaning of every parameter. Set the following parameters:
    - **Input raster or feature source data**: `MyOffice`
    - **Input cost raster**: `TotalCost`

- **Output distance raster**: `ScratchTOPO.gdb\DistancePoint`
- **Maximum distance (optional)**: `<none>`
- **Output backlink raster (optional)**: `ScratchTOPO.gdb\DirectionPoint`

12. Click on **OK**. Running the **Cost Distance** tool will take you around 15 minutes.

    Let's identify the least-cost path from the office to points `8` and `72`:

13. Expand **Spatial Analyst Tools** | **Distance** and double-click on the **Cost Path tool**. Set the following parameters:
    - **Input raster or feature destination data**: `TriangulationPoint`
    - **Destination field (optional)**: `TIP`
    - **Input cost distance raster**: `DistancePoint`
    - **Input cost backlink raster**: `DirectionPoint`
    - **Output raster**: `ScratchTOPO.gdb\LeastCostPath`
    - **Path type (optional)**: `EACH_CELL`

14. Click on **OK**. You should obtain two least-cost paths, as shown in the following screenshot:

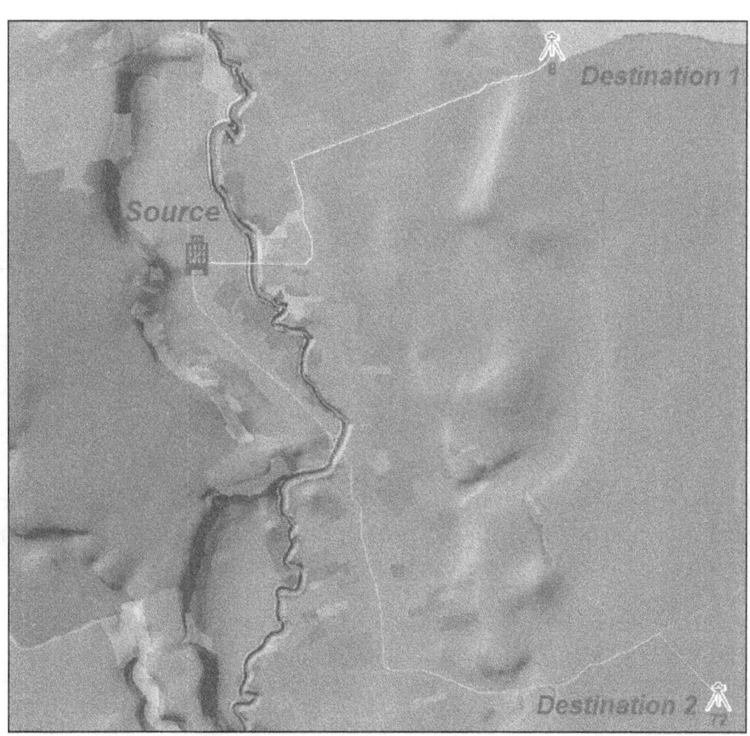

15. Inspect the results. Save your map as `MyLeastCostPath.mxd` and close ArcMap.

You can find the results at `<drive>:\PacktPublishing\Data\SpatialAnalyst \Least-CostPath\Results_LeastCostPath.mxd`.

How it works...

This is how these steps work:

- At step 3, you reclassified cell values for the `LandUse` raster layer. The new value `1` for code `41` means that `Transportation` is the best land use for our path. We will try to avoid crossing the `Hydrography` and `Other Terrains` (for example, built-up area) land use and value `10` indicates the worst choice.
- At step 6, you reclassified the cell values of the `SlopePercent` raster layer using the same scale values as the `LandUse` raster layer. The reclass value `1` corresponds to a shallow slope between `0` and `2` percent. This slope range is the most suitable for our path. Again, value `10` represents the worst slope range for the path.

With the two reclassified layers in a common scale, you created a total cost raster layer at step 9 using the **Plus** tool. At step 11, you calculated *the input for the least-cost path*: cost distance and the cost direction as shown in the following screenshots:

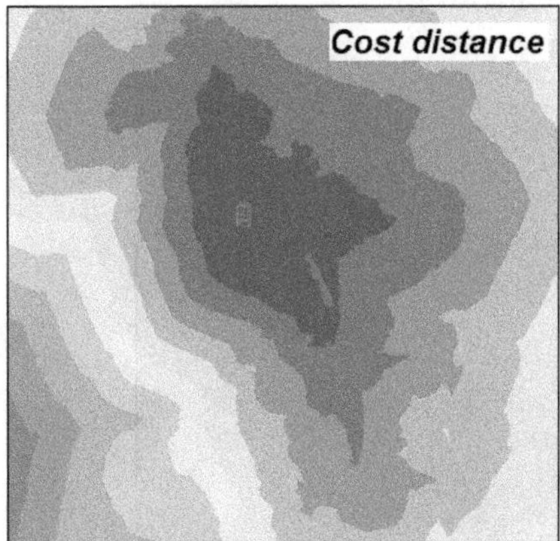

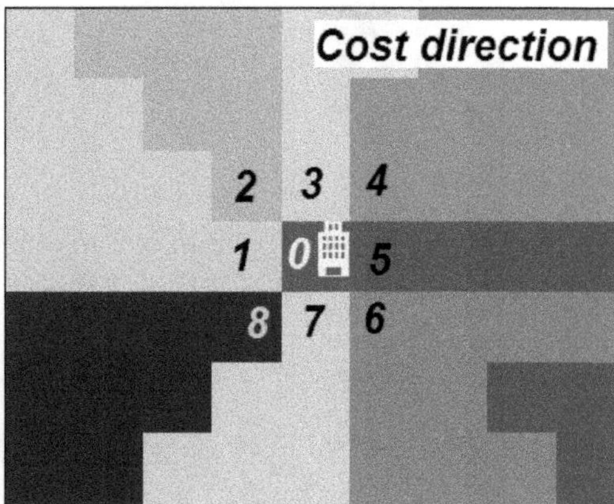

Cost distance values represent accumulated costs and not distance values. Cost direction values range between `1` and `8`. The value `0` represents the source cell. The code from `1` to `8` represents the direction to the least-cost cell back to the source cell.

Going further, you calculated the least-cost path or the shortest path between the `MyOffice` source layer and the `TriangulationPoint` destination layer using the two cost surfaces. The least-cost paths follow the cells with the smallest distance and direction cost, and represent the smallest sum of cell cost between the source and two destinations.

There's more...

Open your previous map document `MyLeastCostPath.mxd`. In the **Catalog** window, go to `Data\SpatialAnalyst\Least-CostPath\TOPO.gdb`. Expand **MyToolbox** and open the **Least Cost Path** model in the **Edit** mode. Explore the model. Notice that the derived raster will be saved at `...SpatialAnalyst\Least-CostPath\TOPO.gdb`. Running the entire model will take you around 20 minutes.

10
Working with 3D Analyst

In this chapter, we will cover the following topics:

- Creating 3D features from 2D features
- Creating a TIN surface
- Creating a terrain dataset
- Creating raster and TIN from a terrain dataset
- Intervisibility
- Creating a profile graph
- Creating an animation

Introduction

The **3D Analyst** extension allows you to display and analyze your 3D data. 3D data means 3D features, rasters, and **Triangulated Irregular Networks** (**TINs**). This 3D data contains a z-value. A z-value refers to an attribute other than the horizontal position (for example, projected *x* and *y* coordinates). In this chapter, your z-value will represent the normal heights (*gravity-related height*) expressed in meters.

The particular term **height** and the general term **elevation** are used interchangeably in this chapter because some ArcGIS instruments and parameters use the term: *elevation*. The term **elevation** means the height above a given level. In particular, our **height** term means the height above the local mean sea level.

According to the international standard ISO 19111:2007 on geographic information—spatial referencing by coordinates:

- A **gravity-related height** is a *height dependent on Earth's gravity field*.
- A **normal height** (H) is a gravity-related height *that approximates the distance of a point above the mean sea level*.
- A **Mean Sea Level** (**MSL**) is the *average level of the surface of the sea over all stages of tide and seasonal variations*. In the local context, MSL is calculated *from observations at one or more points over a given period of time*.

For theory about gravity-related height, please refer to *Jonathan Iliffe* and *Roger Lott, Datums and Map Projections 2nd Edition*; *Whittles Publishing, 2012, pp.30-35 and pp. 135-140.*

In this chapter, you will work in the **ArcScene 3D** environment.

Creating 3D features from 2D features

A 3D feature is a 2D feature that has elevation (height) values (z-values) stored in its geometry. When you are creating a new feature class in your geodatabase, you have from the first panel **Geometry Properties**: **Coordinate include Z values. Used to store 3D data**. In *Chapter 1, Designing Geodatabase*, you created feature classes that store only 2D features.

In this recipe, you will convert the existing simple features (2D) into 3D features using two methods:

- Based on the height values from a raster surface
- Based on the height values from the 2D feature's field attribute

Getting ready

In this recipe, you will work with the **ArcScene** desktop application. ArcScene documents are saved as `.sxd` files. Your scene document will contain:

- A raster surface layer created in *Chapter 9, Working with Spatial Analyst*, in the *Analyzing surfaces* recipe
- Two 2D feature classes: `PointElevation` and `Buildings`

How to do it...

Follow these steps to convert 2D features into 3D features:

1. Start ArcScene and open an existing scene document `3DFeatures.sxd` from `<drive>:\PacktPublishing\Data\3DAnalyst`.
2. From the **Geoprocessing** menu, select **Environments**. Set the geoprocessing environment as follows:
   - **Workspace**:

     Current Workspace: `Data\3DAnalyst\TOPO5000.gdb`

     Scratch Workspace: `Data\3DAnalyst\ScratchTOPO5000.gdb`
   - **Output Coordinates**: **Same as Input**
   - **Processing Extent**: **Same as the Elevation layer**

Let's set base heights for the layers and vertical exaggeration for the scene:

1. In the **Table Of Contents** section, right-click on the `Elevation` raster layer, and navigate to **Properties** | **Base Heights**. Check **Floating on a custom surface**: `Elevation`. Click on **Raster Resolution**. For **Base Surface**, set **Cellsize X** to `2.5` and **Cellsize Y** to `2.5`. Notice that the number of **Rows** and **Columns** is automatically updated. By decreasing the display resolution, you will improve the display performance. Click on **OK**.
2. Click on the **Rendering** tab, check **Render layer at all times**, and type `0.05` seconds for **Draw simpler level of detail if navigation refresh rate exceeds**.

   The ArcScene redraws or renders the layers every time you make a navigation movement. By decreasing the refresh rate, ArcScene will only have time to draw a blue skeleton of your raster layer while you are exploring the scene with the **Navigate** tool. When you stop navigating the scene, the raster layer will render normally.
3. In the **Effects** section, choose `10` for the **Select the drawing priority** option. The **Elevation** layer will have the lowest drawing priority in comparison with other layers from the scene document. The highest priority is value `1`. Click on **Apply** and **OK**.

4. Set the vertical exaggeration for the scene. From the **View** menu, navigate to **Scene Properties** | **General**. Select `2` for **Vertical Exaggeration**. Click on **OK**. Try to navigate around the scene using the **Navigate** tool, as shown in the following screenshot:

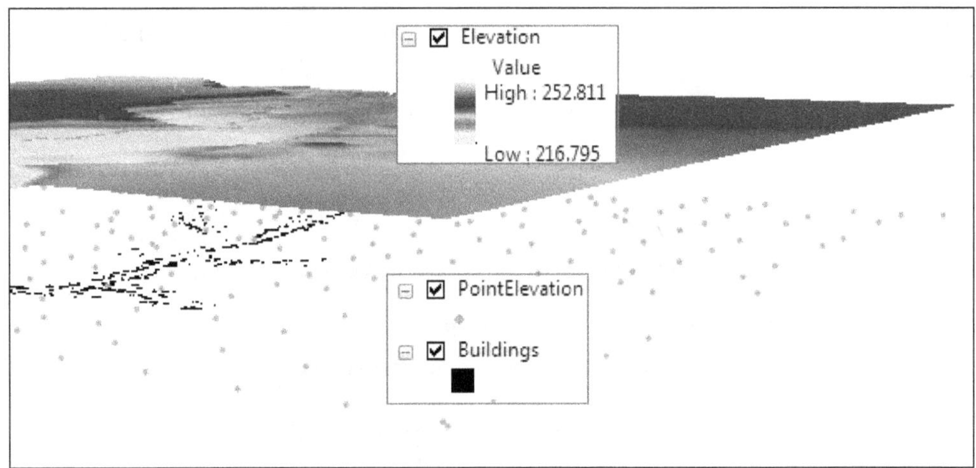

5. In ArcScene, 2D feature layers have their base height set to `0` by default. The `PointElevation` and `Buildings` 2D feature layers are displayed below the `Elevation` layer. Repeat step 3 for the `PointElevation` feature layer using the same surface base heights. For **Select the drawing priority** option, type `1`. Set the base height for the `Buildings` feature layer and for the **Select the drawing priority** option, choose `5`.

   Your 2D feature layers have been added to the `Elevation` surface. Turn off the 2D feature layers. Let's convert the `PointElevation` 2D feature layers to a 3D feature layer:

6. In **ArcToolbox**, expand **3D Analyst Tools** | **3D Features**, and double-click on the **Feature to 3D by Attributes** tool. Click on **Show Help** to see the meaning of every parameter. Set the following parameters:

   - **Input Features**: `TOPO5000.gdb\Relief\PointElevation`
   - **Output Feature Class**: `ScratchTOPO5000.gdb\ PointElevation3D`
   - **Height Field**: `Elevation`
   - **To Height Field (optional)**: `<none>`

7. Click on **OK**. Notice that the `PointElevation3D` feature layer is directly displayed in the 3D space based on the z-value in the point geometry. Open its attribute table. The **Shape** field stores the `Point Z` geometry.

   Let's convert the `Building` 2D polygon feature layers to a 3D feature layer using the `Elevation` surface raster, as follows:

8. In **ArcToolbox**, expand **3D Analyst Tools** | **Functional Surface**, and double-click on the **Interpolate Shape** tool. Click on **Show Help** to see the meaning of every parameter. Set the following parameters:
    - **Input Surface**: `TOPO5000.gdb\Elevation`
    - **Input Feature Class**: `TOPO5000.gdb\Buildings\Buildings`
    - **Output Feature Class**: `...\3DAnalyst\ScratchTOPO5000.gdb\Buildings3D`
    - Check **Interpolate Vertices Only (optional)**
9. Accept the default values for all other parameters. Click on **OK**.
10. Note that the `Buildings3D` feature layer is directly displayed in the 3D space based on the z-value into the polygon geometry. Open the attribute table of the `Buildings3D` feature. The **Shape** field stores the `Polygon Z` geometry.

    Let's improve the 3D appearance of buildings by extruding the polygon features according to the number of levels, as follows:
11. In the **Table Of Contents** section, right-click on the `Buildings3D` layer, and navigate to **Properties** | **Extrusion**. Set the following parameters:
    - Check **Extrude features in layer**
    - **Extrusion value or expression**: `[Stories]*3`
    - **Apply extrusion by**: `adding it to each feature's minimum height`
12. Click on **OK** and explore the results, as shown in the following screenshot:

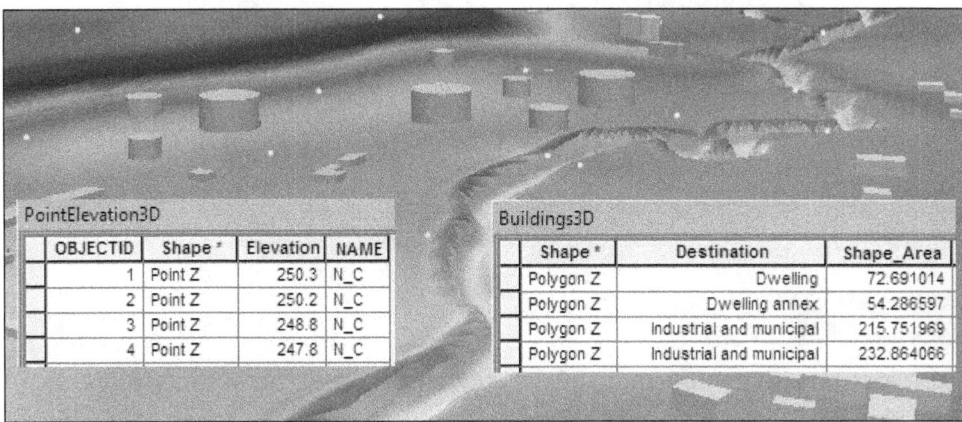

PointElevation3D

OBJECTID	Shape *	Elevation	NAME
1	Point Z	250.3	N_C
2	Point Z	250.2	N_C
3	Point Z	248.8	N_C
4	Point Z	247.8	N_C

Buildings3D

Shape *	Destination	Shape_Area
Polygon Z	Dwelling	72.691014
Polygon Z	Dwelling annex	54.286597
Polygon Z	Industrial and municipal	215.751969
Polygon Z	Industrial and municipal	232.864066

13. Save your scene document as `My3DFeatures.sxd` and close ArcScene.

You can find the final results at `<drive>:\PacktPublishing\Data\3DAnalyst\3DFeatures`.

How it works...

At step 6, you assigned z-values to point features directly from the **Elevation** field. The `Z` from the **Shape** geometry field confirms the 3D feature type.

At step 8, the raster cell values were interpolated to the assigned z-values to every vertex of the 3D polygon feature.

For the `PointElevation3D` and `Buildings3D` layers, it's not necessary anymore to set the base heights (topographic surface). Note also that subtypes and domains were kept in the new 3D feature classes.

At step 11, you raised the building from the ground using the number of levels multiplied by `3`. The value `3` could represent the height of one level expressed in meters. From the **Tools** toolbar, use **Measure tool** | **Measure Height** to measure the reservoir height above the topographic surface, as shown in the following screenshot:

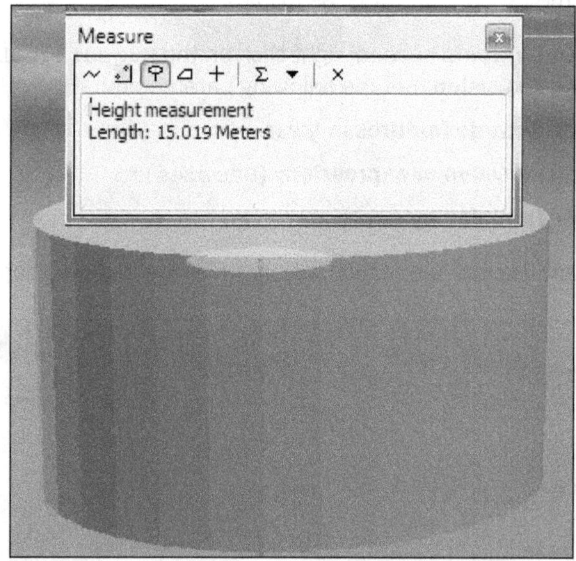

See also

- To create a 3D surface using 2D features, please refer to the *Creating a TIN surface* recipe

Creating a TIN surface

A **Triangulated Irregular Network** (**TIN**) is mesh of non-overlapping triangles that represent a continuous surface. This surface is built by triangulating a set of points that have z-values using the **Delaunay triangulation** method. This method generates triangles that are as near as possible to an equilateral triangle. The set of points represent the primary source of height values (mass points) and incorporates point features or vertices of the polyline and polygon features.

For theoretical aspects about TIN and Delaunay triangulation, please refer to *David O'Sullivan* and *David Unwin, Geographic Information Analysis, John Wiley & Sons, Inc., 2003, page 28 and 215 to 220.*

Getting ready

In this recipe, you will create a TIN surface using 2D feature classes. You will also update the TIN surface with new, upcoming features, such as building footprints and a fish pond.

How to do it...

Follow these steps to create a TIN surface using the ArcScene environment:

1. Start ArcScene and open an existing scene document `TIN.sxd` from `<drive>:\PacktPublishing\Data\3DAnalyst`.

   Keep the geoprocessing environment from the previous recipe. From the **Geoprocessing** menu, select **Geoprocessing Options**. Check whether **Background Processing | Enable** is unchecked to ensure that your process will run in the foreground. By selecting this option, you will not allow other work until the tool stops running.

2. In **ArcToolbox**, expand **3D Analyst Tools | Data Management | TIN**, and double-click on the **Create TIN** tool. Set the following parameters:
   - **Output TIN**: `3DAnalyst\TIN\TOPO5k.gdb`
   - **Coordinate System**: `Pulkovo_1942_Adj_58_Stereo_70`
   - **Input Feature Class** as shown in the following screenshot:

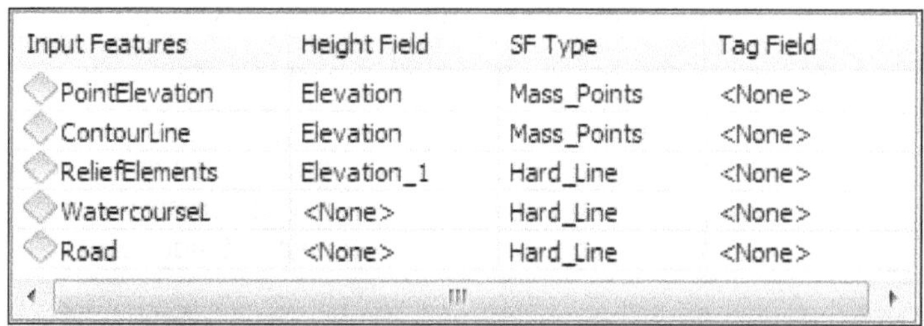

Input Features	Height Field	SF Type	Tag Field
PointElevation	Elevation	Mass_Points	<None>
ContourLine	Elevation	Mass_Points	<None>
ReliefElements	Elevation_1	Hard_Line	<None>
WatercourseL	<None>	Hard_Line	<None>
Road	<None>	Hard_Line	<None>

In the next step, you will create a 3D feature class from the 2D feature class `Buildings` using this time `Topo5k` TIN as the height (elevation) source surface:

3. In **ArcToolbox**, expand **3D Analyst Tools | Functional Surface**, and double-click on the **Interpolate Shape** tool. Set the following parameters:
   - **Input Surface**: `TOPO5k`
   - **Input Feature Class**: `TOPO5000.gdb\Buildings\Buildings`
   - **Output Feature Class**: `ScratchTOPO5000.gdb\Buildings3DFromTIN`
   - **Method (optional)**: `LINEAR`
   - Check **Interpolate Vertices Only (optional)**

4. Click on **OK**. In the **Table Of Contents** section, move the `BuildingsFromTIN` layer above all layers.

   Let's update the TIN with buildings from the `BuildingsFromTIN` layer and with a fish pond from the `Lake` layer.

5. In **ArcToolbox**, expand **3D Analyst Tools | Data Management | TIN**, and double-click on the **Edit TIN** tool. Set the following parameters:
   - **Input TIN**: `3DAnalyst\TIN\TOPO5k.gdb`
   - **Input Feature Class** as shown in the following screenshot:

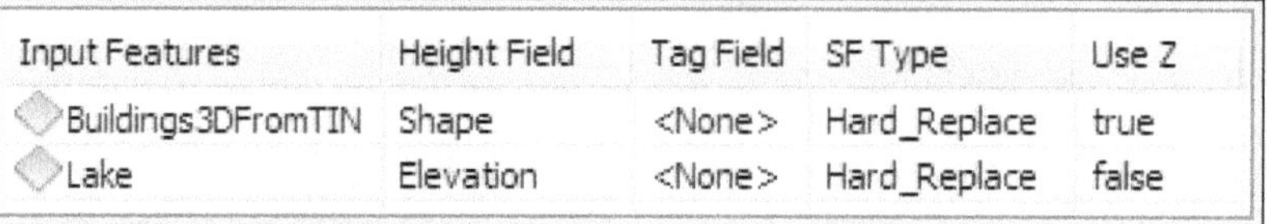

Input Features	Height Field	Tag Field	SF Type	Use Z
Buildings3DFromTIN	Shape	<None>	Hard_Replace	true
Lake	Elevation	<None>	Hard_Replace	false

6. Uncheck **Constrained Delaunay (optional)** and click on **OK**.
7. In the **Table Of Contents** section, turn off all layers except `Topo5k TIN`, `BuildingsFromTIN`, and `Lake`. For the `Lake` feature layer, set `Topo5k` TIN as the base height, and for `Layer` offset, type `0.5`. For the `BuildingsFromTIN` feature layer, set **Extrusion** | **Extrusion value or expression** to `[Stories]*2`. For the `Topo5k` TIN, go to **Layer Properties** | **Symbology**, and uncheck **Edge types**, and for **Elevation**, choose the `Surface` color ramp.
8. From **View** | **Scene Properties** | **General**, set **Vertical Exaggeration** to `2`. You should obtain the following results:

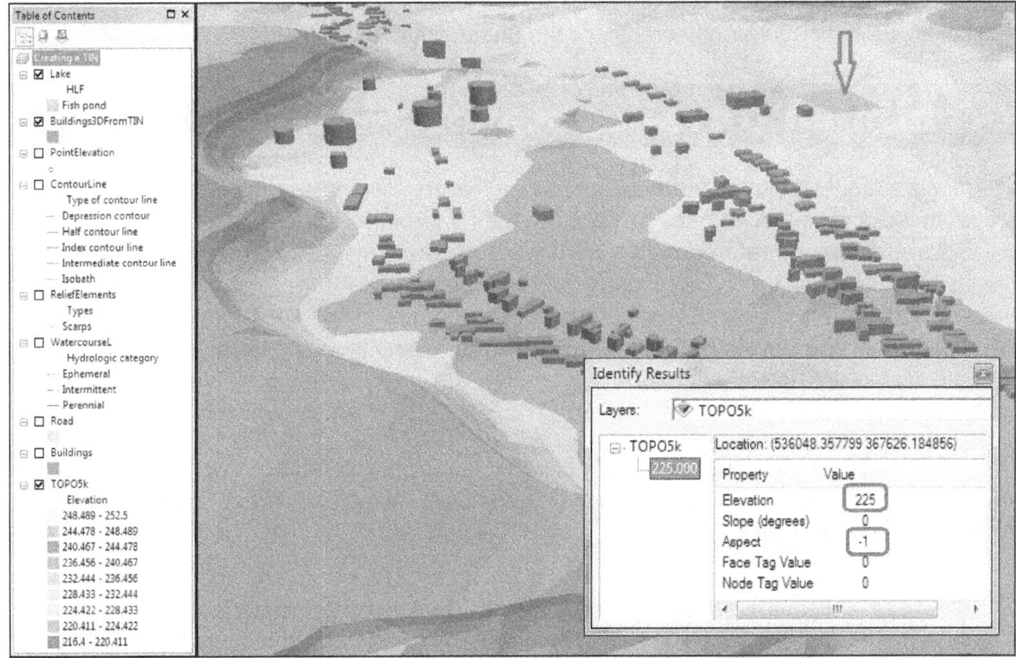

9. Inspect the results. Using the **Identify** tool, click on the fishpond from the `BuildingsFromTIN` layer to get the following information. `Aspect` with values `-1` means a flat area.
10. Save your scene document as `MyTIN.sxd`, and close ArcScene.

You can find the results at `<drive>:\PacktPublishing\Data\3DAnalyst\TIN`.

How it works...

The elevation (height) points and vertices of the contour lines will be used as mass points to define the TIN surface. You have used scarps, water courses, and roads as hard lines; therefore, the TIN triangles will not cross the linear features. Leave unchecked the **Constrained Delaunay (optional)** option; you will allow the Delaunay triangulation process to build multiple triangle edges on the hard lines (breaklines).

At step 5, you updated the TIN surface with flat areas with constant height values. The fishpond from the `Lake` 2D feature layer has the value `225` meters stored in the **Elevation** field. The building footprints from the `BuildingsFromTIN` 3D feature layer have z-values incorporated into the polygon geometry or the **Shape** field.

Creating a terrain dataset

A terrain is a multi-resolution surface stored as a feature class in a feature dataset within a geodatabase. A terrain does not actually store the represented surface unlike TINs or rasters, which do. A terrain only references the participating feature classes and derives TINs at various resolutions on the fly, depending on the display scale. A terrain must be created within a feature dataset together with the feature classes that are used in it. That way, the **terrain**, **terrain surface**, and **terrain dataset** terms are often used interchangeably. You cannot see a terrain surface in the ArcScene environment, but you can use the ArcMap and ArcGlobe applications.

The terrain is used for large amounts of data. In this recipe, you will work with **Light Detection and Ranging** (**LiDAR**) data. LiDAR is an optical remote-sensing technology that provides detailed 3D elevation data.

For further research about LiDAR remote-sensing technology, please refer to *George Vosselman* and *Hand-Gerd Maas, Airborne and Terrestrial Laser Scanning, Whittles Publishing, 2010.*

Getting ready

In this recipe, you will create a terrain using an ASCII text file that stores LiDAR point-cloud data. Your point cloud contains 5 million points. Those points represent a **Digital Terrain Model** (**DTM**) or ground data.

In Windows Explorer, go to the `<drive>:\PacktPublishing\Data\3DAnalyst\LiDAR` folder. Explore the `dtm.xyz` ASCII file with the Notepad application.

How to do it...

Follow these steps to create a terrain dataset:

1. Start ArcCatalog and go to **Customize** | **Extensions** to check the 3D Analyst extension. In **Catalog Tree**, go to `<drive>:\PacktPublishing\Data\3DAnalyst\TOPO5000.gdb`.
2. Your geodatabase already contains an empty feature dataset named `Terrain`. Right-click on the `Terrain` feature dataset and choose **Properties** | **XY Coordinate System**. The current projected coordinate reference system is `Pulkovo_1942_Adj_58_Stereo_70` with the `EPSG` code as `3844`. Click on **Z Coordinate System**. The vertical coordinate reference system is a national gravity-related height system called **Constanta (Black Sea 1975)** with the `EPSG` code as `5781`.

   In the next steps, you will calculate the basic statistics of `dtm.xyz`, and convert mass points from the ASCII file to a multipoint feature class:

3. In **ArcToolbox**, expand **3D Analyst Tools** | **Conversion** | **From File**, and double-click on the **Point File Information** tool. Set the following parameters:
   - **Point Data Browse for Files**: `...\3DAnalyst\LiDAR\dtm.xyz`
   - **Output Feature Class**: `...\3DAnalyst\TOPO5000.gdb\Terrain\DTMInfo`
   - **File Format**: `XYZ`
   - **File Suffix**: `xyz`:

     Coordinate System (optional) | **XY Coordinate System**: **National Grids** | **Europe**: `Pulkovo_1942_Adj_58_Stereo_70`

     Z Coordinate System: **Europe**: `Constanta`
4. Accept the default values for all other parameters. Click on **OK**. You have obtained a polygon feature that defines the XY extent of your LiDAR data.
5. Open ArcMap. Add the `DTMInfo` feature class to your map. Save your map document as `MyTerrain.mxd` in the `...Data\3DAnalyst` folder. Open the `DTMInfo` attribute table to view the following statistical information: `Pt_Count`: `5,722,782` points; `Pt_Spacing`: `1.005`; `Z_Min`: `214.77`, and `Z_Max`: `250.85`. Close the table and turn off the `DTMInfo` layer.
6. In **ArcToolbox**, expand **3D Analyst Tools** | **Conversion** | **From File**, and double-click on the **ASCII 3D to Feature Class** tool. Set the following parameters:
   - **Point Data Browse for Files**: `...\3DAnalyst\LiDAR\dtm.xyz`
   - **File Format**: `XYZ`
   - **Output Feature Class**: `...\3DAnalyst\TOPO5000.gdb\Terrain\MassPoints`

- **Output Feature Class Type**: MULTIPOINT
- **Coordinate System (optional)** | **XY Coordinate System**: **National Grids** | **Europe** | Pulkovo_1942_Adj_58_Stereo_70

  Z Coordinate System: **Europe** | Constanta
- **Average Point Spacing (optional)**: 1.005
- **File Suffix (optional)**: xyz

7. Accept the default values for all other parameters. Click on **OK**. Use the **Zoom In** or the **Magnifier** tool to look closely at the points, as shown in the following screenshot:

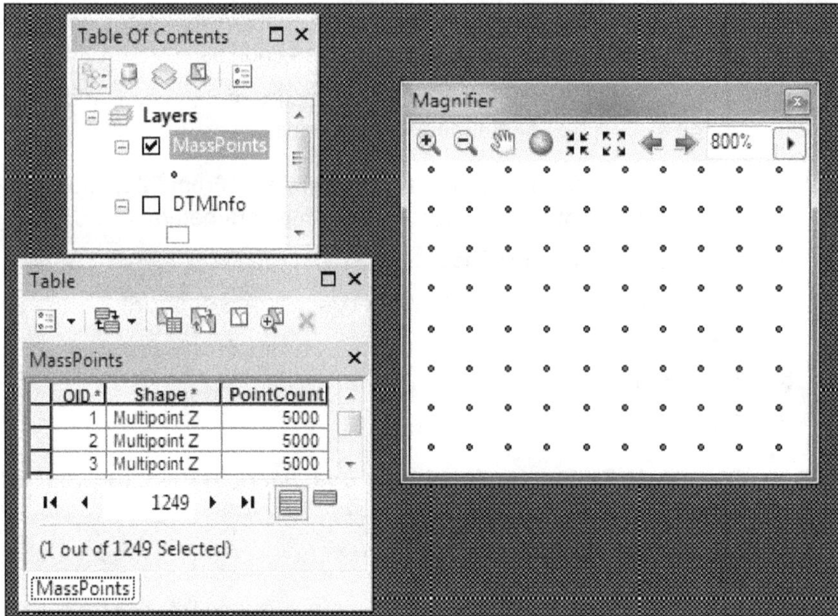

Let's create the terrain surface:

8. In the **Catalog** window, right-click on the TOPO5000.gdb\Terrain feature dataset and navigate to **New** | **Terrain**. Set the following parameters:
   - **Enter a name for your terrain**: DTM
   - Check MassPoints and DTMInfo
   - **Approximate point spacing**: 1.005

9. Click on **Next**. Click on **Advanced**. Set the **Embedded** option to **Yes**. Notice the terrain fields. Click on **Next** and set the following parameters:
    - **Point selection method**: `Z Mean`
    - **Secondary thinning method**: `Moderate`
    - **Secondary thinning threshold**: `0.5`
10. Click on **Next.** Click on **Calculate Pyramid Properties**. **Adjust Terrain Pyramid Levels**, as shown in the following screenshot:

No.	Window Size	Maximum Scale
1	1	1000
2	5	2500
3	10	5000
4	20	10000

11. Click on **Next**. Inspect the parameter's summary. Click on **Finish**. Click on **Yes** to build the terrain. Add the `DTM` terrain to your map. Let's explore the terrain at different scales. Add the `Lake` feature class in your map document and **Zoom In** to the fishpond. Turn off the `Lake` layer to see the terrain surface. Display and explore the area of interest at different scales, as shown in the following screenshots:

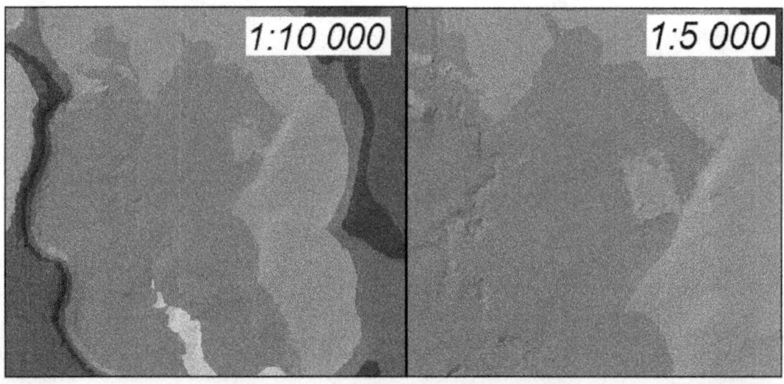

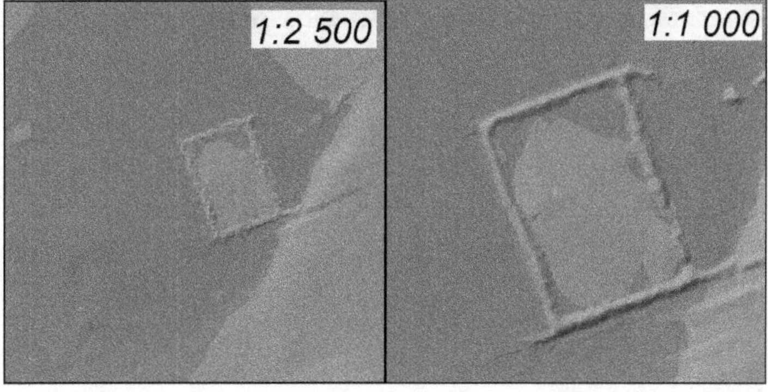

12. Let's examine the properties of the `DTM` terrain layer. Right-click on the `DTM` layer and choose **Properties** | **Display**. Let's reduce the default number of points used in the rendering process. From the **Apply point limit** value, erase a zero to obtain the point `80,000`. If you want to see the height when you point to the map, check **Show Map Tips**. Click on **OK**. Display and explore the area of interest at the scale `5,000` and `2,500`. Apparently, there is no difference. But how about a `1:1,000` scale?
13. Let's display the contour lines for the `DTM` terrain. Right-click on the `DTM` layer and choose **Properties** | **Symbology**. In the **Show** section, click on the **Add** button, and select **Contour with the same symbol**. Click on **Add** and **Dismiss**. Click on the **Import** button to import the symbol and parameter values from `...\3D Analyst\Terrain\DTM.lyr`. You should obtain the following results:

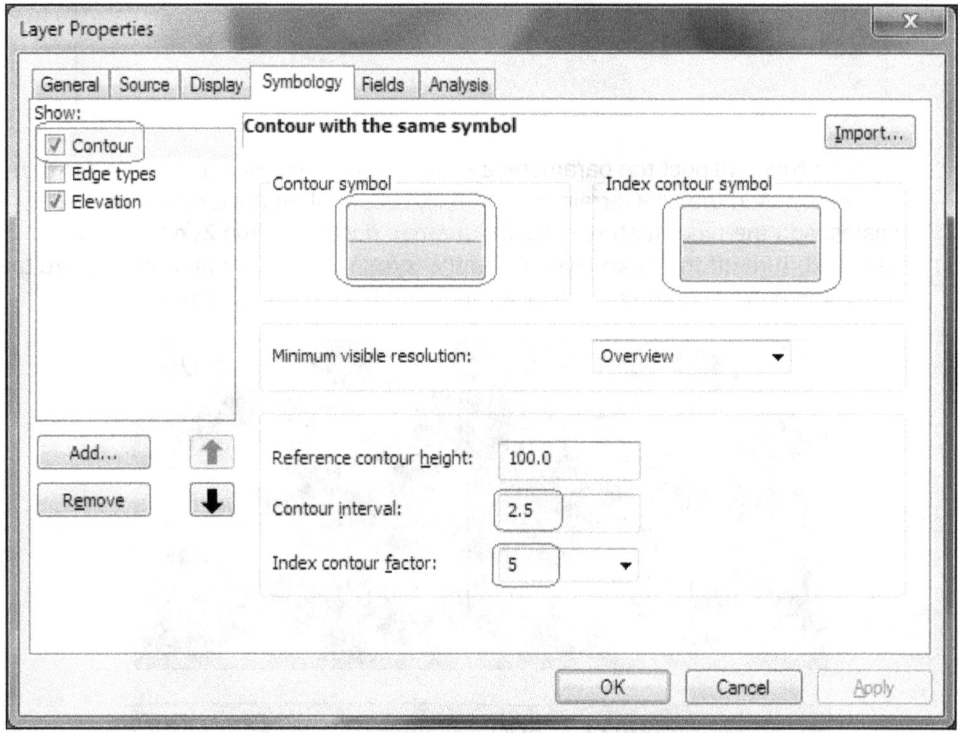

14. Click on **OK**. Inspect the results. If you add the `ContourLine` 2D feature class, you can explore the difference between the topographic map surface at the scale `1:5,000` and a new terrain surface, as shown in the following screenshot:

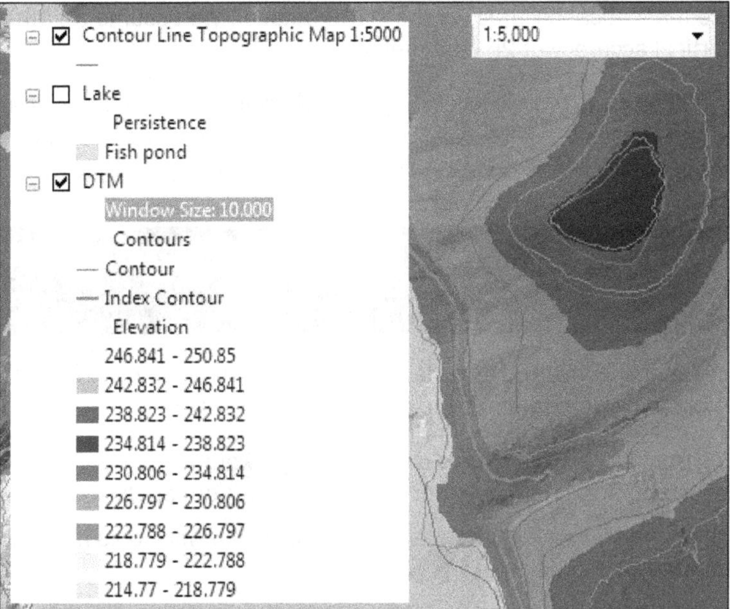

15. In the **Table Of Contents** section, note that the **Window Size** label values is updated when you change the display scale. Save your map and close ArcMap.

You can find the results at `...\3DAnalyst\Terrain\Terrain.mxd`.

How it works...

Before starting to load your LiDAR data into your geodatabase, you evaluated the point density. Because your ACSII file contains around five million records, you grouped the LiDAR points into *multipoint* features. The multipoint geometry reduces the size of the `MassPoints` output feature class and the number of rows in the attribute table.

The **Window Size** pyramid will divide the data extent into equal windows and will select the representative points with `Z Mean` (a z-value closest to the mean). You choose this option because you want to avoid the extreme values. The resultant surface will be integrated in our product, which is a topographic map at the scale `1:5,000`.

From steps 9 to 10, you defined the rules for display terrain (scalability). The **Moderate** secondary thinning method will eliminate irrelevant points but will preserve the linear discontinuities from your surface. If z-values in the window area are within 0.5 meters, then it will be considered a flat area and will result in a more generalized surface. The 0.5 meters represents twice the vertical accuracy of the LiDAR dataset. We assume that the vendor mentioned a vertical accuracy of 25 centimeters.

At step 10, you defined four pyramid levels. Those levels define the vertical accuracy of the terrain, through four ranges of scales. Note that **Level 1** of **Terrain Pyramid** has a **Window Size** of 1 meter, which means that the full surface resolution will be displayed up to scale `1:1,000`.

Creating raster and TIN from a terrain dataset

A terrain dataset can be converted to rasters or TINs using the geoprocessing tools from the **3D Analyst** toolbar. Why would we need to convert terrain to raster or TIN?

- Starting from large-scale source data, you might want to obtain derivative vector or raster surfaces for small-scale (for example `1:25,000`) applications.
- You might need a raster surface to use statistical analysis tools from the **Spatial Analyst** extension.
- You might want to refine and update the surface with new polygon or polyline features having a precision corresponding to the `1:3,000` scale. You will need a TIN with a vertical accuracy corresponding to that scale.
- You might want to visualize your surfaces in ArcScene.
- You might want to store the raster and TIN directly onto disk.

Getting ready

In this recipe, you will generate a raster and a TIN based on the `DTM` terrain surface created in the previous recipe.

You can start from your map document `MyTerrain.mxd` and skip step 1. Otherwise, you will use the terrain surface from `...\Data\3DAnalyst\Terrain\TOPO5000.gdb`.

How to do it...

Follow these steps to convert a terrain dataset to raster and TIN using ArcMap:

1. Start ArcMap and open an existing map document, `Terrain_ToRasterTIN.mxd`, from `...\3DAnalyst\Terrain`.
2. For the geoprocessing environment, set **Current Workspace** and **Scratch Workspace**: `TOPO5000.gdb\Terrain`. For **Processing Extent**, choose **Union of Inputs**.
3. In **ArcToolbox**, expand **3D Analyst Tools** | **Conversion** | **From Terrain**, and double-click on the **Terrain to Raster** tool. Set the parameters, as shown in the following screenshot:

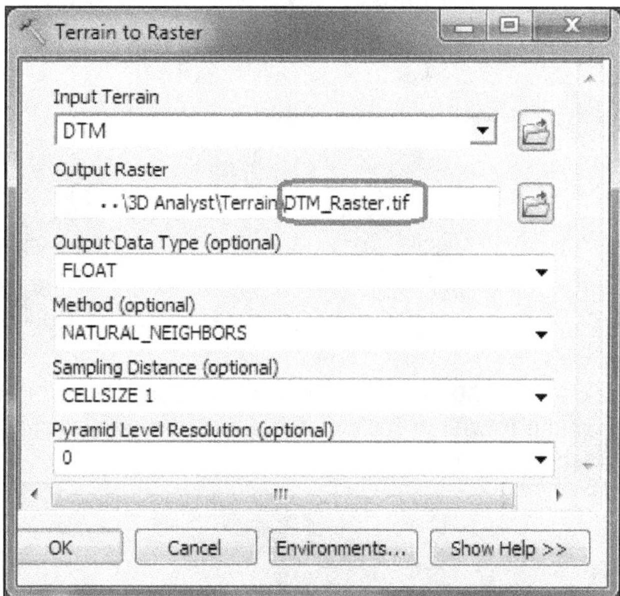

4. Click on **OK**. In the **Table Of Contents** section, drag the `DTM_Raster.tif` surface above the `DTM` terrain surface. Change the symbology from `Black to White` to `Elevation#1 Color Ramp`. In the **Effects** toolbar, select `DTM_Raster.tif` as the layer drawing. Inspect the result using the **Swipe** tool, as shown in the following screenshot:

5. In **ArcToolbox**, expand **3D Analyst Tools** | **Conversion** | **From Terrain**, and double-click on the **Terrain to TIN** tool. Set the parameters, as shown in the following screenshot:

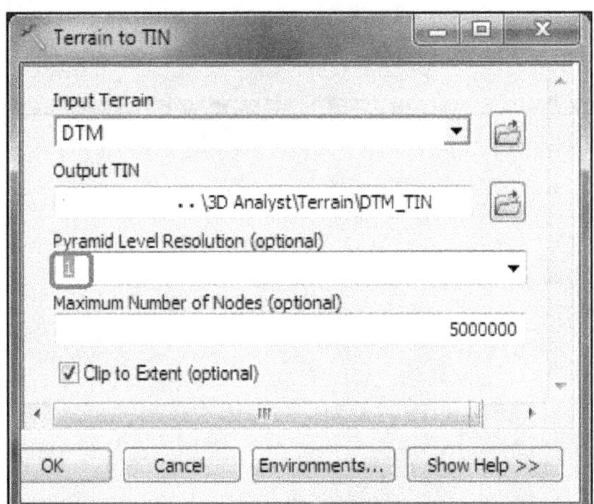

6. Click on **OK**. In the **Table Of Contents** section, drag the `DEM_TIN` surface above the `DTM` raster surface. Inspect the result in the same manner as the step 4 screenshot.
7. Save your map as `MyTerrain_ToRasterTIN.mxd` and close ArcMap.

You can find the final results at `...\Data\3DAnalyst\Terrain\ ResultsTerrain_ToRasterTIN.mxd`.

How it works...

At step 3, you converted the terrain surface to a raster surface in the TIFF format. You used the **natural neighbors** interpolation method to obtain a smoother raster surface. Notice that you have to mention the `.tif` file extension to create a raster in the TIFF format. To define the cell size of your output raster, you mentioned **CELLSIZE** as `1`. You will not improve the accuracy having a cell size smaller than the average point spacing of your initial LiDAR data. You have kept the full resolution of the input terrain by accepting the value `0` for the **Pyramid Level Resolution** parameter.

At step 5, you converted the terrain surface to a TIN surface. If you keep the value `0` for **Pyramid Level Resolution**, your geoprocessing tool will fail and display the message, **The TIN will be too big**. This is because of the allowed number of mass points in a TIN—up to `5,000,000`. You chose **Level 1** and accepted the default number of nodes.

Intervisibility

Intervisibility refers to the analysis of a *sight line* between an observer and a target feature *through potential obstructions* such as 3D surfaces or features.

Getting ready

We continue to plan the topographic survey campaign mentioned in *Chapter 9, Working with Spatial Analyst*, in the *Analyzing surfaces* and *Analyzing the least-cost path* recipes. In this recipe, you will analyze the visibility between three geodetic points and buildings (reservoirs and two churches).

`TriangulationPoint3D` has an attribute field named **SurveyInstruments** that stores the normal height (H) of the instrument calculated as the sum of the **Height Black Sea'75** field and a constant value of `1.5` meters (height of the survey instrument).

`Buildings3D` has a field named **H_Buildings** that stores the height calculated as the sum of the **Z_Mean**, the normal height (H) field, and the approximate height of the building. The **Z_Mean** field has been calculated using the **Add Z Informative** tool.

The `Buildings3D` feature and the `Elevation` surface will be used as potential obstructions.

How to do it...

Follow these steps to analyze the visibility between the 3D triangulation points and the polygon buildings, as shown in the following screenshot:

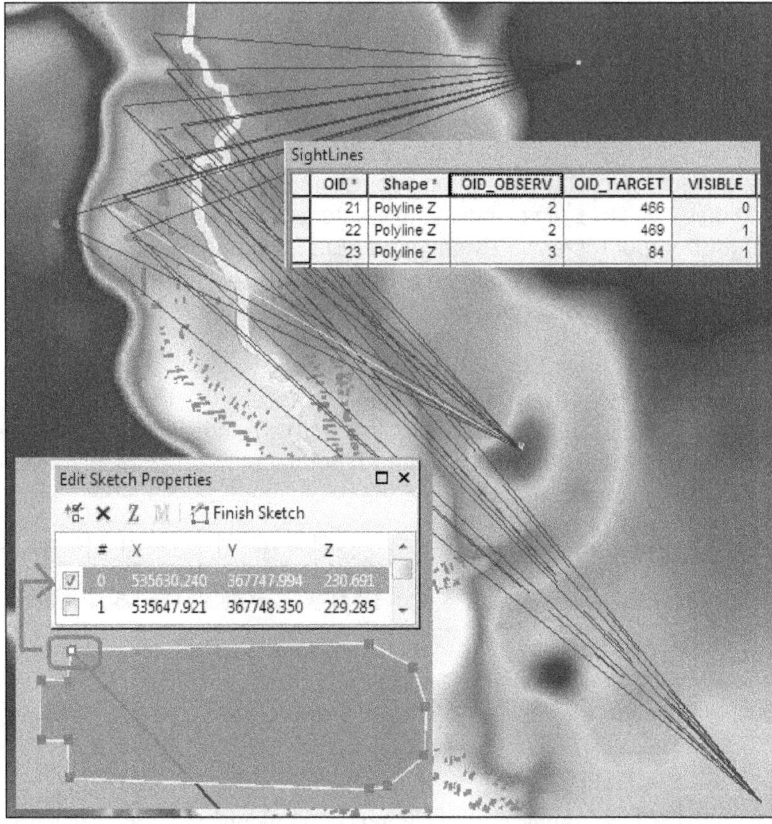

1. Start ArcScene and open the scene document, `Intervisibility.sxd`, from `<drive>:\PacktPublishing\Data\3DAnalyst\Intervisibility`.
2. Firstly, you will select the reservoirs and two churches. Open the `Buildings3D` attribute table. Open the **Select By Attributes** tool and type the expression: `CFC = 319 OR CFC = 27`. Click on **Apply**.
3. Only selected features will be used as target polygon features. In **ArcToolbox**, go to **3D Analyst Tools | Visibility**, and double-click on the **Construct Sight Lines** tool. Set the following parameters:
    - **Observer Points**: `TriangulationPoint3D`
    - **Target Features**: `Buildings3D`

- **Output**: ...\Intervisibility\TOPO5000.gdb\SightLines
- **Observer Height Field (optional)**: SurveyInstrument
- **Target Height Field (optional)**: H_Building
- **Join Field (optional)**: <none>
- **Sampling Distance (optional)**: 100
- Check **Output the Direction (optional)**

4. Click on **OK**. Open the SightLines attribute table to inspect the fields. Based on those 33 sight lines, you will analyze the visibility. In **ArcToolbox**, go to **3D Analyst Tools** | **Visibility**, and double-click on the **Intervisibility** tool. This tool will modify the input SightLines layer. Set the following parameters:
   - **Sight Lines**: SightLines
   - **Obstructions**: For this, add the Buildings3D and Elevation layers
5. Accept the default values for **Visible Field Name (optional)**. Click on **OK**. The SightLines attribute table has been updated with a new field: **VISIBLE**.

   There are 7 sight lines with **VISIBLE** value 0, which means they are obstructed by the elevation surface or buildings. The remaining 26 sight lines with value 1 are not obstructed, and there is visibility between the observer and target features.

6. Save your scene as MyIntervisibility.sxd and close ArcScene.

You can find the final results at <drive>:\PacktPublishing\Data\3DAnalyst \Intervisibility\Results_Intervisibility.sxd.

How it works...

Please note that the result of the **Construct Sight Lines** tool is a 3D polyline layer. By setting the value 100 for **Sampling Distance (optional)**, we kept only one sight line between the triangulation points and the *first vertex* of every selected building feature, as shown in the previous image. If you type 0, a sight line will be created for every vertex of a building polygon.

Creating a profile graph

A profile graph shows the height changes along a given line. A profile graph can be created from:

- 3D line graphics (created with the **Steepest path** and **Line of Sight** tools)
- 3D polyline features (for example, geodatabase feature class and shapefile)

Getting ready

In this recipe, you will learn how to generate a profile graph using:

- 3D polylines from the `SightLines3D` feature class
- `Elevation` raster as a topographic surface

To generate a profile graph, you have to use the ArcMap environment and the **3D Analyst** toolbar.

How to do it...

Follow these steps to generate profiles using the `SightLines3D` 3D polyline layers, as shown in the following screenshot:

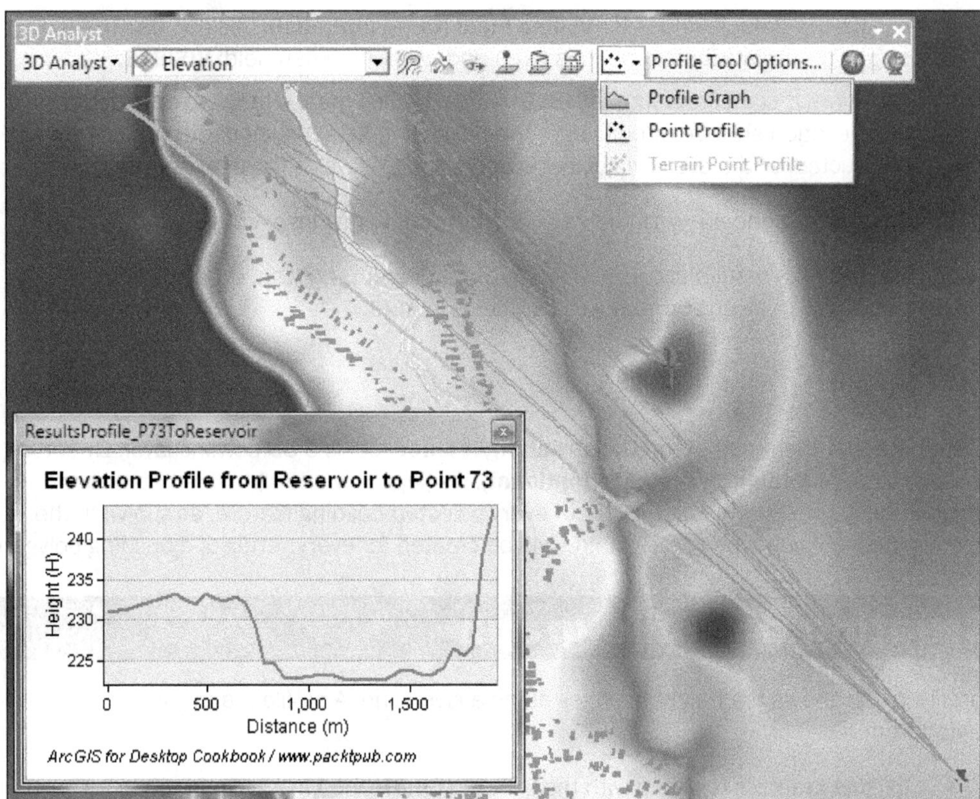

1. Start ArcMap and open an existing map document, `Profile.mxd`, from `<drive>:\PacktPublishing\Data\3DAnalyst\ElevationProfile`.
2. Load the **3D Analyst** extension and add the `3D Analyst` toolbar.
3. With the **Select Feature** tool, select a polyline feature from the `SightLines3D` layer. Click on the **Profile Graph** tool from the **3D Analyst** toolbar. The z-value is represented along the *Y* axis and distance is represented along the *X* axis.

If you select more polyline features, you will obtain more profiles in the same graph.

4. Right-click on the profile graph, select the **Properties** option, and select the **Appearance** tab. For **General graph properties**, edit **Title** and **Footer**, and for **Axis properties**, add **Title** for **Left** and **Bottom**, as shown in the preceding screenshot. Click on **OK**.
5. Right-click on the profile graph, and select the **Save** option to save your graph as the `MyProfile_P73ToReservoir.grf` graph file to the `...\3DAnalyst\ElevationProfile` folder.
6. Right-click on the profile graph and select the **Export** option. In the **Picture** frame, select the **as PDF** format and accept the default parameters. Click on **Save** and go to the previous location. Save the PDF file as `MyElevationProfile.pdf`.
7. Select the **Data** tab, check **Excel Format**, and accept the default parameters. Click on **Save** and repeat the previous step. Inspect the results with the Windows Explorer application.
8. In ArcMap, close the graph window. Go to **View** | **Graphs** and select **Manage Graphs**. If you right-click on `ResultsProfile_P73ToReservoir`, you will find the same options from steps 5 and 6. To reopen your graph, double-click on the selected graph. Explore by yourself **Advanced Properties**.
9. Save your map document as `MyProfile.mxd` and close ArcMap.

You can find the final results at `<drive>:\PacktPublishing\Data\3DAnalyst\ElevationProfile\ResultsProfile.mxd`. In the `ResultsProfile.mxd` map document, please go to **View** | **Graphs**, and explore all five profile graphs.

How it works...

If you want to use the sight lines features created in the *Intervisibility* recipe, you have to go through the following steps:

1. Recreate the `SightLines` layer using the **Construct Sight Lines** tool, with **Observer Height Field (optional)** with the value `Shape.Z` and **Target Height Field (optional)** with the value `Shape.Z`. You will have a sight line at surface level. Notice that your sight lines have only two vertices along the polyline feature. Only two vertices are not enough for a profile graph.
2. Use the **Densify** tool with a **DISTANCE** of `50`—unit as **Meters**—to add more vertices to the polyline features.
3. Use the **Interpolate Shape** tool with **Interpolate Vertices Only** checked. Your new vertices will have interpolated Z-values based on the `Elevation` surface.

The `Profile_P73ToReservoir` graph reflects the direction of polyline features. To change the direction of your path graphic (from point `73` to the reservoir), please follow the next steps: start an edit session, select the polyline with the **Edit Vertices** tool, right-click on the selected feature to change the polyline direction with **Flip**, and stop the edit session. Select the polyline and generate a new profile graph. Note that the direction of the graphic path has been changed.

There's more...

You can create profiles from a terrain surface, as shown in the following screenshot:

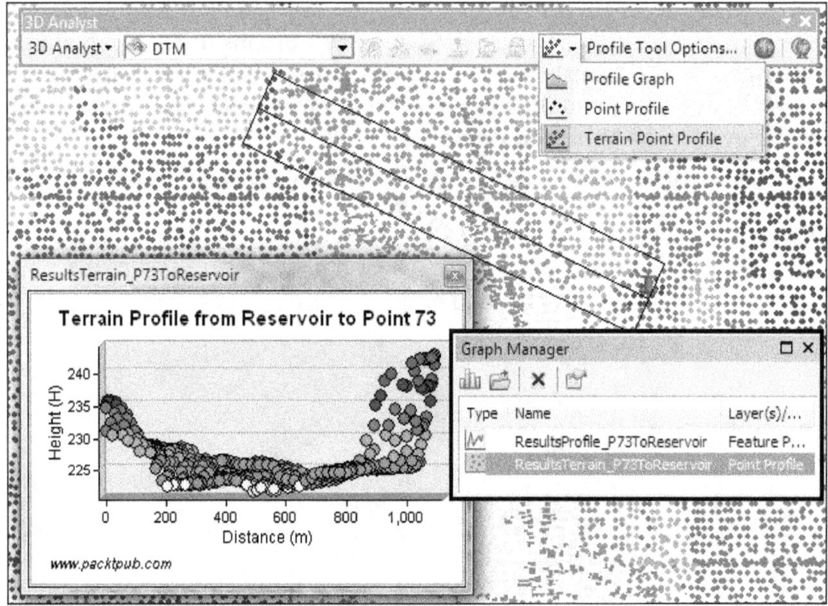

Open your map document `MyProfile.mxd`. Add the `DTM` terrain surface from `...3DAnalyst\ ElevationProfile\TOPO5000.mdb\Terrain`.

To activate the **Terrain Point Profile** tool, you have to symbolize the terrain surface with one of the three `Terrain point` renders. For the `DTM` terrain layer, choose the `Terrain point elevation with graduated color ramp` render.

In the **3D Analyst** toolbar, choose `DTM` as the surface layer, select the **Terrain Point Profile** tool, and click on two locations—the starting and ending points. Change the width for the black selection box, as shown in the preceding image. A new profile graph will be displayed. From **Properties** | **Appearance** of the profile graph, check the option **Graph in 3D view**.

Save and export your terrain point profile using the **View** | **Graphs** | **Manage Graphs** dialog box.

Creating an animation

There are two ways to create an animation:

- Capturing keyframes
- Recording your navigation through the scene

Keyframes are individual views. When you create an animation by capturing keyframes, you can use the **Capture View** tool or the **Animation Manager** dialog from the **Animation** toolbar. The rest of the views from the animation file will be interpolated based on those keyframes to result in a smooth animated picture.

Getting ready

In this recipe, you will create an animated tour of your 3D feature in ArcScene and export it to a video file. Before you start working, please watch the movies `3DFeatures_ByKeyFrames.avi` and `3DFeatures_ByRecording.avi` from the `...\Data\3DAnalyst\Animation` folder. These are the two results you should obtain at the end of the exercise. You can use a media player such as **Windows Media Player** or **VLC**.

You can start with your scene document `My3DFeatures.sxd` from the *Creating 3D features from 2D features* recipe and skip step 1. Otherwise, use a scene document from the `Animation` folder.

How to do it...

Follow these steps to create an ArcScene animation:

1. Start ArcScene and open an existing scene document `3DFeatures.sxd` from `...\Data\3DAnalyst\Animation`.

   In the next steps, you will create an animation by capturing keyframes:

2. Firstly, you will load seven bookmarks that will help you in capturing camera keyframes. Go to **Bookmarks** | **Manage Bookmarks** and click on the **Load** button to add the `KeyFrames_Animation.dat` file from `...\3DAnalyst\Animation`. You will load seven bookmarks, as shown in the following screenshot:

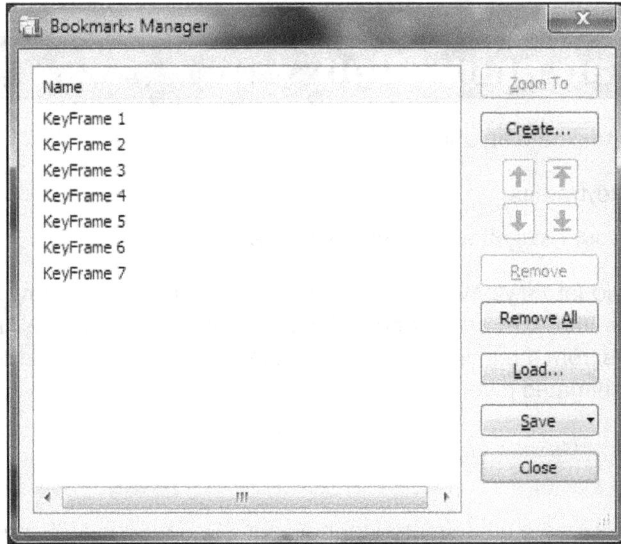

3. Close the **Bookmarks Manager** dialog. Explore all seven bookmarks and finish by choosing the first bookmark.

4. From **Customize** | **Toolbars**, add the **Animation** toolbar. From the **Animation** toolbar, select the **Animation** | **Animation Manager** | **Keyframes** tab. Move the **Animation Manager** dialog to the right-hand side of the scene.

   Let's capture some keyframes for your animation.

5. From the **Animation Manager** dialog window, click on the **Create** button, and set the parameters as shown in the following screenshot:

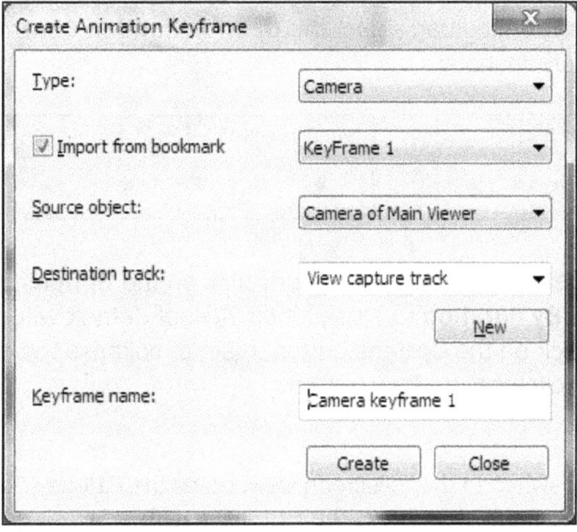

6. Check the **Import from bookmark** option and select the first bookmark named `KeyFrame 1`. Click on the **Create** button. Note that the **Animation Manager** dialog window has been updated with the `Camera keyframe 1`.
7. Repeat step 6 for the remaining bookmarks (from `KeyFrame 2` to `KeyFrame 7`). You have to capture all seven keyframes of the animation, as shown in the following screenshot:

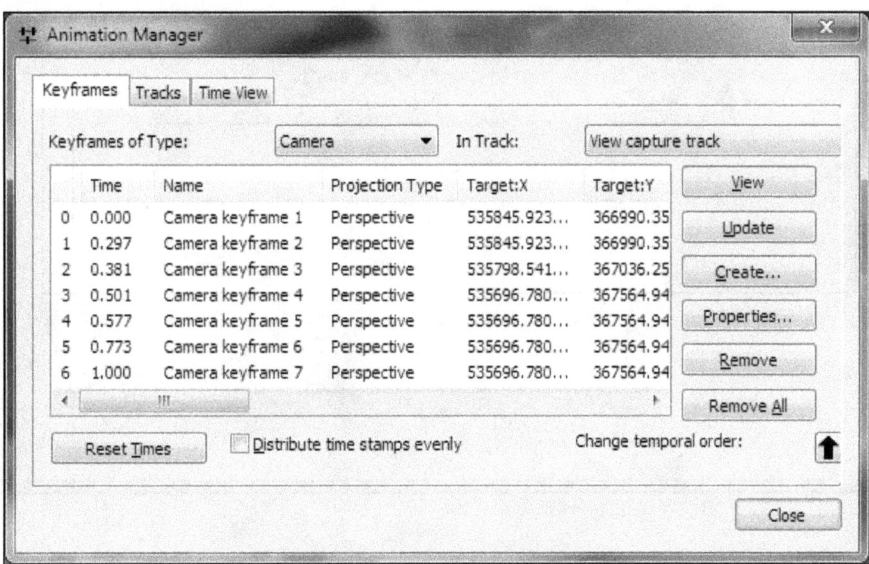

8. Close the **Create animation Keyframe** window. Explore all the columns of `Keyframes`. Select the **Time View** tab to see the capture track line of your animation. Leave open the **Animation Manager** dialog window.
9. From the **Animation** toolbar, select the **Open Animation Control** button, as shown in the following screenshots:

10. To adjust the default animation duration, click on the **Options** button, and change **Play Options** | **By duration** to `15` seconds. Accept default values for all other parameters. Click on the **Options** button again to collapse the dialog box. To watch the animation, click on the **Play** button.

To switch to the fullscreen view, press the *F11* key. To leave the fullscreen view, press the *F11* key again.

11. While you are watching your animation, notice the capture track line from the **Animation Manager** dialog window, as shown in the following screenshot:

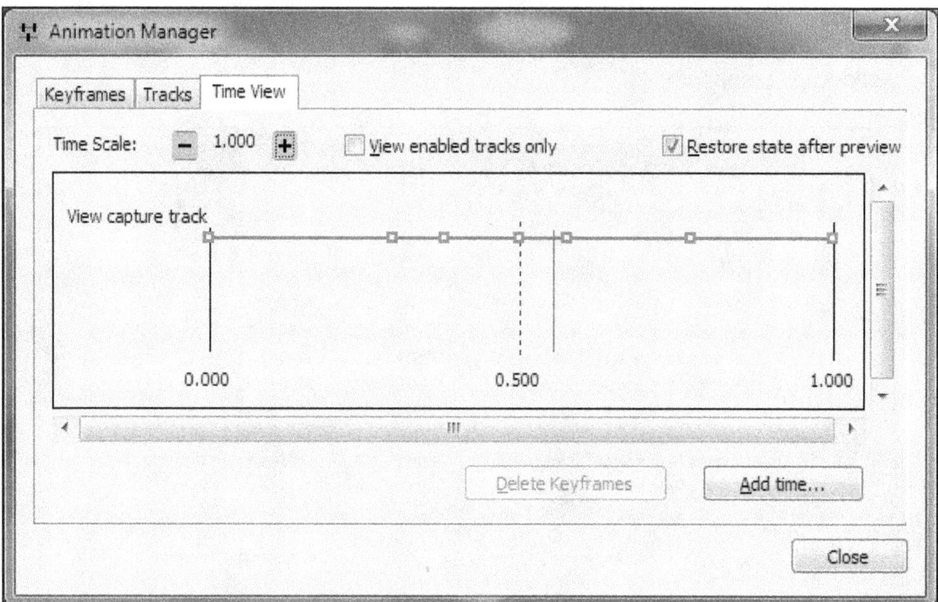

12. Let's save the animation into an animation file with the `.asa` extension. From the **Animation** toolbar, select **Save Animation File**. Save your animation as `MyKeyFrames.asa` in `...Data\3DAnalyst\Animation`. Select the **Export Animation** option to export your animation file to an AVI file `MyKeyFrames.avi`. In the **Video Compression** dialog window, accept the default **Compressor** as `Microsoft Video 1`, and click on **OK**.

    In the next steps, you will create an animation by recording a flying through the scene:

13. In the **Animation Manager** dialog window, select the **Keyframes** tab, and click on **Remove All**. From the **Animation** toolbar, select the **Open Animation Control** button. From the **Bookmarks** menu, select `KeyFrame 2`.
14. To start recording, perform the following steps:
    - Select the **Fly** tool from the **Tools** toolbar.
    - Click on the **Record** button from **Open Animation Control** to start recording.
    - Left-click to start navigating as a virtual airplane. Notice the speed values in the lower-left corner of the scene, as shown in the following screenshot:

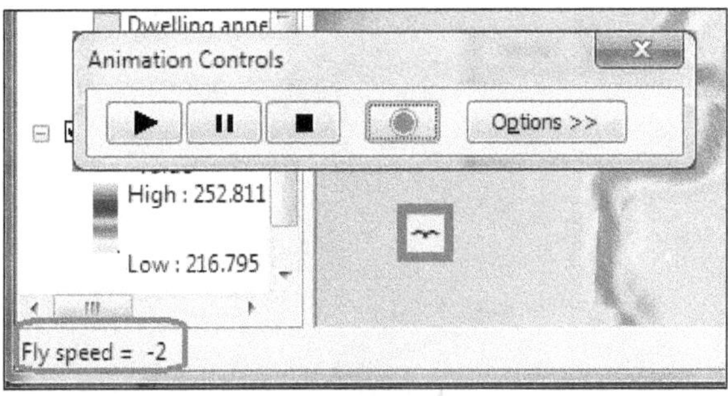

> To control the flight, press the *Shift* key to maintain a constant altitude, use the left-click to increase the speed and right-click to decrease the speed. A negative speed value (for example, `-1`, `-2`) means backwards flight.

    - After 5-10 seconds, stop flying by pressing the *Esc* key or the mouse wheel.

15. Click on the **Record** button again to stop recording.
16. The **Animation Manager** | **Keyframes** dialog window has been updated with the corresponding keyframes when you have to stop recording. Check also the **Tracks** and **Time View** tabs. Watch your recorded navigation by pressing **Play** from **Animation Control**.

17. If you are not satisfied with the result, in **Animation Manager** | **Keyframes**, click on the **Remove All** button and start recording once again.
18. Repeat step 30 to save your animation as `MyRecording.asa`, and to export it as `MyRecording.avi`. Close the dialog windows and the ArcScene application.

Watch `MyKeyFrames.avi` and `MyRecording.avi` using a video player.

How it works...

In the **Animation Manager** | **Keyframes** dialog window, you can erase and replace any keyframe from your animation by selecting its `ID` and using the **Remove** button. Also, you can change the order of keyframes using the **Change temporal order** arrows.

At step 11, you can change the position or timing of every keyframe (small green square) along the capture-track red line. If your keyframes are too close, your animation will proceed rapidly. If your keyframes are too distant, your animation will move slowly. To improve the timing, click on the **Keyframes** tab, check **Distribute time stamps evenly**, and click on **Reset Times**. These steps will be reflected in the **Time View** window and will smoothen the animation.

11
Working with Data Interoperability

In this chapter, we will cover the following topics:

- Exporting file geodatabase to different data formats
- Importing data from the XML format
- Importing vector data
- Importing raster data
- Spatial ETL

Introduction

Interoperability means the capability of two or more systems to exchange information and use/interpret the information shared without knowledge of the data structure. According to the online ESRI GIS Dictionary:

> *The interoperability is required for a GIS user using software from one vendor to study data compiled with GIS software from a different provider.*

There are two important aspects of data interoperability in GIS:

- Spatial data type (for example, GIS, CAD, raster, XML, and LiDAR)
- Spatial data format (for example, shapefile, DWG, ECW, GML, and XYZ)

Data standards support you in dealing with data interoperability issues. Standards refer to formats, data models, or both.

Here you have some examples of international and national spatial standards:

- ISO/TC 211 Geographic information/Geomatics: `http://www.iso.org`
- Open Geospatial Consortium: `http://www.opengeospatial.org`
- Ordnance Survey MasterMap: `http://www.ordnancesurvey.co.uk`

The **ArcGIS Data Interoperability** extension is a spatial **Extract/Transform/Load** (**ETL**) toolset that use the **Feature Manipulation Engine** (**FME**) technology from Safe Software Inc. Spatial ETL allows you to read (extract), transform, and write (load) over 70 spatial data formats.

For more details about data interoperability, please refer to:

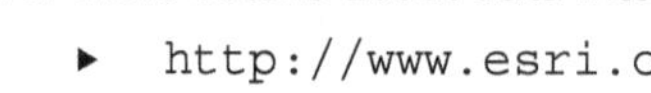

- `http://www.esri.com/products/technology-topics/standards`
- Online *ArcGIS Help (10.2)* by navigating to **Extensions** | **Data Interoperability** at `http://resources.arcgis.com/en/help/main/10.2`

Exporting file geodatabase to different data formats

In **ArcToolbox** | **Conversion Tools**, you will find a lot of tools to convert geodatabase data to another format. With the **Search** tool, you can find four tool types: **Script** (for example, Python file—`.py`), **Built-in**, **Model**, and **Specialized** (for example, **Spatial ETL Tool**). Use the **Search** window for a quick built-in tool search with the **Tools** filter and the **Local Search** index option, as shown in the following screenshots:

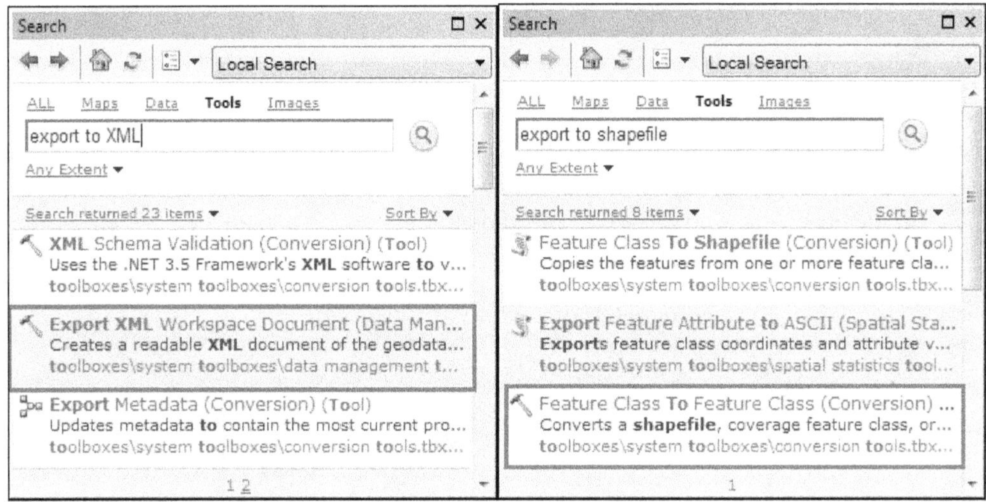

Getting ready

In this recipe, you will migrate a geodatabase into the **Extensible Markup Language** (**XML**) interchange format and the Esri shapefile format using the ArcCatalog context menu.

The XML workspace document stores the specification of the geodatabase and the data stored in it. It is a data exchange format used for:

- Sharing geodatabase data and schema to other applications
- Adding more data, domains, objects, and rules to an existing geodatabase

You will exclude data and will export only the schema of an existing file geodatabase. In the second part of this exercise, you will transfer the subtype and domain descriptions as additional field attributes in the shapefile format.

How to do it...

Follow these steps to export a file geodatabase using ArcCatalog:

1. Start ArcCatalog click on the **Connect To Folder** tool from the **Standard** toolbar, and navigate to `<drive>:\PacktPublishing\Data\Interoperability`. Click on **OK**.

   In the next steps, you will export the geodatabase schema to an XML file, as shown in the following screenshot:

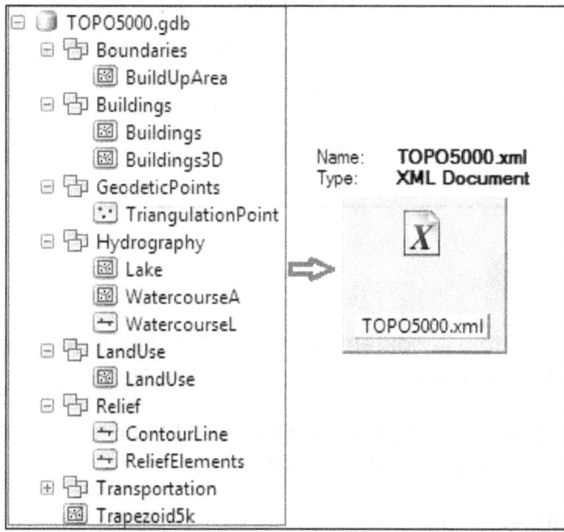

2. In **Catalog Tree**, go to `Interoperability\SourceData`, and expand `TOPO5000.gdb` to explore the contents. Right-click on `TOPO5000.gdb` and navigate to **Export** | **XML Workspace Document**.
3. You will export only the schema of the geodatabase without data. Check the **Schema Only** option.
4. For **Specify the output XML file**, navigate to `...\Data\Interoperability\MyResults`, and type `TOPO5000`. Click on **Save**. As you will not export data, uncheck the **Export Metadata** option. You will edit your own metadata information in the next recipe.

5. Click on **Next**. In the second panel, you can exclude one or more feature classes from the exported schema, as shown in the following screenshot:

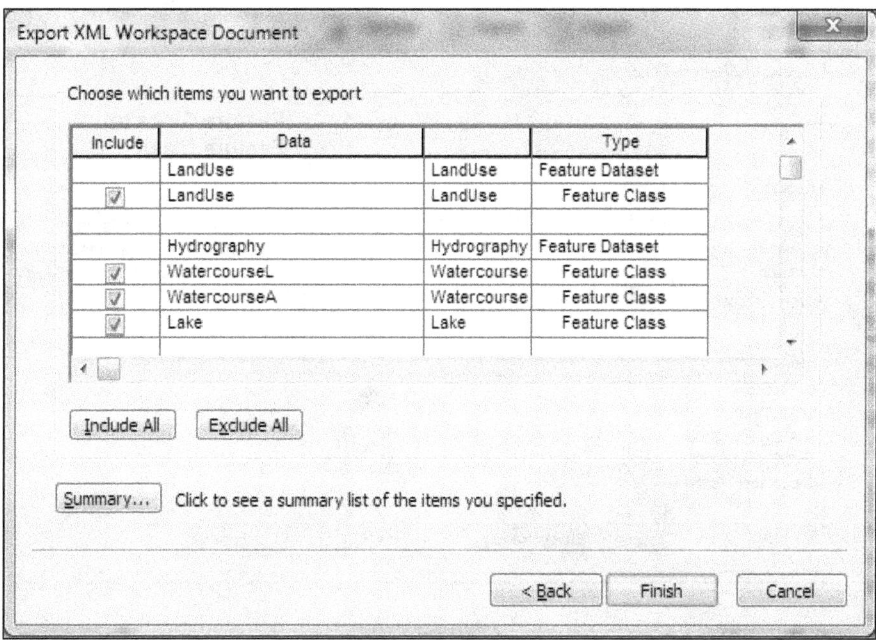

6. After you have finished exploring all the listed feature classes, click on the **Include All** button to ensure that all data is checked. Click on **Finish**. You can now send the geodatabase schema by e-mail to your colleagues.

   In the next steps, you will export the geodatabase data to the shapefile format:

7. Open the **Results** window to track the geo-processing tools. This will help you out in the next steps.

8. Your geodatabase has a lot of descriptions for subtypes and domains. Right-click on the `TOPO5000.gdb\LandUse\LandUse` feature class and navigate to **Export** | **To Shapefile (single)**. Set the parameters as shown in the following screenshot:

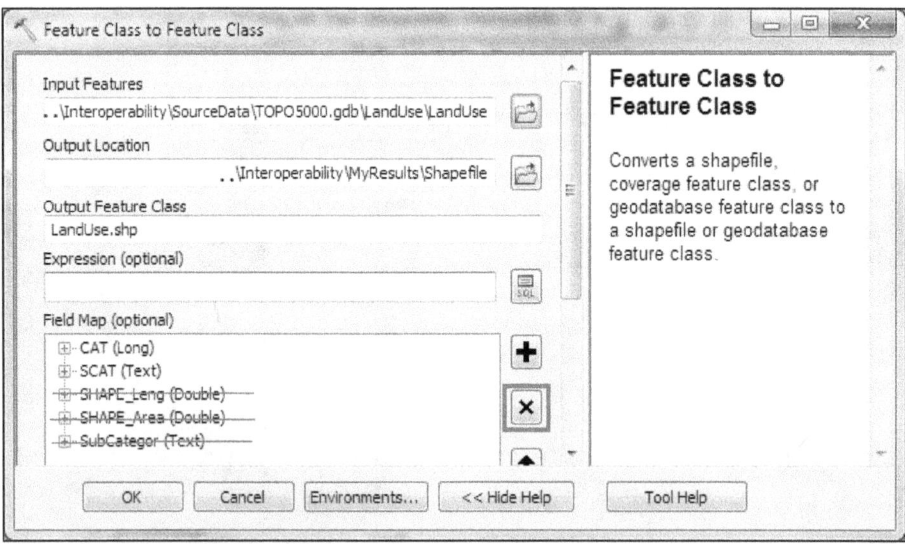

9. Delete the **SHAPE_Leng (Double)**, **SHAPE_Area (Double)**, and **SubCategor (Text)** fields using the **Delete** button. Click on the **Environments** button, and go to the **Fields** parameter. Check **Transfer field domain descriptions**. This option will transfer to you the subtype and domain descriptions. You have worked with those descriptions in *Chapter 1, Designing Geodatabase*. Click on **OK** twice to run the tool.
10. In ArcCatalog, inspect the output shapefile in the **Preview**: **Table** mode view. Your shapefile now has the codes and descriptions for subtypes and domains.
11. Export the remaining feature classes at once by navigating to **Export** | **To Shapefile (multiple)**.
12. From **Results** | **Current Session**, double-click on **Feature Class** to reopen the tool. Change only the **Input Features** parameter and **Output Feature Class**. The **Output Location** and the **Environment** parameter will remain unchanged. Click on **OK** to run the tool again. Explore the results.

How it works...

At step 4, even if you uncheck all feature classes from a feature dataset, the exported XML workspace document will keep the empty feature dataset. To see the XML file, you can use the Notepad++ free text editor.

For more details about the XML schema of your geodatabase, please refer to the Esri white paper, `XML Schema of the Geodatabase.pdf`.

At the feature class level, you can save the data into an XML file using **Export XML Recordset Document**.

When you are sharing your schema, it's good practice to document the XML workspace. You should keep the metadata information or create supplementary material (for example, a PDF containing summary information about the lineage of data, responsible party, coordinate reference system, and spatial resolution).

There's more...

In *Chapter 10, Working with 3D Analyst*, you worked with 3D features. If you want to see those buildings in the **Google Earth** application, you have to export the `Buildings3D` feature class to a **Keyhole Markup Language** (**KML**) format. For the next exercise, you need to install Google Earth for desktop from `https://www.google.com/earth`.

For more details about the KML format, please refer to:

- Online *ArcGIS Help (10.2)* navigating to **Geodata** | **Data types** | **KML** at `http://resources.arcgis.com/en/help/main/10.2`
- Online Google documentation at `http://developers.google.com/kml`

Follow these steps to import a feature class into Google Earth:

1. Start ArcCatalog. In **Catalog Tree**, go to `...\SourceData\TOPO5000.gdb`. Firstly, prepare the `Buildings3D` feature class.
2. In **ArcToolbox**, expand **Data Management Tools** | **Projections and Transformation**, and double-click on the **Project** tool. Set the following parameters:
    - **Input Dataset**: `SourceData\TOPO5000.gdb\Buildings\Buildings3D`
    - **Output Dataset**: `SourceData\TOPO5000.gdb\ Buildings3D_UTM35N`
    - **Output Coordinate System**: `Projected Coordinate Systems\UTM\WGS 1984\Northern Hemisphere\WGS 1984 UTM Zone 35N`
    - **Geographic Transformation (optional)**: `Pulkovo_1942_Adj_1958_To_WGS_1984_19`

3. In ArcScene, add and symbolize `Buildings3D_UTM35N`. Extrude features in layers based on the **[Stories]** field.
4. In **ArcToolbox**, expand **Conversion Tools** | **To KML** and double-click on the **Layer To KML** tool. Set the following parameters:
    - **Layer**: `Buildings3D_UTM35N`
    - **Output File**: `MyResults\Buildings3D_UTM35N.kmz` (the KMZ file format is a compressed KML)
    - Check **Clamped feature to ground (optional)**. Click on **OK**.
5. Open the Google Earth application. Go to the **View** menu and check **Grid**. Go to **Tools** | **3D View** and check **Universal Transverse Mercator**. Click on **OK**.
6. Go to **File** | **Open** and browse to `Buildings3D_UTM35N.kmz`. Explore the results.

You can find the results at `<drive>:\PacktPublishing\Data\ Interoperability\DataResults`.

See also

- If you want to see how your colleagues will generate a geodatabase schema from the XML workspace document received from you, please go to the next recipe *Importing data from the XML format*

Importing data from the XML format

An XML file can transport and store geodatabase schema and spatial data. Geodatabase XML is Esri's XML-based interchange format used to share data (*workspace data*) and geodatabase schema in full or in part (*workspace definition*) between geodatabases or other software systems.

Getting ready

In this recipe, you will create a file geodatabase schema by importing an XML workspace document that contains only the workspace definition as shown in the following screenshot:

You will use the `TOPO5000.XML` file created in the previous exercise.

How to do it...

Follow these steps to create a file geodatabase using as a template an ArcGIS XML workspace document:

1. In **Catalog Tree**, go to `Interoperability\MyResults`. Right-click on the `MyResults` folder and navigate to **New** | **New File Geodatabase**. Rename **New File Geodatabase.gdb** to `MyTOPO`.

2. Right-click on `TOPO5000.gdb` and navigate to **Import** | **XML Workspace Document**. Check **Schema Only** and specify the XML source to import by selecting the `TOPO5000.XML` file you created in the previous recipe. Click on **Next** and **Finish** to start importing schema as shown in the following screenshots:

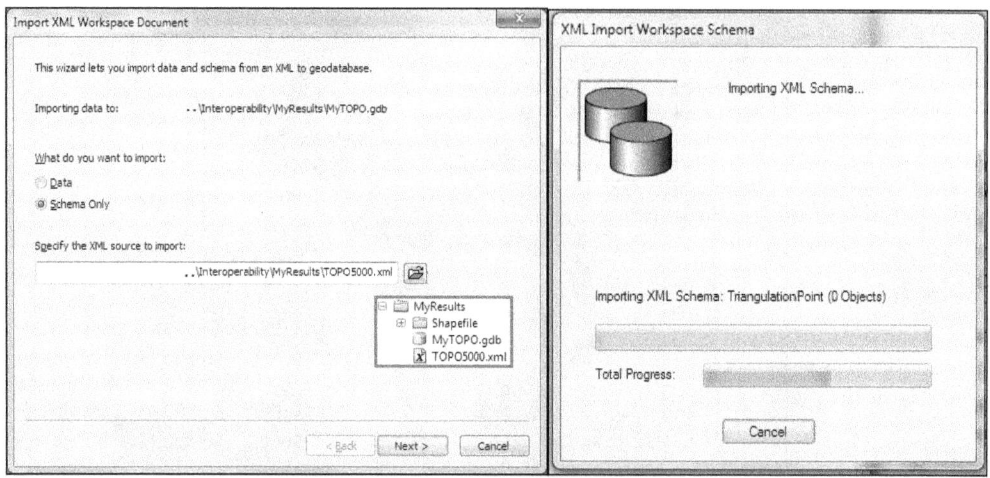

3. Expand `MyTOPO.gdb` to explore the contents.

You can find the results at `<drive>:\PacktPublishing\Data\ Interoperability\DataResults\MyTOPO_SchemaOnly.gdb`.

How it works...

By importing an XML workspace document that stores a geodatabase schema, you saved a lot of time in defining the geodatabase structure. The feature datasets, feature classes, attribute domains, and subtypes are now built into the `MyTOPO` geodatabase.

There are lots of ArcGIS template data models at `http://solutions.arcgis.com`. Use your Esri global account created in *Chapter 7, Exporting Your Maps*, to download a sample data model from **ArcGIS for Local Government** | **Land Records** | **Aggregate Authoritative Data**.

There's more...

Let's add some metadata information to your geodatabase from an XML file:

1. In ArcCatalog, go to **Customize** | **ArcCatalog Options**, and select **Metadata Style: ISO 19139 Metadata Implementation Specification**. Click on **OK**.

2. In **Catalog Tree**, go to `Interoperability\MyResults`, and select `MyTOPO.gdb`. Select the **Description** tab and click on **Import**. For **Source Metadata**, select `MyTOPO_ISO19139.XML` from the `SourceData` folder, as shown in the following screenshot:

3. Click on **OK**. Your metadata content has been updated with new information. Explore the contents.

While you are populating `MyTOPO.gdb` with data, it's good practice to create and update your metadata for every feature dataset and feature class from your geodatabase.

When you use ArcMap, you can view and edit metadata information from the **Catalog** context menu: right-click on the selected feature class and choose **Item Description**. Before starting to edit and export metadata elements, select the metadata style from the **Customize** | **ArcMap Options** | **Metadata** tab. The core metadata elements and the required elements and subelements might vary slightly depending on the metadata standard used (for example, ISO 19115, INSPIRE, FGDC CSDGM Metadata).

See also

- After you have built the structure of file geodatabase, you should populate it with the existing vector and raster datasets. To see the ArcGIS options, please explore the next recipes *Importing vector data* and *Importing raster data*.

Importing vector data

When you are importing existing vector data in your geodatabase, you have to inspect the following characteristics of your data:

- Source and data format (for example, internal/external, reliable, supported)
- Spatial reference (for example, maybe you will need to transform or project the data)
- Attributes (for example, field data type: short integer, text, date)
- Corresponding scale of the source data
- Metadata information (for example, quality of data, access, use and legal constraints)

You have already added data to a file geodatabase:

- Using **Load Data** to add a feature class in the *Editing features in a geodatabase* recipe of *Chapter 2, Editing Data*
- Using **Import** from the ArcCatalog context menu to add a feature class in the *Projecting vector data* recipe of *Chapter 3, Working with CRS*
- Using **Load Data** and **Export** from the ArcCatalog context menu to add a DXF file from the external source in the *Spatial joining features* recipe of *Chapter 4, Geoprocessing*

Getting ready

In this recipe, you will convert a `.csv` and a `.xlsx` file to a new feature class. In addition, you will import an XML file and a shapefile data file in the `MyTOPO` geodatabase, as shown in the following screenshot:

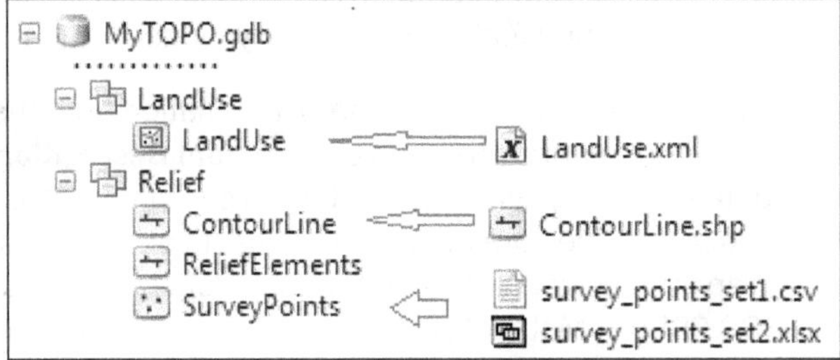

All those four formats are stored in the `...\Interoperability\SourceData` folder.

How to do it...

Follow these steps to convert the `.csv` and `.xlsx` file formats to a feature class:

1. Start ArcMap and open an existing map document `ImportData.mxd` from `<drive>:\PacktPublishing\Data\Interoperability\SourceData`.
2. From the **File** menu, navigate to **Add Data** | **Add XY Data**. Set the following parameters:
   - **Choose a table from the map or browse for another table**: `...\SourceData\survey_points_set1.csv`
   - **X Field**: `East (S42)`
   - **Y Field**: `North (S42)`
   - **Z Field**: `Elevation (H)`
   - **Coordinate System of Input Coordinates**: This parameter will inherit the coordinate system of the active data frame
3. Accept the default **Coordinate System** and click on **OK**. Read the warning message and click on **OK**.
4. Select the **Add Data** button and go to `...\SourceData`. Double-click on `survey_points_set2.xlxs` to see the Excel worksheet `survey_points`. Click on **Add**. To see a nonspatial table, select **List By Source** in the **Table Of Contents** section. Open the attribute table of the `survey_points` worksheet to inspect the fields.
5. Right-click on the `survey_points` table and select **Display XY Data**. For **Specify the fields for the X, Y and Z coordinates**, set the following parameters:
   - **X Field**: `East`
   - **Y Field**: `North`
   - **Z Field**: `H`
6. Accept the default **Coordinate System** and click on **OK**. Read the warning message and click on **OK**.

   Let's export the point events to feature classes:
7. Right-click on **SurveyPoints_Set1.csv Events** and select **Data** | **Export Data**. Set **Output feature class**: `...\Interoperability\MyResults\MyTOPO.gdb\SurveyPoints_Set1`. Click on **OK**.
8. Repeat step 7 for the `survey_points Events` layer and save it as `SurveyPoints_Set2`. Inspect the attribute fields of both feature classes.

9. In **ArcToolbox**, expand **Data Management Tools | General**, and double-click on the **Merge** tool. Click on **Show Help** to see the meaning of every parameter. Set the following parameters:
    - **Input Datasets**: `SurveyPoints_Set1 and SurveyPoints_Set2`
    - **Output Feature Class**: `MyTOPO.gdb\Relief\SurveyPoints`
10. Accept all fields and click on **OK**. Open the attribute table of the `SurveyPoints` layer.
11. Notice that you have duplicated fields for the `East`, `North`, and `H` coordinates.

    Before you start to use the **Merge** tool, you should check and rename the fields of your feature classes. Let's correct this:
12. Use **Select by Attribute** to select all features from `set1` with **ID IS NULL**. Right-click on the **East** field, select **Field Calculator**, and double-click on the **[East__S42_]** field to build the following expression: `East= [East__S42_]`. Click on **OK**.
13. Repeat the previous step for the **North** and **H** fields. Erase the **East_S42**, **North _S42**, **Elevation_H**, and **ID** fields, as shown in the following screenshot:

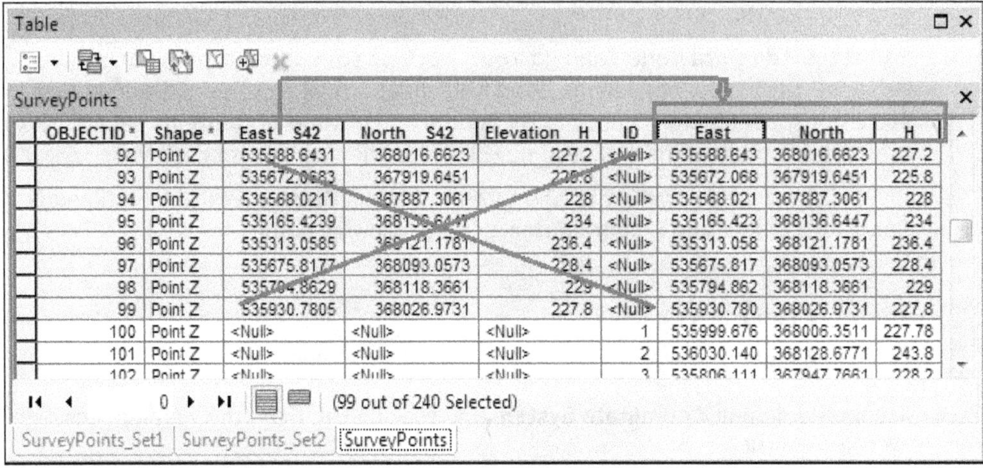

14. Save your map document as `MyImportData.mxd` and close ArcMap.

    In the next steps, you will continue to load a shapefile into the `MyTOPO` geodatabase using the ArcCatalog context menu:

15. In **Catalog Tree**, go to your `Interoperability` folder connection, and expand the `SourceData` folder to explore the contents. If you inspect **Properties** | **Fields** of `ContourLine.shp`, you will notice that you don't have the information related to the types of contour lines. Inspect also the **Fields** of the `ContourLine` feature class. You have a subtype field called `CNT` that stores this information. You will load all polyline features into the `Unknown` subtype, and later on, you can add more information.
16. Navigate to `...\MyResults\MyTOPO\Relief` and right-click on the empty `ContourLine` feature class. Select **Load** | **Load Data** and set the following parameters from the panels:
    - **Input data**: `...\SourceData\ContourLine.shp`
    - Click on **Add**
    - **I want to load all features into a subtype**: `Unknown`
    - For field mapping, accept **Target** and **Matching Source Field**: `Elevation [double]`
    - Check **Load all of the source data**
17. Click on **Finish**. In the **Preview** panel, inspect the geometry and attribute table.

    In the next step, you will load data from an XML recordset document:
18. In **Catalog Tree**, navigate to `MyTOPO\LandUse`, and right-click on the empty `LandUse` feature class. Select **Load** | **Load XML Recordset Document** and set the following parameters from the panels:
    - **Specify the XML source to load**: `LandUse.xml`
    - Inspect and accept the fields for **Target Field** and **Matching Source Field**
19. Click on **Finish**. Inspect the result in the **Preview** panel.

You can find the results at `...\Interoperability\DataResults\MyTOPO.gdb`.

How it works...

At steps 2 and 5, the survey points have been added in ArcMap as *event tables layers*. Those points are saved in map documents, but they are not permanent features. To save them, you exported them as a feature class in the `MyTOPO.gdb` geodatabase.

You can create a point feature class from the `survey_points_set1.csv` file using the ArcCatalog context menu. Right-click on the file and select **Create Feature Class** | **From XY Table**.

Before starting to load data from any format, it's important to evaluate the existing data. By carefully planning the load steps, you will avoid losing data in the field-mapping step, generating duplicate/residual data or failing the entire load process. Here, you have some supplementary steps:

- Change the data type of the attribute fields for the editable source data format (for example, shapefile, feature class)
- Export source data in an intermediary editable format (for example, from the *DXF* format to a *file geodatabase feature class*, or from an *Excel* table to a *dBASE* or *file geodatabase table*)
- Change the geometry of the source data according to the destination feature class

The field-mapping process can be a critical step if the data types of the target and source fields do not match properly.

Importing raster data

You can store one or more rasters in a geodatabase as:

- Individual raster datasets
- Mosaic dataset
- Raster catalog

You have already worked with the raster dataset in *Chapter 9, Working with Spatial Analyst* and *Chapter 10, Working with 3D Analyst*.

In this recipe, you will work with the mosaic dataset and the raster catalog.

A mosaic dataset creates a seamless view of two or more adjacent or non-adjacent images. In a mosaic dataset, all rasters have the same cell size, format, and coordinate system.

A raster catalog is a table that groups two or more raster datasets. In a catalog, rasters may have different cell sizes and different formats, but they must have the same coordinate system.

Getting ready

In this recipe, you will create a mosaic dataset using four rasters in the **Enhanced Compression Wavelet** (**ECW**) file format.

For more details about the ECW raster format, refer to `http://www.hexagongeospatial.com/products/data-management-compression/ecw`.

All those four rasters are stored in the `...\Interoperability\SourceData\ECW` folder.

How to do it...

Follow these steps to create a *raster catalog* and a *mosaic dataset* using the ArcCatalog context menu and the **ArcToolbox** tools:

1. Start ArcCatalog. In the **Catalog** window, go to your `Interoperability` folder connection, and expand the `SourceData\ECW` folder to explore its contents. Those four rasters cover four small adjacent areas.
2. Right-click on the `orthoimage_1.ecw` raster, and select **Properties** | **General**. Inspect the properties of the raster, such as **Number of Bands**, **Pixel Depth**, and **Pyramids resampling**. Click on **OK** to close the dialog window.
3. From the **Geoprocessing** menu, select **Environments**. Set the geoprocessing environment as follows:
    - **Workspace** | **Current and Scratch Workspace**: `...\Interoperability\MyResults\MyTopo.gdb`
    - **Output Coordinates** | **As Specified Below**: Import from `<Feature Dataset>`
    - **Raster Storage** | **Pyramid levels**: `10`
    - **Raster Storage** | **Compression**: `LZ77`; click on **OK**.

    In the next step, you will learn how to manage your four rasters in a raster catalog:
4. In the `MyResults` folder, right-click on the `MyTOPO` geodatabase, and navigate to **New** | **Raster Catalog**. Set the following parameters:
    - **Output Location**: `...\MyResults\MyTopo.gdb`
    - **Raster Catalog Name**: `MyRasters`
    - **Raster Management Type (optional)**: `MANAGED`
5. Accept the default values for all other parameters. Click on **OK**.
6. Let's load four ECW rasters. Right-click on the `MyRasters` raster catalog and navigate to **Load | Load from Workspace**. For **Input Workspace**, navigate to the `...\SourceData\ECW` folder. Leave the two remaining options unchecked. Click on **OK**.

7. Explore the raster catalog as shown in the following screenshot:

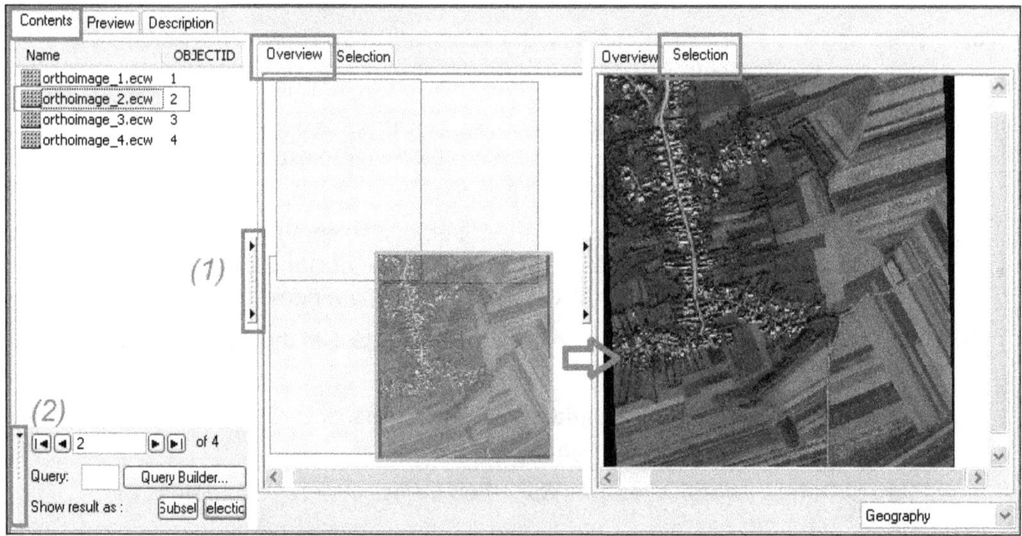

8. Use the **Expand Window** button **(1)** to open the raster catalog browser and the **Show Advanced Tools** button **(2)** to query a raster catalog.
9. The **Overview** panel displays the footprints of rasters, and the **Selection** panel displays the selected raster.

   In the next step, you will learn to manage your four rasters in a raster mosaic:

10. Let's create first an empty container for rasters. In **Catalog Tree**, navigate to the `MyResults` folder, and right-click on the `MyTOPO` geodatabase. Navigate to **New** | **Mosaic Dataset** and set the following parameters:
    - **Output Location**: `..\MyResults\MyTopo.gdb`
    - **Mosaic Dataset Name**: `OrthophotoMap`
    - **Coordinate System**: **Import** from `MyTopo.gdb\LandUse`
    - **Product Definition (optional)**: `NATURAL_COLOR_RGB`
11. Accept the default values for all other parameters. Click on **OK**.
12. Let's add the rasters. In **ArcToolbox**, expand **Data Management Tools** | **Raster** | **Mosaic Datasets**, and double-click on the **Add Rasters To Mosaic Dataset** tool. Click on **Show Help** to see the meaning of every parameter. Set the following parameters:
    - **Mosaic Dataset**: `MyTopo.gdb\OrthophotoMap`
    - **Raster Type**: `Raster Dataset`
    - **Input Data**: `Workspace`

- Navigate to `...\SourceData\ECW`
- Check **Updates Overviews**

13. Accept the default values for all other parameters. Click on **OK**.
14. In ArcMap, open the `MyImportData.mxd` map document, and load your `OrthophotoMap` mosaic dataset, as shown in the following screenshots:

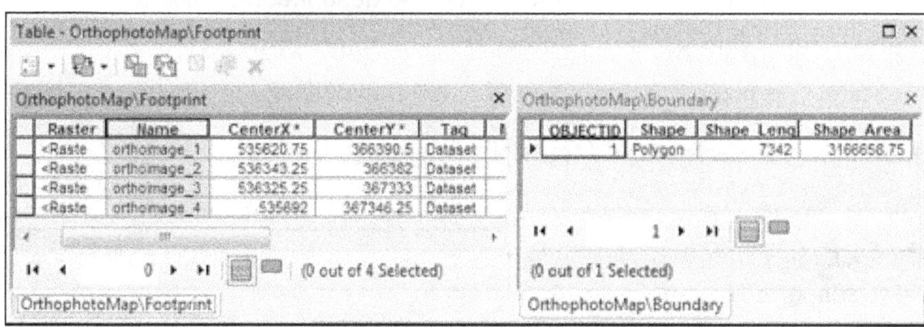

15. Open and inspect the attribute table for the `Boundary` and `Footprint` layers.
16. Save your map and close ArcMap.

How it works...

At step 4, we chose **Raster Management Type (optional)** value as `MANAGED`. The raster catalog will be managed by the geodatabase. This means that your rasters will be stored in `MyTopo.gdb`. Each ECW raster will be stored as a separate raster dataset.

The `UNMANAGED` option will not convert the rasters, and the catalog will store only the path to the source rasters. You will not be able to see those rasters if the geodatabase or the referenced rasters are moved to another location.

A mosaic dataset will treat the four raster datasets as a single one. At step 12, the `Footprint` layer represents the extent of each individual raster from mosaic. The `Boundary` layer represents the edge of the combined rasters. If you want to change the display order of the rasters, right-click on the `Image` layer, and navigate to **Properties** | **Mosaic**.

Spatial ETL

In the previous recipes, you used the traditional data translation between two data types and formats. The ArcGIS Data Interoperability uses the FME functionality to extend the formats you can read, transform, and translate.

The **transformation** process allows you to manipulate the structure and content of your data. The transformation process occurs during the format translation process. **Translation** refers to the process of converting a data format into another data format using:

- **Workspace**: This is a container that stores the translation definition
- **Reader and writer**: These are the source and destination datasets
- **Feature type**: This is a feature class, layer, or object class
- **Feature**: This is the base component of the translation

You have three tools to translate and transform interoperability data:

- **Quick Import**: This imports any FME-supported format into a new geodatabase
- **Quick Export**: This exports any feature class and interoperability data to any FME-supported format
- **Spatial ETL Tool**: This restructures and transforms Esri formats or other FME-supported formats

A **Spatial ETL Tool** is created, edited, and run by the user. A **Spatial ETL Tool** uses a graphic application called **workbench**, as shown in the following screenshot:

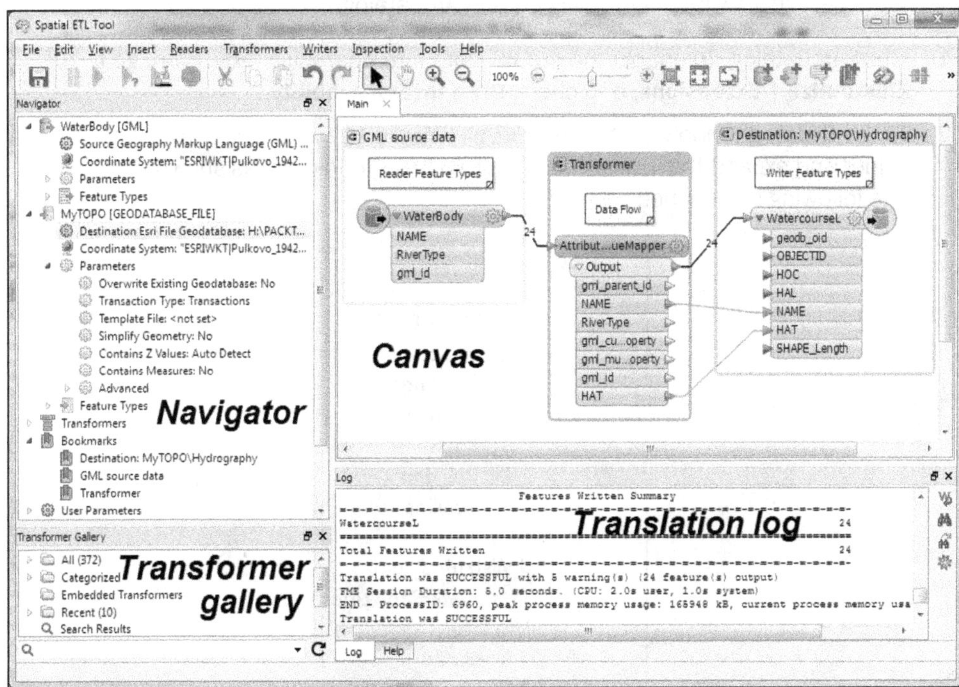

In workbench, you can manipulate the geometry and attributes of the datasets using **transformers**. Transformers allow you to restructure and re-project the data source before loading it into the destination data format. There are two types of transformers: geometric and attribute. When you are using a transformer, you have to set all mandatory parameters.

Getting ready

In this recipe, you will explore three data formats: **DWG**, **GML**, and **MIF/MID**. After you finish exploring the interoperability data, you will use the **Spatial ETL Tool** to translate the GML file into a feature class by transforming the data content.

- For more details about the GML format, please refer to `http://www.opengeospatial.org/standards/gml`
- For more details regarding the MIF/MID data interchange format, please go to `http://www.mapinfo.com/support/Product Documentation` and search by `MIF/MID`
- For more details regarding the FME, please refer to `http://www.safe.com/support/support-resources`

How to do it...

Follow these steps to view interoperability data in ArcCatalog:

1. Start ArcCatalog. To enable the ArcGIS Data Interoperability extension, go to **Customize** | **Extensions**, and check **Data Interoperability**.
2. In the **Catalog** window, go to the `Interoperability\SourceData` folder and expand the `+DWG`, `+GML`, and `+MapInfo` folders, as shown in the following screenshot:

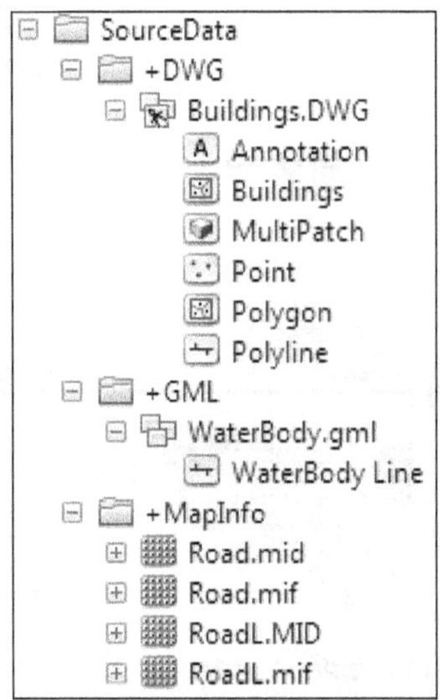

3. You can directly read a DWG file. Expand the `Buildings.dwg` file. There are displayed all types of geometry in six CAD feature classes. Navigate to **Buildings.DWG** | **Polyline** to see the features in the **Preview** pane. Switch **Preview** from the **Geography** mode to the **Table** mode. Explore the attribute fields and note the layers stored in **Layer** field. This field will help us to extract the features based on the `dwg` layers.
4. Let's explore the GML format. Expand the `WaterBody.gml` interoperability feature class. Use the **Identify** tool to examine the attributes.

   To extend the functionality of the direct-read data formats, you can create your own link to data source (interoperability connection) through the `Interoperability Connections` folder in **Catalog Tree**.

5. In the **Catalog Tree** window, go to `Interoperability Connections`, and double-click on the **Add Interoperability Connection** tool, as shown in the following screenshots:

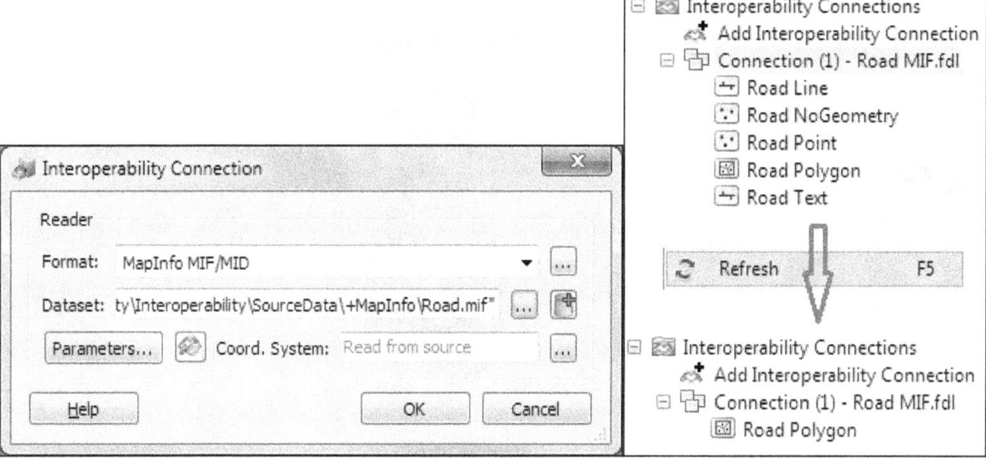

6. To specify **Format**, click on **Browse the gallery** to open the **FME Reader Gallery** window. Search for `MapInfo`, and double-click on `MapInfo MIF/MID` to select the format and exit from the reader gallery.
7. Set **Dataset** by navigating to `Interoperability\SourceData\+MapInfo\Road.mif`. Click on **OK**. The MIF format is directly read by ArcGIS. Expand `Connection (1) - Road MIF.fdl` to display the feature classes. Here again are listed all types of geometry in five interoperability feature classes.
8. Select the `Road Polygon` feature class and select the **Preview** pane. All features will be indexed and displayed. As shown in the previous screenshot, press the *F5* key to remove all empty feature classes.
9. Repeat the steps 5 to 8 to create an interoperability connection for the `RoadL MIF/MID` dataset.

   Continue to follow the steps to import the GML file into the `WatercourseL` feature class from `MyTOPO.gdb` geodatabase created in the *Importing data from the XML format* recipe using the Spatial ETL tool.

10. In the **ArcToolbox** window, right-click on **ArcToolbox**, select **Add Toolbox**, and click on the **New Toolbox** tool. Click on **Open**. Rename the newly created toolbox as `MyInteroperability`.
11. Right-click on `MyInteroperability` and navigate to **New** | **Spatial ETL Tool**.

12. In the wizard window, define your source format as **GML (Geography Markup Language)**. Click on **Next**. Browse to the `SourceData\+GML` folder and select the `WaterBody.gml` file, as shown in the following screenshots:

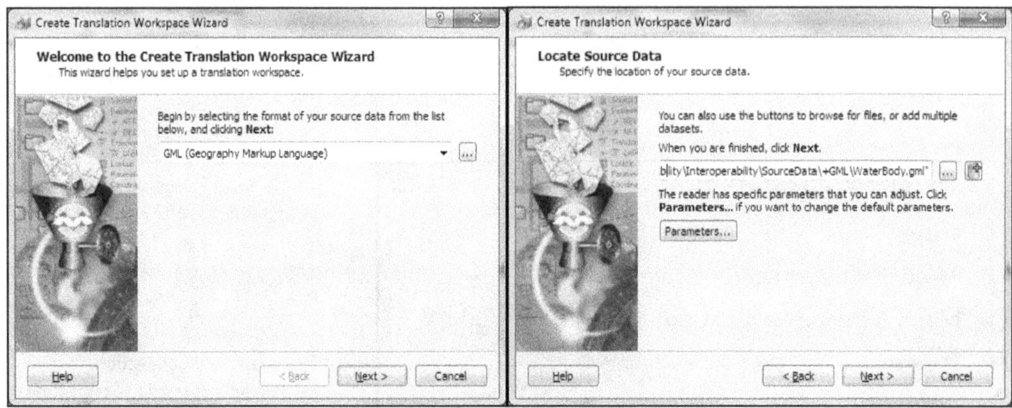

13. Go to the next panel and define **Destination Format**: `Esri Geodatabase (File Geodb ArcObjects)`. Go to the next two panels, accept the default parameters, select **Static Schema**, and click on **Finish** to create the translation workspace. The workbench has opened.

    Firstly, you will add a new writer to the workspace:

14. Navigate to **Writers** | **Import Feature Types**, and for **Dataset**, navigate to `Interoperability\MyResults\MyTOPO.gdb`. Click on **OK**. In the **Select Feature Types** window, uncheck **Select all**. Check only `WatercourseL` and click on **OK**.

15. Erase the `Waterbody` writer, and drag a connection from the yellow arrow of the `WaterBody` reader to the red arrow of the `WatercourseL` writer, as shown in the following screenshot:

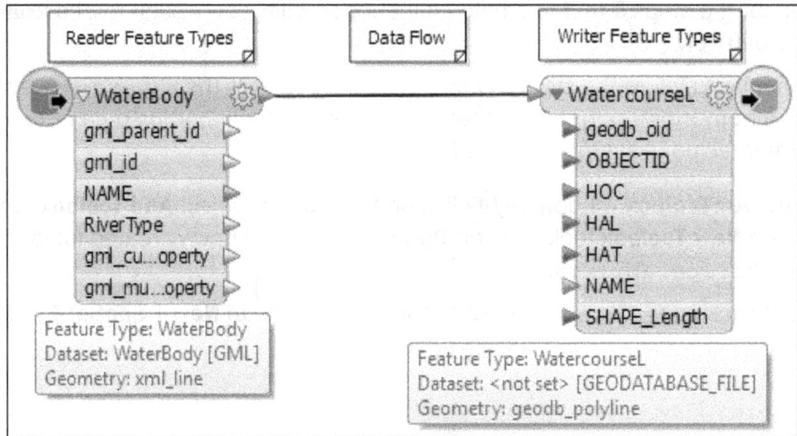

In the next step, you will set the destination parameter for **Writer**:

16. In the **Navigator** window, double-click on **Destination Esri File Geodatabase** of `<notset> [GEODABASE_FILE]`, and navigate to `MyResults\MyTOPO.gdb`. Click on **OK**.

    In the next step, you will set the properties for **Reader**:

17. In workbench canvas, double-click on `WaterBody`, and select the **User Attributes** tab. Uncheck all attribute fields except `NAME` and `RiverType`. This option will extract from the GML file only those two fields. Select the **Format Attributes** tab and uncheck `gml_id`. You should have all format attributes unchecked. Click on **OK**.

    In the next step, you will set the properties for **Writer**:

18. Double-click on `WatercourseL` and select the **Format Parameters** tab. Select **Yes** for **Truncate Table First** and type `Hydrography` for **Feature Dataset**, as shown in the following screenshot:

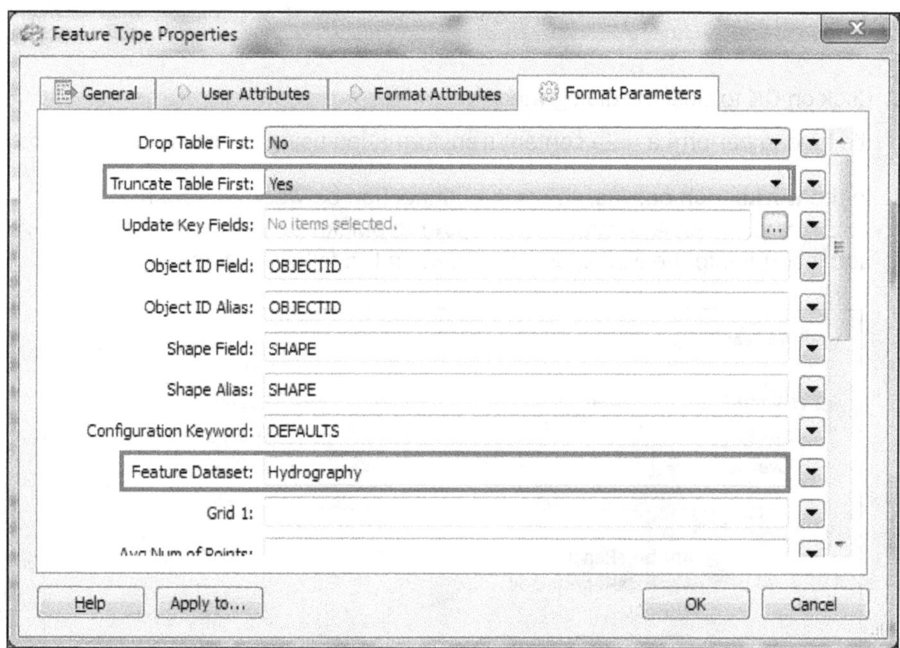

19. Select the **User Attribute**s tab. Select the **Name** field and move above the **HAT** field with the **Moves the current row up** arrow. This will avoid overlapping links between features in canvas. For the **HAT** field, select the **Type**: `subtype`. Click on the **Edit** button and define the subtype code, as shown in the following screenshots:

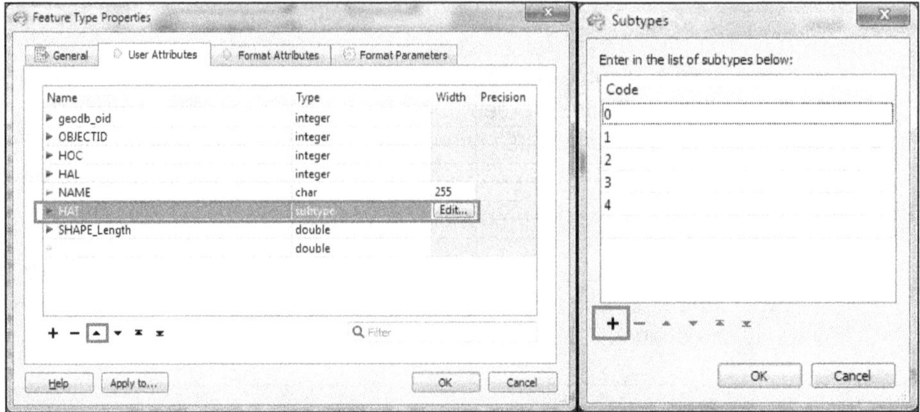

20. Click on **OK** to close all dialog windows.

    It's time to perform a data content transformation using an attribute transformer:

21. In the **workbench canvas**, select the connection between **Reader** and **Writer**. Type `attri` text, select **AttributeValueMapper** to quickly access the transformers gallery, and insert it into the workspace, as shown in the following screenshot:

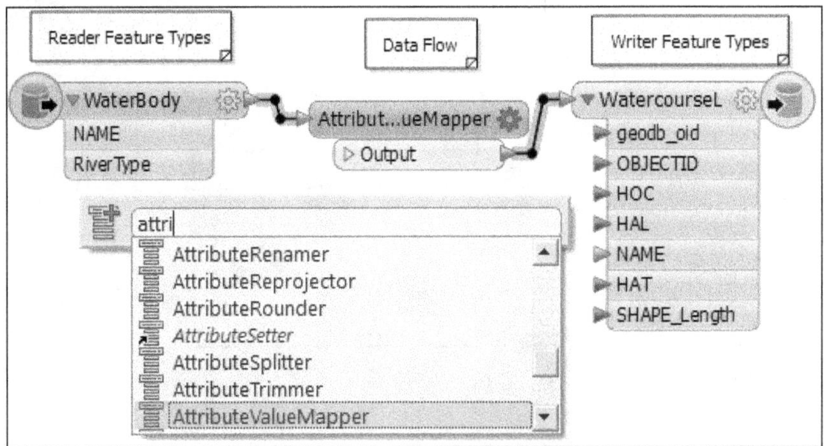

22. Double-click on the **AttributeValueMapper** transformer to open the **Parameters** dialog. For **Source Attribute**, select the `RiverType` field. For **Destination Attribute**, type `HAT`, which classifies the water features.

23. For **Source Value**, navigate to **Import** | **Source Value** to import the values from the `RiverType` field. Select the `WaterBody` GML file and click on **Next** twice. In the **Select Attributes for "Source Value"** panel, select the `RiverType` field, and click on **Next** and **Import**, as shown in the following screenshots:

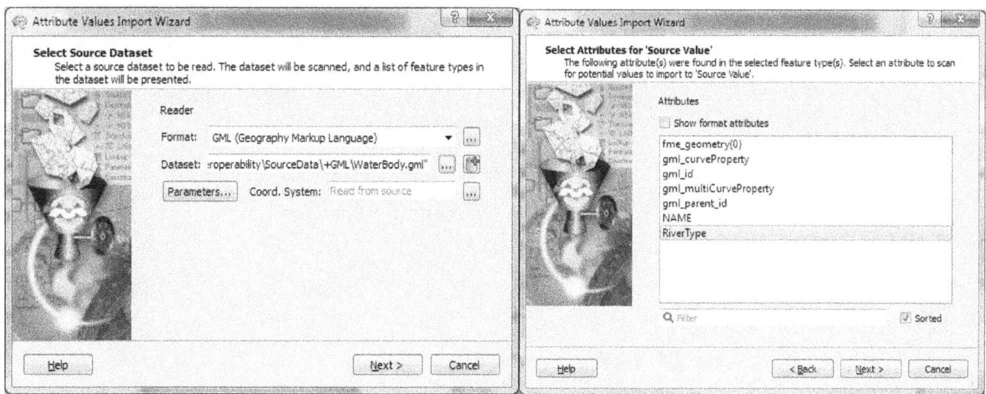

24. For the **Destination Value**, type the corresponding code, as shown in the following screenshot:

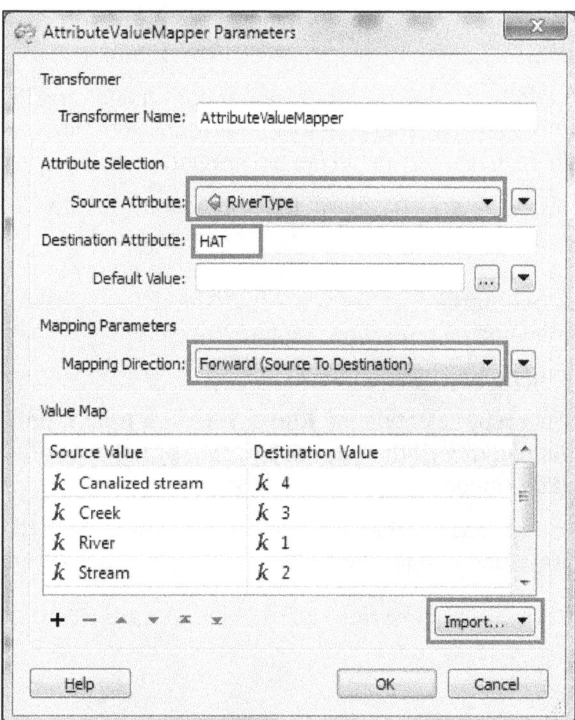

25. Click on **OK**. In the workbench canvas, create the connection between the **Reader** and **Writer** fields, as shown in the following screenshot:

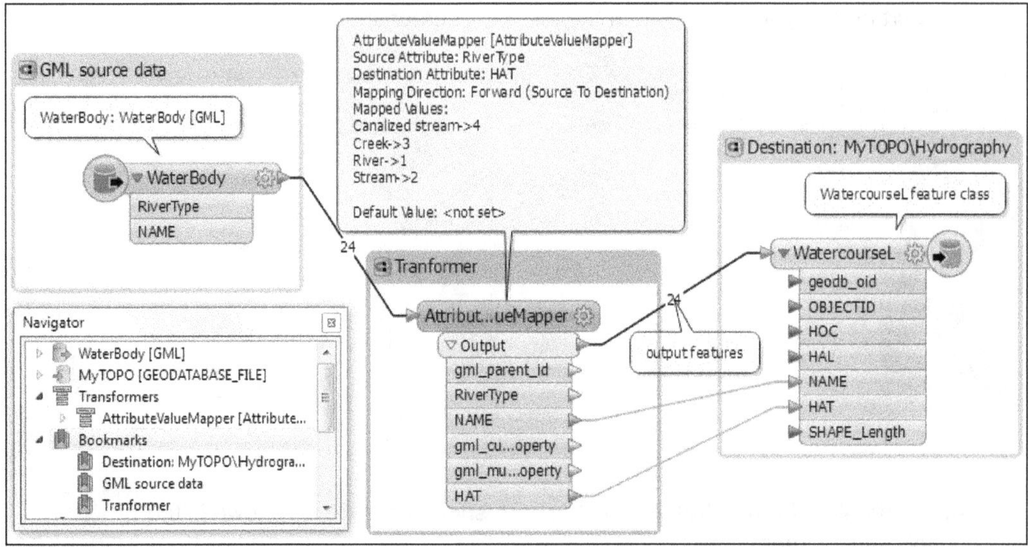

26. Let's add some annotations. Right-click on `WaterBody` and select **Add Annotation**. Explore the results and edit the default text.
27. Let's see the setting of the **AttributeValueMapper** transformer. Right-click on the transformer and select **Show Summary Annotation**. Note that the summary annotation cannot be edited and is colored blue to differentiate from the user annotation.
28. Let's add some bookmarks as shown in the preceding image. With bookmarks, you can underline different parts of your workspace, and you have easy access to the different parts of your workspace. Right-click on the empty space on the canvas and select **Insert Bookmarks**. The appearance of the default bookmark can be changed: text, color, dimension, and position. Explore the options.
29. Go to the **File** menu and click on **Save**.
30. Run the translation by selecting the **Run** or **Resume Translation** tool. Exit the workspace. You can modify the **Spatial ETL Tool** by right-clicking on the tool and selecting the **Edit** option.
31. Inspect the `WatercourseL` feature class in ArcCatalog. Note that subtype descriptions were updated for the `HAT (Classification)` field.

You can find the results at `...\Interoperability\DataResults\SpatialETLTool`.

How it works...

The datasets explored in ArcCatalog from steps 2 to 9 can be used in ArcMap like any other Esri data. Those datasets are read-only and cannot be edited. Use the **Quick Import** tool to quickly translate the non-Esri formats to a geodatabase. This tool will save the output in a new geodatabase.

At step 10, we started to create the **Spatial ETL Tool** to import a GML file into your existing geodatabase. Before running a tool, be sure that **Coordinate System** for **Reader** and **Writer** are specified.

From steps 14 to 17, we added **Writer** by importing an existing feature class from the `MyTOPO` geodatabase.

At step 18, we set the **Writer** properties. To keep the schema of the existing feature class table, we set **Drop Table First** to **No**. To make sure that the destination table was empty, we set **Truncate Table First** to **Yes**. We also mentioned that the `WatercourseL` output will be a feature class in the `Hydrography` feature dataset.

At step 19, we added the subtype definition for the `WatercourseL` feature class output.

From steps 21 to 24, we worked with the **AttributeValueMapper** parameter. It sets up value mapping between the `RiverType` text field (source) and the subtype codes of the feature class (destination). **Source Values** were imported from the `WaterBody` GML file.

Here are some best practices to create an easy-to-understand workspace:

- Avoiding overlapping links
- Adding annotation comments and bookmarks to the workspace
- Renaming the transformers
- Disabling/enabling the links (by right-clicking on the link)
- Using the **Translation Log** tool

Index

Symbols

A

B

G

H

I

J

V

W

X

Z

About Packt Publishing

Packt, pronounced 'packed', published its first book, *Mastering phpMyAdmin for Effective MySQL Management*, in April 2004, and subsequently continued to specialize in publishing highly focused books on specific technologies and solutions.

Our books and publications share the experiences of your fellow IT professionals in adapting and customizing today's systems, applications, and frameworks. Our solution-based books give you the knowledge and power to customize the software and technologies you're using to get the job done. Packt books are more specific and less general than the IT books you have seen in the past. Our unique business model allows us to bring you more focused information, giving you more of what you need to know, and less of what you don't.

Packt is a modern yet unique publishing company that focuses on producing quality, cutting-edge books for communities of developers, administrators, and newbies alike. For more information, please visit our website at `www.packtpub.com`.

Writing for Packt

We welcome all inquiries from people who are interested in authoring. Book proposals should be sent to `author@packtpub.com`. If your book idea is still at an early stage and you would like to discuss it first before writing a formal book proposal, then please contact us; one of our commissioning editors will get in touch with you.

We're not just looking for published authors; if you have strong technical skills but no writing experience, our experienced editors can help you develop a writing career, or simply get some additional reward for your expertise.

Building Web and Mobile ArcGIS Server Applications with JavaScript

ISBN: 978-1-84969-796-5 Paperback: 274 pages

Master the ArcGIS API for JavaScript, and build exciting, custom web and mobile GIS applications with the ArcGIS Server

1. Develop ArcGIS Server applications with JavaScript, both for traditional web browsers as well as the mobile platform.
2. Acquire in-demand GIS skills sought by many employers.

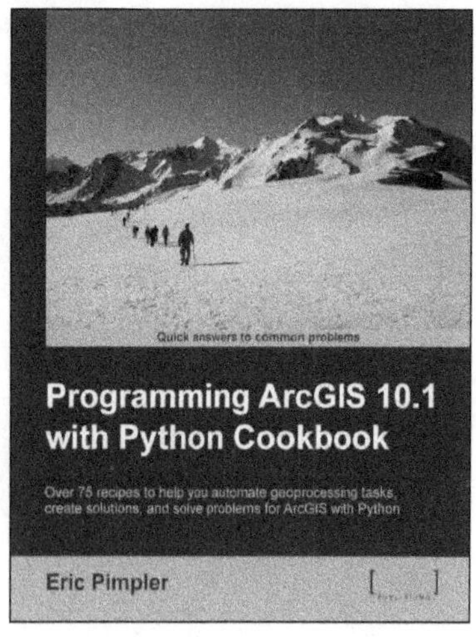

Programming ArcGIS 10.1 with Python Cookbook

ISBN: 978-1-84969-444-5 Paperback: 304 pages

Over 75 recipes to help you automate geoprocessing tasks, create solutions, and solve problems for ArcGIS with Python

1. Learn how to create geoprocessing scripts with ArcPy.
2. Customize and modify ArcGIS with Python.
3. Create time-saving tools and scripts for ArcGIS.

Please check **www.PacktPub.com** for information on our titles

Google Maps JavaScript API Cookbook

ISBN: 978-1-84969-882-5 Paperback: 316 pages

Over 50 recipes to help you create web maps and GIS web applications using the Google Maps JavaScript API

1. Add to your website's functionality by utilizing Google Maps' power.
2. Full of code examples and screenshots for practical and efficient learning.
3. Empowers you to build your own mapping application from the ground up.

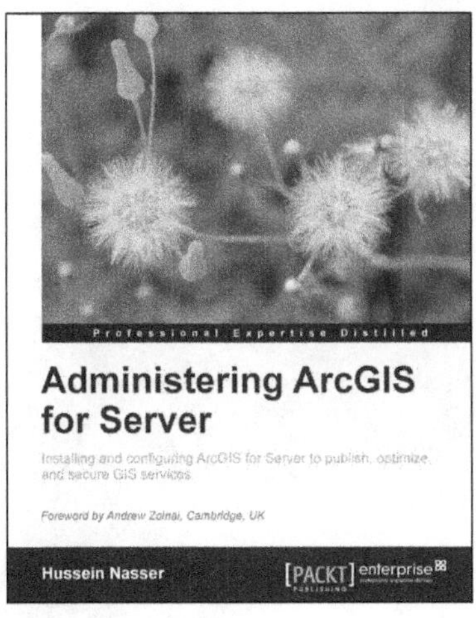

Administering ArcGIS for Server

ISBN: 978-1-78217-736-4 Paperback: 246 pages

Installing and configuring ArcGIS for Server to publish, optimize, and secure GIS services

1. Configure ArcGIS for Server to achieve maximum performance and response time.
2. Understand the product mechanics to build up good troubleshooting skills.
3. Filled with practical exercises, examples, and code snippets to help facilitate your learning.

Please check **www.PacktPub.com** for information on our titles

www.ingramcontent.com/pod-product-compliance
Lightning Source LLC
Chambersburg PA
CBHW062050050326
40690CB00016B/3037

* 9 7 8 1 7 8 3 5 5 9 5 0 3 *